Lecture Notes in Computer Science   14901

Founding Editors

Gerhard Goos

Juris Hartmanis

AF147900

The series Lecture Notes in Computer Science (LNCS), including its subseries Lecture Notes in Artificial Intelligence (LNAI) and Lecture Notes in Bioinformatics (LNBI), has established itself as a medium for the publication of new developments in computer science and information technology research, teaching, and education.

LNCS enjoys close cooperation with the computer science R & D community, the series counts many renowned academics among its volume editors and paper authors, and collaborates with prestigious societies. Its mission is to serve this international community by providing an invaluable service, mainly focused on the publication of conference and workshop proceedings and postproceedings. LNCS commenced publication in 1973.

Anna Lisa Ferrara · Ram Krishnan
Editors

Data and Applications Security and Privacy XXXVIII

38th Annual IFIP WG 11.3 Conference, DBSec 2024
San Jose, CA, USA, July 15–17, 2024
Proceedings

 Springer

Editors
Anna Lisa Ferrara 🆔
University of Molise
Campobasso, Italy

Ram Krishnan 🆔
The University of Texas at San Antonio
San Antonio, TX, USA

ISSN 0302-9743   ISSN 1611-3349 (electronic)
Lecture Notes in Computer Science
ISBN 978-3-031-65171-7   ISBN 978-3-031-65172-4 (eBook)
https://doi.org/10.1007/978-3-031-65172-4

This Springer imprint is published by the registered company Springer Nature Switzerland AG
The registered company address is: Gewerbestrasse 11, 6330 Cham, Switzerland

If disposing of this product, please recycle the paper.

Preface

This volume contains the papers presented at the 38th Annual IFIP WG 11.3 Conference on Data and Applications Security and Privacy (DBSec 2024), San Jose, USA – July 15–17, 2024.

In response to the call for papers for this edition, we received 39 submissions. Each submission underwent a thorough single-blind peer review process, evaluating its significance, novelty, and technical quality. The Program Committee, consisting of 49 members, performed an outstanding job in reviewing all the submitted works, with additional assistance from external reviewers. Each submission received an average of three reviews. Of the submitted papers, 14 full papers and 6 short papers were selected for presentation at the conference.

The success of DBSec 2024 was made possible by the dedicated efforts of many individuals, and we would like to extend our sincere appreciation to all. We are grateful to the members of the Program Committee and the external reviewers for their hard work in evaluating the papers and promptly engaging in the discussion and selection process. We also thank everyone who provided assistance and ensured a smooth organization process, particularly Ashish Kundu and Haibing Lu for their efforts as General Chairs. Additionally, we acknowledge Sara Foresti (IFIP WG 11.3 chair) and Jaideep Vaidya for their guidance and support, Valentina Piantadosi (Publicity Chair) for her assistance with publicity, Roberta Cimorelli Belfiore (Web Chair) for managing the conference web page, and the local committee for contributing to other conference arrangements.

The use of EasyChair greatly facilitated the conference review and publication process, ensuring its smooth execution.

Last but certainly not least, we extend our gratitude to all the authors who submitted their contributions and to all the conference attendees. We hope that the proceedings of DBSec 2024 will be interesting, stimulating, and inspiring for your future research endeavors.

July 2024

Anna Lisa Ferrara
Ram Krishnan

Organization

General Chairs

Ashish Kundu   Cisco Research, USA
Haibing Lu   Santa Clara University, USA

Program Committee Chairs

Anna Lisa Ferrara   University of Molise, Italy
Ram Krishnan   University of Texas at San Antonio, USA

IFIP WG 11.3 Chair

Sara Foresti   Università degli Studi di Milano, Italy

Publicity Chair

Valentina Piantadosi   University of Molise, Italy

Web Chair

Roberta Cimorelli Belfiore   University of Molise, Italy

Program Committee

Ayesha Afzal   Air University, Pakistan
Hafiz Asif   Rutgers University, USA
Vijay Atluri   Rutgers University, USA
Francesco Buccafurri   Università Mediterranea di Reggio Calabria, Italy
Frédéric Cuppens   Polytechnique Montréal, Canada
Nora Cuppens-Boulahia   Polytechnique Montréal, Canada
Sabrina De Capitani di Vimercati   Università degli Studi di Milano, Italy
Fausto Fasano   University of Molise, Italy

Additional Reviewers

Abbadini, Marco
Amour, Shmuel
Berlato, Stefano
Bisegna, Andrea
Cimato, Stelvio
Cimorelli Belfiore, Roberta
De Angelis, Vincenzo
Galdi, Clemente
Glas, Magdalena
Guerra, Michele

Lazzaro, Sara
Milanese, Roberto
Morales, Daniel
Mueller, Mathis
Muñoz, Antonio
Neudert, Raphael
Rios, Ruben
Sharif, Amir
Sinha, Arunesh

Biometrics: Challenges and Opportunities (Keynote)

Vincenzo Piuri ⓘ

Computer Science Department, Università degli Studi di Milano
vincenzo.piuri@unimi.it

Biometric technologies and applications are pervasively permeating our everyday life.

Biometrics for user authentication, once typically used for restricting access to critical (physical or digital) environments, are now making their ways in everyday life applications and services, from the emerging fast-paced adoption of biometric passports and face boarding at the airports to the widespread use of face and fingerprint recognition for unlocking smart phones.

Biometrics are also seamlessly at the basis of many of the services and applications of today's smart society, from governmental and business services to leisure. Video cameras are increasingly being adopted in public places to respond to security needs or providing services in smart cities. Smart cars monitor the drivers facial expressions and movements to react in case of sleepy drivers. IoT and smart assistant devices and applications rely on voice recognition for user interactions. Advanced entertainment and e-commerce applications can acquire and process user facial or body postures for emotion and sentiment analysis. Adding to the examples, which show the richness of applications relying on biometrics when interacting with users, biometrics technologies are also behind other services operating offline and downstream, such as the longly debated automatic user tagging from images in social network applications.

The widespread adoption of biometrics, the enormous amount of biometrics data gathered, collected and processed, as well as advancements in artificial intelligence open new challenges and opportunities in the field of biometrics and biometric data processing. The importance and criticality of biometrics are also testified by their explicit mention and particular consideration in the Artificial Intelligence Act just approved by the European Parliament.

These advancements in applications call for novel biometric solutions, able to operate in new and emerging scenarios seamlessly and balancing the need of catering advanced services based on biometrics with the rightful desire for an ethical, secure, and privacy-respectful use of biometrics.

In this talk, I will illustrate the main biometrics techniques discussing their characteristics (e.g., universality, uniqueness, permanence, performance, and circumvention), strengths, limitations, and applications. I will also discuss challenges and research directions, with particular focus on opportunities from the application of AI.

Keywords: Biometrics · Privacy · Artificial Intelligence

Acknowledgements. This work was supported in part by the EC under projects EdgeAI (101097300) and GLACIATION (101070141), and by project SERICS (PE00000014) under the MUR NRRP funded by the EU - NGEU. Views and opinions expressed are however those of the authors only and do not necessarily reflect those of the European Union or the Italian MUR. Neither the European Union nor the Italian MUR can be held responsible for them.

References

1. Abukmeil, M., Ferrari, S., Genovese, A., Piuri, V., Scotti, F.: A survey of unsupervised generative models for exploratory data analysis and representation learning. ACM Comput. Surv. **54**(5), 1–40 (2021)
2. Chen, C.F., Moriarty, B., Hu, S., et al.: Model-agnostic utility-preserving biometric information anonymization. Int. J. Inf. Secur. (2024). https://doi.org/10.1007/s10207-024-00862-8
3. Donida Labati, R., et al.: Biometric recognition in automated border control: a survey. ACM Comput. Surv. **49**(2), 24:1–24:39 (2016)
4. Donida Labati, R., Genovese, A., Piuri, V., Scotti, F., Vishwakarma, S.: I-SOCIAL-DB: a labeled database of images collected from websites and social media for iris recognition. Image Vis. Comput. **105**(104058), 1–9 (2021)
5. Genovese, A., Muñoz, E., Piuri, V., Scotti, F.: Advanced biometric technologies: emerging scenarios and research trends. In: Samarati, P., Ray, I., Ray, I. (eds.) From Database to Cyber Security. LNCS, vol. 11170, pp. 324–352. Springer, Cham (2018). https://doi.org/10.1007/978-3-030-04834-1_17

Contents

Attack

ML Attack, Vulnerability

Security User Studies

Differential Privacy

Access Control

A Graph-Based Framework for ABAC Policy Enforcement and Analysis

Mian Yang[1(✉)], Vijayalakshmi Atluri[1], Shamik Sural[2], and Jaideep Vaidya[1]

[1] Rutgers University, Newark, USA
{mian.yang,atluri}@rutgers.edu, jsvaidya@business.rutgers.edu
[2] Indian Institute of Technology Kharagpur, Kharagpur, India
shamik@cse.iitkgp.ac.in

Abstract. In the realm of access control mechanisms, Attribute-Based Access Control (ABAC) stands out for its dynamic and fine-grained approach, enabling permissions to be allocated based on attributes of subjects, objects, and the environment. This paper introduces a graph model for ABAC, named G_{ABAC}. The G_{ABAC} leverages directional flow capacities to enforce access control policies, mapping the potential pathways between a subject and an object to ascertain access rights. Furthermore, graph based modeling of ABAC enables the utilization of readily available commercial graph database systems to implement ABAC. As a result, enforcement and analyses of ABAC can be accomplished simply through graph queries. In particular, we demonstrate this using the Neo4j graph database and present the performance of executing enforcement and different analyses queries.

1 Introduction

Attribute-Based Access Control (ABAC) is the next generation access control model that is increasingly becoming popular due to its high flexibility, scalability, and portability across enterprises. Unlike traditional Role-Based Access Control (RBAC), where accesses are granted based on a user's role [10], in ABAC, the access control decisions are made based on attributes associated with the user (subject), the resource (object) being accessed, and the environment. This model allows for the dynamic nature of access control, meaning that it can adapt to changes in the environment and policies in real time. Additionally, ABAC is identity-less, which means that it uses user and resource attributes rather than their identities, which lends itself to be portable to cloud environments [17]. These features make ABAC an attractive choice for organizations for enforcing effective access control measures that can be easily managed and adapted to changing environments.

A flow graph is a specialized type of graph used to model the flow of some entity through a graph. This model comprises vertices connected by directed edges, each of which has a defined capacity that limits the flow between the vertices it connects. Here capacity denotes the maximum amount of the entity that can traverse a particular point in the graph. By applying graph algorithms

A. L. Ferrara and R. Krishnan (Eds.): DBSec 2024, LNCS 14901, pp. 3–23, 2024.
https://doi.org/10.1007/978-3-031-65172-4_1

that leverage such flow capacities, it becomes feasible to ascertain whether a subject s has the permission p to access an object o within the framework of access control policies. In this paper, we model ABAC as a flow graph, where the flow capacities act as constraints similar to those that satisfy user and object attribute conditions.

According to the guidelines developed by NIST [11], enterprises are recognizing the need for a robust mechanism for reviewing and analyzing access control policies, which emphasizes the importance of having the ability to review attributes associated with subjects (or users) and objects, as well as access rules. Specifically, organizations typically want to know the exact capabilities associated with user attributes, as well as the access control regulations associated with object attributes. This process not only enhances security by identifying potential risks and preventing unauthorized access but also ensures compliance with regulatory standards like GDPR, HIPAA, and SOX. Our model leverages graph algorithms to facilitate these user-centric and object-centric analyses effectively, offering a practical tool for detailed policy assessment and compliance assurance.

An additional benefit of graph based modeling of ABAC is that, it enables utilization of readily available commercial graph database systems. In particular, in this paper we demonstrate how a Neo4j graph database, which is the pioneer in the graph market [3], can be used to represent the ABAC policy translated into a graph and how access control evaluation and other analyses can be carried out by simply issuing queries to Neo4j.

Our specific contributions in this paper include:

- We propose a graphical model, named G_{ABAC}, that represents all the elements of an ABAC system. G_{ABAC} utilizes the flow capacity in graphs in such a way that it lends itself for ABAC enforcement and analyses.
- We introduce an approach for access request evaluation that employs traversing the G_{ABAC}. We theoretically prove that our approach results in access control evaluations that are consistent with those of the original ABAC system. In other words, we prove that an access request will be evaluated by G_{ABAC} to be "permit (or deny)" if and only if it is permitted (or correspondingly denied) under the ABAC system.
- We develop approaches to perform user-centric and object-centric evaluation of ABAC, again by querying/traversing the G_{ABAC}. Essentially, user-centric evaluation gives users' capabilities on objects and object-centric evaluation results in objects' accessibility by users. It is important to note that such analyses is not trivial, and even if it is feasible through certain implementations, only one of them would be easier, but not both.
- We implement the G_{ABAC} in Neo4j and formulated queries to facilitate both access control evaluation, and user- and object-centric analyses. We carry out performance analysis to study the execution cost of access request evaluation in different settings.

The rest of the paper is organized as follows. In Sect. 2, we review the foundational concepts and components of ABAC. In Sect. 3, we present details of the methodology for modeling the ABAC components within a graph structure and provide illustrative examples to elucidate the graph modeling process. In Sect. 4, we present our approach and algorithm for enforcing ABAC utilizing G_{ABAC}, provide proof of correctness for the algorithm and analyze the runtime complexity of the algorithm. In Sect. 5, we introduce the algorithms for checking user capability and object accessibility within G_{ABAC}. In Sect. 6, we present the implementation details of G_{ABAC} in Neo4j database and the design of the graph queries based on the algorithms to achieve access request evaluation of ABAC. We present our experimental results with respect to the execution time with our graph modeling. In Sect. 7, we review related work. Finally, in Sect. 8, we present our concluding remarks and future research directions.

2 ABAC Preliminaries

In this section, we review the ABAC model. The basic components of ABAC [11,18] are as follows:

$Users(U)$: Represents a set of authorized users or subjects. For each member u_i in the set, $1 \leq i \leq |U|$.

$Objects(O)$: Represents a set of resources that require to satisfy some condition to be accessed. For each member o_i in the set, $1 \leq i \leq |O|$.

$Environments(E)$: Represents a set of environment conditions, independent of users and objects. For each member e_i in the set, $1 \leq i \leq |E|$.

U_A, O_A, E_A: Represent the sets of user attribute names, object attribute names, and environment attribute names in the system, respectively.

P: A set of all possible permissions/operations that the system allows U to O. For example, if $read$ and $write$ are the only two possible operations on objects, then $P = \{read, write\}$. For each member p_i in the set, $1 \leq i \leq |P|$

U_C: Represents a set of all possible user attribute conditions denoted as uc_j, for $1 \leq j \leq |U_C|$. Members of this set are represented as equalities of the form $n = c$, where n is a user attribute name and c is either a constant or any.

O_C: Represents a set of all possible object attribute conditions denoted as oc_k, for $1 \leq k \leq |O_C|$. Members of this set are represented as equalities of the form $n = c$, where n is an object attribute name and c is either a constant or any.

UA: User attribute relation $UA \subseteq U \times U_C$ is a many-to-many mapping of users and user attribute conditions. We use UA_i to denote the set of all the user attribute conditions possessed by u_i.

OA: Object attribute relation $OA \subseteq O \times O_C$ is a many-to-many mapping of objects and the set of all attributes conditions. We use OA_i to denote the set of all the object attribute conditions possessed by o_i.

Π: Represents a set of access rules called the rule base of the ABAC system. Each member of this set is denoted as r_i, for $1 \leq i \leq |\Pi|$. A rule r_i in ABAC is of the form $\langle [uc]_i^*, [oc]_i^*, p_i \rangle$ where $[uc]_i^*$ and $[oc]_i^*$ represent the set of user and object attribute conditions, respectively. Specifically, $[uc]_i^* = \{uc_{i1}, \ldots, uc_{im}\}$ such that every $uc_{ij} \in U_C$, and $[oc]_i^* = \{oc_{i1}, \ldots, oc_{in}\}$ such that every $oc_{ij} \in O_C$.

For the sake of simplicity, in this paper, we assume the environment condition to be any.

3   Modeling ABAC as a Graph

In this section, we introduce how ABAC can be modeled as a graph, where vertices symbolize the entities involved in access control, and edges represent the relationships or flow between these entities. Flow signifies the movement or transmission of data or resources, depicted through directed edges with capacities defining their maximum supportable flow. Applying this foundational concept to ABAC, the flow within our graph encapsulates the dynamic process by which specific conditions are met and contribute to the decision-making in access control.

To map each element of ABAC into this model, we need to consider not only the entities and their relations but also how the participation of each condition affects rule satisfaction. To quantify the participation of user and object attribute conditions ($[uc]_i^*$ and $[oc]_i^*$), we measure their cardinalities $|[uc]_i^*|$ and $|[oc]_i^*|$, respectively. This measurement serves as a foundation for defining the fractional contribution ($fract$) on each condition, where for every $uc_{ij} \in [uc]_i^*$, $fract(uc_{ij}) = \frac{1}{|[uc]_i^*|}$, and analogously for $oc_{ij} \in [oc]_i^*$, $fract(oc_{ij}) = \frac{1}{|[oc]_i^*|}$. Such a fractional contribution is pivotal in our graph model, representing the flow capacity of uc_{ij} and oc_{ij} through the edges, thus accurately reflecting each condition's weight towards achieving rule satisfaction. It is noteworthy that variations to this weighting scheme may be considered. Such variations, though not standard within ABAC, could allow for differential weighting of conditions based on their perceived importance, potentially offering a more nuanced and customizable approach to access control.

In the following, we define how the graph G_{ABAC} can be constructed from the given ABAC system.

Definition 1. *Considering the ABAC system comprising of U, O, UA, OA, U_c, O_c and Π, the ABAC graph $G_{ABAC} = (V, E)$ is defined as follows, where V is the set of vertices and E is the set of edges.*

> *for every $u_i \in U$ there exists a vertex $u_i \in V$.*
> *for every $o_i \in O$ there exists a vertex $o_i \in V$.*
> *for every $uc_i \in U_C$ there exists a vertex $uc_i \in V$.*
> *for every $oc_i \in O_C$ there exists a vertex $oc_i \in V$.*
> *for every u_i to uc_j mapping in UA, there exists an edge $(u_i \rightarrow uc_j) \in E$.*
> *for every o_i to oc_j mapping in OA, there exists an edge $(o_i \rightarrow oc_j) \in E$.*
> *for every $r_i = \langle [uc]_i^*, [oc]_i^*, p_i \rangle \in \Pi$:*
>> *there exists a vertex $AND^u_{[uc]_i^*}$ and a vertex $AND^o_{[oc]_i^*}$,*
>> *there exists an edge $AND^u_{[uc]_i^*} \xrightarrow{p} AND^o_{[oc]_i^*}$,*
>> *for every $uc_{ij} \in [uc]_i^*$ there exists an edge $uc_{ij} \xrightarrow{fract(uc_{ij})} AND^u_{[uc]_i^*} \in E$.*
>> *for every $oc_{ij} \in [oc]_i^*$ there exists an edge $oc_{ij} \xrightarrow{fract(oc_{ij})} AND^o_{[oc]_i^*} \in E$*

Essentially, all users, objects, user and object attribute conditions are represented as nodes in G_{ABAC}. All mappings from user to their respective user-attribute conditions, and mappings from object to their object-attribute conditions are represented as edges in G_{ABAC}.

For every $r_i \in \Pi$, the permission p is granted only when all $uc_{ij} \in [uc]_i^*$ and $oc_{ij} \in [oc]_i^*$ are fulfilled. To model this, vertices $AND_{[uc]_i^*}^u$ and $AND_{[oc]_i^*}^o$ are introduced for each rule. These vertices function as logical $'AND'$ gates: $AND_{[uc]_i^*}^u$ for aggregating user attribute conditions and $AND_{[oc]_i^*}^o$ for object attribute conditions, ensuring that all required attributes are met before granting permission p. To articulate these requirements, edges are drawn from each uc_{ij} and oc_{ij} to their respective AND vertices, with labels $fract(uc_{ij})$ and $fract(oc_{ij})$ signifying the flow capacity. Such a mechanism ensures that the total inflow to both $AND_{[uc]_i^*}^u$ and $AND_{[oc]_i^*}^o$ vertices equals 1, effectively mirroring the logical requirement that each and every specified attribute condition for both user and object must be satisfied for access to be granted.

It is important to note that $AND_{[uc]_i^*}^u$ is an AND^u node with specific set of user attribute conditions. If another rule contains the same set of user attribute conditions, we need not create another AND^u node for that rule. Specifically, if two rules r_i and r_j are such that $[uc]_i^* = [uc]_j^*$ then the graph contains only one AND^u node corresponding to both these two rules. Similar is the case with the AND^o nodes.

Example 1. Consider a scenario involving a set of users $U = \{\text{Alice}, \text{Bob}, \text{Cindy}\}$ within a company. Alice serves as an Assistant in the HR department and is involved with the Pioneer team. Bob holds a Manager position without specific assignments to any department or project. Cindy is also a Manager but within the HR department. In this context, we define the user attribute conditions as follows: uc_1 : {Department = HR}, uc_2 : {Role = Assistant}, uc_3 : {Role = Manager}, and uc_4 : {Team = Pioneer}. Thus, $UA_{\text{Alice}} = \{uc_1, uc_2, uc_4\}$, $UA_{\text{Bob}} = \{uc_3\}$, and $UA_{\text{Cindy}} = \{uc_1, uc_3\}$.

The scenario also includes two objects, where $O = \{\text{FileX}, \text{FileY}\}$, both associated with the Hope Project but differing in their related field: FileX is related to NameList, and FileY is related to Budget. Thus, we establish object attribute conditions as: oc_1 : {Related = NameList}, oc_2 : {Related = Budget}, and oc_3 : {Project = Hope}. Accordingly, $OA_{\text{FileX}} = \{oc_1, oc_3\}$, and $OA_{\text{FileY}} = \{oc_2, oc_3\}$.

Consider the following rules in the rule set Π: The first rule: Assistant in HR department within Pioneer team can read file in Hope project related to Name List. This rule can be denoted as $r_1 = \langle [uc_1, uc_2, uc_4], [oc_1, oc_3], read \rangle$, where $r_1 \in \Pi$. In G_{ABAC}, for r_1, there exists a vertex AND_I^u where $I = [uc_1, uc_2, uc_4]$ and AND_J^o where $J = [oc_1, oc_3]$. Given that $[uc]_1^*$ consists of the $[uc_1, uc_2, uc_4]$, its cardinality $|[uc]_1^*| = 3$. Consequently, the capacity of each incoming edge from the $uc_{1j} \in [uc]_1^*$ to AND_I^u is $\frac{1}{3}$. Similarly, $[oc]_1^*$, which includes the attributes $[oc_1, oc_3]$, consists of two elements, so its cardinality $|[oc]_1^*| = 2$. Therefore, the capacity of each incoming edge from the $oc_{1j} \in [oc]_1^*$ to AND_J^o is $\frac{1}{2}$.

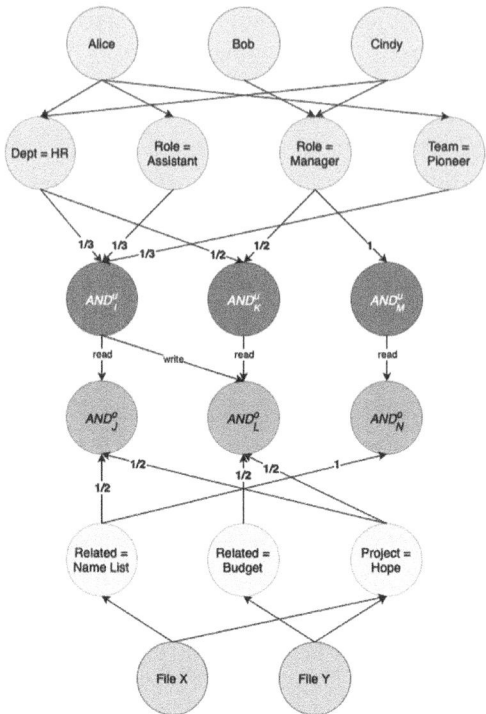

Fig. 1. G_{ABAC} of Example 1

The second rule: **Manager** in the **HR** department can read file in **Hope** project related to **Budget**. This rule $r_2 = \langle [uc_1, uc_3], [oc_2, oc_3], read \rangle$. In G_{ABAC}, for r_2, there exists a vertex AND_K^u where $K = [uc_1, uc_3]$ and AND_L^o where $L = [oc_2, oc_3]$. As $|[uc]_2^*| = 2$, and $|[oc]_2^*| = 2$, the capacity of each incoming edge from the $uc_{2j} \in [uc]_2^*$ to AND_K^u is $\frac{1}{2}$ and from the $oc_{2j} \in [oc]_2^*$ to AND_L^o is $\frac{1}{2}$.

The third rule: **Manager** can read file related to **Name List**, which can be denoted as $r_3 = \langle [uc_3], [oc_1], read \rangle$. Since $[uc_3]$ is the only $uc_{3j} \in [uc]_3^*$, the capacity of incoming edge from uc_3 to AND_M^u where $M = [uc_3]$ is 1. Similarly, the capacity of the incoming edge from oc_1 to AND_N^o where $N = [oc_1]$ is 1.

The fourth rule: **Assistant** in **HR** department within **Pioneer** team can write file in **Hope** project related to **Budget**. This rule $r_4 = \langle [uc_1, uc_2, uc_4], [oc_2, oc_3], write \rangle$. Given that $[uc]_{4j} = [uc]_{1j}$, all $[uc]_{4j} \in [uc]_4^*$ connect to AND_I^u with $\frac{1}{3}$ without the need to create another AND^u. In other words, both r_1 an r_4 have the same AND^u node. Similarly, since $[oc]_{4j} = [oc]_{2j}$, all $[oc]_{4j} \in [oc]_4^*$ connected to AND_L^o with $\frac{1}{2}$, again avoiding the creation of an addition AND^o. Figure 1 illustrates the G_{ABAC} graph constructed from the information provided above. The graph uses **Department** as **Dept** for brevity.

4 Enforcement of ABAC Using G_{ABAC}

In this section, we propose our graph traversal algorithm that ensures how an access request can be *correctly* evaluated using G_{ABAC}, provide its proof of correctness and Speaking informally, and discuss its complexity.

4.1 Algorithm for Access Request Evaluation

When an access request $\langle u_i, o_j, p \rangle$ is received, the traversal begins with navigating the graph starting from u_i, examining all uc_j linked to u_i and assessing the inflow capacity of their connected AND^u vertices. A similar process is conducted for the object o_j, reviewing all oc_k and their connection to AND^o vertices. Access request $\langle u_i, o_j, p \rangle$ is granted if there exists both an AND^u and an AND^o vertex, each with an inflow capacity of 100%, indicating that all necessary user and object attribute conditions are met, and there is a direct edge between AND^u and AND^o labeled with p. If any of these criteria are not satisfied, i.e., either the inflow capacities of any AND nodes is not 100% or the requisite p labeled edge between AND^u and AND^o is not present, the access request is denied.

Algorithm 1 Access Request Evaluation in G_{ABAC}

Input: $\langle u_i, o_j, p \rangle$, G_{ABAC}
Output: "Permit" or "Deny"
 1: **Compute** UA_{ig} such that $uc_k \in UA_{ig}$ where there exists $u_i \rightarrow uc_k$
 2: **Compute** OA_{jg} such that $oc_l \in OA_{jg}$ where there exists $o_j \rightarrow oc_l$
 3: **for** each AND^u with inflow capacity of 1 such that the set of incoming nodes to AND^u is a subset of UA_{ig} **do**
 4:     **if** there exist $AND^u \xrightarrow{p} AND^o$ **then**
 5:         **if** inflow capacity of $AND^o = 1$ such that the set of incoming nodes to AND^o is a subset of OA_{jg} **then return** permit
 6:         **end if**
 7:     **end if**
 8: **end for**
 9: **return Deny**

Algorithm 1 outlines the process of ABAC enforcement within G_{ABAC}. The algorithm begins by identifying the UA_{ig} for user u_i and the OA_{jg} for object o_j. The algorithm then iterates over AND^u and first searches for any AND^u vertices that are connected to u_i through the subset of UA_{ig} with an inflow capacity of 1. If no such AND^u satisfies this condition, access is immediately denied, as it indicates that the user does not meet the necessary attribute conditions for access. If an AND^u fulfilling the condition is found, the algorithm proceeds to check if there exists a connection such that $AND^u \xrightarrow{p} AND^o$. Even after iterating over all the AND^u no such AND^o can be found, the access is denied. If an AND^o is found, the algorithm further checks whether it connects to o_j

through the subset of OA_{ig} with an inflow capacity of 1. If these conditions are met, access is permitted. Otherwise, if no suitable AND^o is found after checking all possibilities, the access request is denied.

Example 2. Using the G_{ABAC} illustrated in Fig. 1 as a reference, let us consider a scenario where Cindy requests access to read FileX. The algorithm initiates by identifying the attribute conditions associated with both the user Cindy, and the object FileX. Cindy is associated with the HR Department and her role as a Manager ($UA_{\text{Cindy}} : \{\text{Dept} = \text{HR}, \text{Role} = \text{Manager}\}$). FileX is related to NameList and its association with Project Hope ($OA_{\text{FileX}} : \{\text{Related} = \text{NameList}, \text{Project} = \text{Hope}\}$).

The algorithm searches for any AND^u vertex that gathers 100% inflow capacity from a subset of UA_{Cindy}. Two vertices, AND^u_K and AND^u_M, meet this criterion. AND^u_K achieves its required inflow equally from the Dept = HR and Role = Manager vertices, with each contributing half of the necessary inflow. AND^u_M receives a complete inflow directly from the Role = Manager vertex. Subsequently, the algorithm first identifies AND^u_K. For AND^u_K, the algorithm identifies AND^o_L as being connected via a *read* relationship. The algorithm then assesses if AND^o_L can achieve full inflow capacity from the subset of the OA_{FileX}. However, it is observed that AND^o_L only receives half of its required inflow from the Project = Hope vertex, failing to meet the necessary 100% inflow capacity for granting access. The algorithm continues its search because not all satisfied AND^u vertices have been explored.

Next, the algorithm examines AND^u_M, finding AND^o_M connected via an edge labeled *read*. This time, the assessment reveals that AND^o_M obtains 100% inflow capacity from the Related = NameList vertex, which is subset of OA_{FileX}. Since this pathway satisfies the necessary condition, the algorithm grants Permit for Cindy to read FileX.

4.2   Proof of Correctness

Theorem 1. *Given an access request $\langle u_i, o_j, p \rangle$, Algorithm 1 when evaluated on G_{ABAC} constructed according to Definition 1 returns permit (deny) iff it evaluates to permit (deny) under the ABAC policy.*

Proof. **The if Case:** We prove that given an access request $\langle u_i, o_j, p \rangle$, if it evaluates to permit (deny) under the ABAC policy, then Algorithm 1 when evaluated on G_{ABAC} constructed according to Definition 1 also returns permit (deny).

We begin by assuming that the ABAC policy dictates a permit decision for the given access request. This implies that in ABAC, there exists a rule $r_i = \langle [uc]^*_i, [oc]^*_i, p \rangle$ such that $[uc]^*_i \subseteq UA_i$ for u_i and $[oc]^*_i \subseteq OA_j$ for o_j, with a permission p. Note that in G_{ABAC}, according to Definition 1, this rule is represented as: for every r_i, its $[uc]^*_i$ is represented as a vertex $AND^u_{[uc]^*_i}$ with inflow capacity of 1 from every $uc_{ij} \in [uc]^*_i$, its $[oc]^*_i$ is represented as a vertex $AND^o_{[oc]^*_i}$ with inflow capacity of 1 from every $oc_{ij} \in [oc]^*_i$ and p is represented as $AND_i^u \xrightarrow{p} AND_i^o$. Algorithm 1 will then detect these fully satisfied conditions and return a "permit".

Conversely, assume that the access request is evaluated as deny by the ABAC policy. This means that there does not exist a rule that satisfies any of the conditions, i.e., either $[uc]_i^* \nsubseteq UA_i$, $[oc]_i^* \nsubseteq OA_j$, or the permission is not p. According to our graph construction, no such fully connected configuration can exist in the graph. Consequently, the algorithm will find either incomplete inflow capacities or missing p labelled edges, resulting in a "deny".

The Onlyif Case: We prove that given an access control request $\langle u_i, o_j, p \rangle$, if Algorithm 1 when evaluated on G_{ABAC} constructed according to Definition 1 returns permit (deny), then the request evaluates to permit (deny) under the ABAC policy.

Algorithm 1 permits access only when G_{ABAC} contains AND^u and AND^o vertices connected by a p labeled edge, each with full inflow capacity of 1. That is, every $[uc]_i^* \subseteq UA_i$ for u_i should be connected to the AND^u and every $[oc]_i^* \subseteq OA_j$ for o_j should be connected to the AND^o. Since AND^u represents $[uc]^*$, and AND^o represents $[oc]^*$ of the rule, by checking the full capacity requirement of the incoming nodes of these AND^u and AND^o nodes, the algorithm checks for the presence of all the attribute conditions in the rule. Thus, if Algorithm 1 returns 'permit', it must be the case that ABAC policy evaluation is also a permit, as our algorithm does not add any extraneous permits.

On the other hand, if the algorithm denies access, it means that either the inflow capacities for AND^u or AND^o are not met or a required p connection is missing. Such a scenario implies that the conditions for any rule r_i are not completely met, corresponding directly to a policy decision to deny access. □

4.3 Time Complexity of Algorithm 1

The worst case arises when the algorithm must conduct an exhaustive search through G_{ABAC} to determine access permissions. This situation is primarily driven by the density of the graph, specifically the number of user attribute conditions and object attribute conditions and the complex interconnections between AND^u and AND^o vertices representing the aggregation of these conditions. In essence, the scenario occurs when every possible path through the graph must be explored to ascertain whether a user u can perform an operation p on an object o. The complexity and depth of this exhaustive search are underscored by two main factors: the number of attribute conditions connected to the entities involved and the structure of the AND vertices' connections. For each user u and object o, the algorithm first identifies all relevant attribute conditions. It then assesses AND^u vertices to find any that signify the complete satisfaction of user attribute conditions through an inflow capacity of 1. Similarly, it checks for AND^o vertices connected to satisfied AND^u vertices under the specified permission p, also requiring a full inflow capacity. The worst case occurs when the algorithm has to iterate over all vertices and edges in the graph, a condition met when the final vertices evaluated are those that determine the access decision.

Thus, the algorithmic complexity is then $O(|UCS|+|OCS|+|AND^u|\times|AND^o|)$, where $|UCS|$ and $|OCS|$ represent the counts of user and object attribute conditions respectively, and $|AND^u|\times|AND^o|$ reflects the exhaustive combinations of vertex connections evaluated.

5   User/Object-Centric Analysis of ABAC Using G_{ABAC}

In this section, we show how G_{ABAC} can answer (graph) queries that list all the objects to which a user u_i has access, and list all the subjects by whom an object o_j can be accessed. As discussed in Sect. 1, these two situations align with the guidelines developed by NIST, which addresses the need to review privileges and monitor authorizations. Algorithm 2 is designed to identify all objects that a specific user u_i is authorized to access under a given permission p_k within G_{ABAC}. The process begins by initializing an empty set AOP (Accessible Objects with Permission), which holds all the permissions (o_j, p_k) that u_i possesses.

The algorithm identifies UA_{ig} linked to the user u_i. Next, the algorithm evaluates whether there exists an AND^u vertex with an inflow capacity of 1 from the subset of UA_{ig}. If no such AND^u vertex exists, the algorithm concludes that u_i does not have the requisite attribute conditions to access any object under p and returns the empty AOP set. For each AND^u that meets the inflow criterion, the algorithm then searches a AND^o with connection such that $AND^u \xrightarrow{p_k} AND^o$. If such AND^o is founded, the algorithm then identifies o_j connected to each satisfied AND^o where the inflow capacity from the subset of OA_{ij} to AND^o equals 1. Each such o_j and p_k is added to AOP set, signifying that it is accessible by user u_i under p_k.

Algorithm 2 Evaluating User Capabilities in G_{ABAC}

Input: u_i, G_{ABAC}
Output: the set of (o_j, p_k) where $o_j \in O$ and $p_k \in P$
 1: initialize an empty set AOP (Accessible Objects with Permission)
 2: **Compute** UA_{ig} such that $uc_k \in UA_{ig}$ where there exists $u_i \rightarrow uc_k$
 3: **for** each AND^u with inflow capacity of 1 such that the set of incoming nodes to AND^u is a subset of UA_{ig} **do**
 4:     **for** each $AND^u \xrightarrow{p_k} AND^o$ **do**
 5:         **if** AND^o has inflow capacity $=1$ from OA_{jg} such that $oc_l \in OA_{jg}$ where there exists $o_j \rightarrow oc_l$ **then**
 6:             Add (o_j, p_k) to the set AOP
 7:         **end if**
 8:     **end for**
 9: **end for**
10: **return** AOP

Example 3. If we want to know all objects that can be *read* by *Alice* in Fig. 1, the Algorithm 2 identifies the $UA_{\texttt{Alice}}$ vertices connected to Alice, which is Dept = HR, Role = Assistant, and Team = Pioneer. Then the algorithm proceeds to search any AND^u vertex that gathers 100% inflow capacity from a subset of UA_{Alice}. In this scenario, AND^u_I satisfied the inflow requirement and connected to AND^o_J in edge labeled with *read*.

Given that AND^o_J secures an inflow capacity of 1 from Related = NameList and Project = Hope, object FileX, which are linked to those two object attributes, is considered accessible for Alice to read and thus (FileX, read) added to the AOP set. Since AND^o_J is the only satisfied AND^o connected with AND^u_I and there are no other AND^u satisfied, the algorithm return the set $AOP = \{(\text{FileX}, \text{read})\}$.

The evaluation of object accessibility within G_{ABAC} serves as the converse to accessing user capabilities. The prior discussion on user capabilities emphasized identifying objects a user can access under specific permission. Conversely, evaluating object accessibility focuses on pinpointing which users are granted access to a particular object, effectively presenting the dual side of the access equation. Algorithm 3 is designed to methodically identify the users who are authorized to access a specific object o_i under a given permission p_k.

Algorithm 3 Evaluating Object Accessibility in G_{ABAC}

Input: o_i, G_{ABAC}
Output: the set of (u_j, p_k) where $u_j \in U$ and $p_k \in P$
1: initialize an empty set AUP (Accessible Users with Permission)
2: **Compute** OA_{ig} such that $oc_k \in OA_{ig}$ where there exists $o_i \rightarrow oc_k$
3: **for** each AND^o with inflow capacity of 1 such that the set of incoming nodes to AND^o is a subset of OA_{ig} **do**
4:     **for** each $AND^u \xrightarrow{p_k} AND^o$ **do**
5:         **if** AND^u has inflow capacity =1 from UA_{jg} such that $uc_l \in UA_{jg}$ where there exists $u_j \rightarrow uc_l$ **then**
6:             Add (u_j, p_k) to the set AUP
7:         **end if**
8:     **end for**
9: **end for**
10: **return** AUP

Example 4. To compute the set of all users allowed to access FileX in Fig. 1, Algorithm 3 begins by identifying the OA_{FileX} vertices, which is Related = NameList and Project = Hope. Next, it searches for AND^o vertices that have 100% inflow capacity from a subset of OA_{FileX}. In this case, both AND^o_J and AND^o_N satisfy the inflow capacity requirement. Moving forward, the algorithm then examines whether there exist AND^u vertices connected to these AND^o vertices with a p_k label on its edge. For AND^o_J, it finds AND^u_I which satisfies the read permission and subject Alice satisfies the inflow capacity of 100% to AND^u_I, making her eligible to read FileX. So (Alice, read) is added to AUP. Since no other subject meets the criteria, and no other AND^u connected with AND^o_J, we proceed to AND^o_N. Here AND^u_M connected with AND^o_N in read permission and both Bob and Cindy contribute 100% inflow to AND^u_M due to their Role = Manager. So here (Bob, read) and (Cindy, read) are added to

AUP. As there are no other relevant AND^u vertices connected to AND_N^o, and no additional AND^o meets the inflow constraint, the algorithm concludes by returning the set of $AU = \{(\texttt{Alice}, \texttt{read}), (\texttt{Bob}, \texttt{read}), (\texttt{Cindy}, \texttt{read})\}$.

6   Implementation Using Neo4j

Neo4j is a native graph database designed specifically for storing, managing, and querying highly connected data efficiently using its graph-specific storage mechanisms and Cypher query language [4]. In this section, we present how G_{ABAC} has been implemented in the graph database system, Neo4j.

6.1   Changes to the Graph

In the process of implementing our G_{ABAC} using Neo4j, we encounter a specific requirement to adjust the model, particularly concerning the assignment of $fract$ to edges. Neo4j does not natively support direct representation of fractions in the format defined in Sect. 3. To maintain the integrity and accuracy of the our ABAC graph model within the constraints imposed by Neo4j, a modification is necessary. Instead of assigning the fractional value directly as the edge weight, we adopt an alternative approach:

1. We set the $fract$ of edges between every $uc_{ij} \in [uc]_i^*$ and its corresponding $AND_{[uc]_i^*}^u$, and between every $oc_{ij} \in [oc]_i^*$ and its corresponding $AND_{[oc]_i^*}^o$, to a fixed value of 1.
2. To compensate for this adjustment, we manage the total inflow for the $AND_{[uc]_i^*}^u$ vertex to be equal to $|[uc]_i^*|$. A similar approach is applied for the AND_i^o vertex, where the total inflow is set to match $|[oc]_i^*|$. For example, if there are three incoming edges for a $AND_{[uc]_i^*}^u$ node, the label on these edges has been changed to 1 instead of 1/3. And the capacity requirement has been changed to 3 instead of 1. Similar adjustments are made to the capacity of $AND_{[oc]_i^*}^u$ nodes.

By implementing this strategy, we effectively preserve the functional aspects of G_{ABAC} within Neo4j, ensuring that it remains consistent with the original design intent, despite the limitations in representing fractional weights directly.

6.2   Query for Enforcement

To tailor Neo4j's Cypher query capabilities to our need, we integrate Python for enhanced control. While Cypher excels in returning all paths that satisfy a given search criterion, Algorithm 1 requires an immediate halt upon discovering the first pair of AND^u and AND^o that match the capacity requirement as well as connected with the right permission. However, since it is a database query, the Cypher query execution does not terminate, but continues to find all the pairs that satisfy the above requirement. This integration allows us to leverage Neo4j for data storage and searching, with Python managing the logical termination of the search based on specific conditions.

The detailed code is included in Appendix A. A Python function *access_check* operationalizes Algorithm 1. Initially, it processes input variables - user, object and permission types - through sanitation to remove illegal characters for input and interact with the graph database. The core of both the algorithm and the Python function involves querying the graph to evaluate access requests based on predefined user and object attributes conditions, along with permissions. The function first identifies relevant AND^u connected to the specified user. This mirrors the algorithm's step of finding AND^u vertices with a 100% inflow pattern from its user attribute conditions. If no such AND^u vertices are found, the function immediately denies access, adhering to the algorithm's directive to deny access when the initial conditions fail. For each qualifying AND^u vertex, the function further explores connected AND^o through specified permission label. The decision to permit or deny access hinges on finding at least one AND^o vertex that satisfies the final condition - having a 100% inflow of relationships from the object's object attribute conditions.

6.3 Experimental Evaluation

The experimental evaluation is designed to assess the effectiveness of the proposed approach across various settings and to analyze the impact of different factors. Due to the difficulty in acquiring suitable real datasets, we have created synthetic datasets with specific parameter values. In order to manage the rule size, we set the maximum $|[uc]_i^*|$ and $|[oc]_i^*|$ for each rule r_i to be 5. Every G_{ABAC} is constructed using synthetic data generated with consistent random seeds, but varying sets of parameter values. The key parameters are $|U|$, $|O|$, $|\Pi|$, $|U_C|$, $|O_C|$, participation of $|uc_j|$ in $[uc]^*$, and $|oc_k|$ in $[oc]^*$.

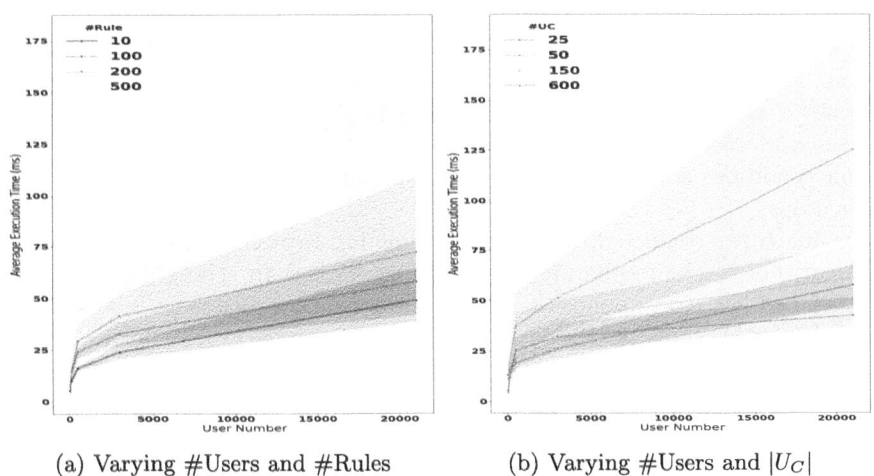

(a) Varying #Users and #Rules   (b) Varying #Users and $|U_C|$

Fig. 2. Access Request Evaluation with varying the size of rules and attribute conditions

We randomly generate 100 access requests to test the average execution time of G_{ABAC}, with each request being processed five times to determine the average execution time over these iterations. In average execution time of G_{ABAC} result, we ignore the first experimental time because this includes the time to connect the Neo4j database. All the average execution times are recorded in milliseconds.

Figure 2a illustrates how the average execution time increases with the number of users under different $|\Pi|$ and different configurations of $|U_C|$. Each line on the Fig. 2a represents one of the rule size, and the shaded area around each line depicts the variability in execution times for different $|U_C|$ settings, namely $|U_C| = [25, 50, 150, 600]$. This shaded area demonstrates the upper and lower bounds of execution times at various points, highlighting how performance fluctuates under different $|U_C|$ configurations, with wider shaded areas showing greater variability. Similarly, Fig. 2b explores how the average execution time changes with the number of users for different $|U_C|$, and the surrounding shaded areas represent the variability in average execution times under varying $|\Pi|$. Moreover, the Fig. 2b shows that having more $|U_C|$ does not necessarily correlate with longer execution times. Interestingly, the graph reveals that when $|U_C| = 50$, the execution time displays the most variability across different number of rules. In contrast, other levels of $|U_C|$ show relatively consistent execution times.

We further analyze three scenarios: (i) all access request evaluate to grant, (ii) all access request evaluate to deny, and (iii) equal mixture of both. These are illustrated in Fig. 3a. In the deny case, which takes longer execution time, the algorithm exhaustively iterates through all AND^u and AND^o without finding any that satisfy the given constraints. Conversely, in the grant case, since the algorithm terminates as soon as a match is found, it always takes less time than that of the deny case. The case considering both grant and deny scenarios takes an intermediate amount of time due to the mix of immediate terminations and exhaustive checks.

To further investigate the impact the density of the graph on execution times, we varied the percentage of participation of the user (and object) attribute conditions in the rules, where we kept $|U_C| = 50$. The results, shown in Fig. 3b, indicate a clear trend: higher percentages of uc participation correlate with increased average execution times, suggesting that the involvement of more uc in rule conditions adds to the computational load and thus the time required for evaluations.

Similarly, we evaluated the execution time for user-centric analysis outlined in Algorithm 2. The results are shown in Figs. 4a and 4b. The findings suggest that a larger $|U_C|$ corresponds to a reduced average execution time. The width of the shaded area represents the variability of $|\Pi|$ in average execution time for the same $|U|$ and $|U_C|$ in Fig. 4a and the variability of $|U_C|$ in average execution time for the same $|U|$ and $|\Pi|$ in Fig. 4b. Notably, at $|U_C| = 25$, there is a significant variability in the execution time across various rule sizes, with a broader shaded area in the figure indicating increased variability across different number of rules in execution times under the same conditions. Additionally, Fig. 4b illustrates that an increase in the number of rules leads to longer average execution times.

7 Related Work

A number of researchers have modeled access control systems as graphs. Specifically, Alves et al. [7] present a novel framework for analyzing access control policies using graph-based methods. It focuses on CBAC policies and demonstrates how these can be visually and formally represented and analyzed using graph theory. Berolissi et al. [8] introduce a graph-based language designed for

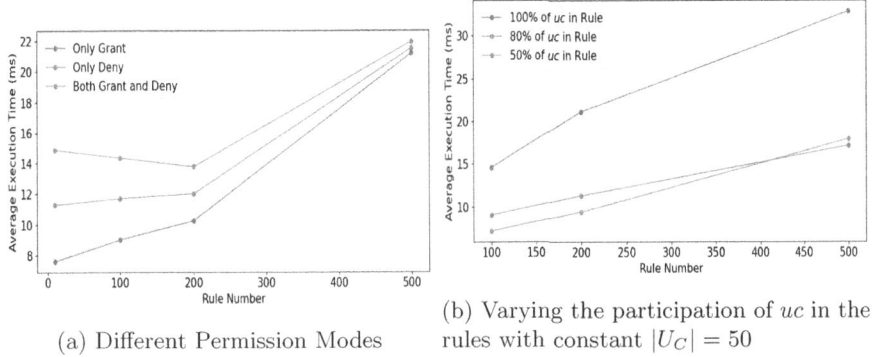

(a) Different Permission Modes

(b) Varying the participation of uc in the rules with constant $|U_C| = 50$

Fig. 3. Access Request Evaluation with varying access control outcomes and the % of attribute condition participation in rules

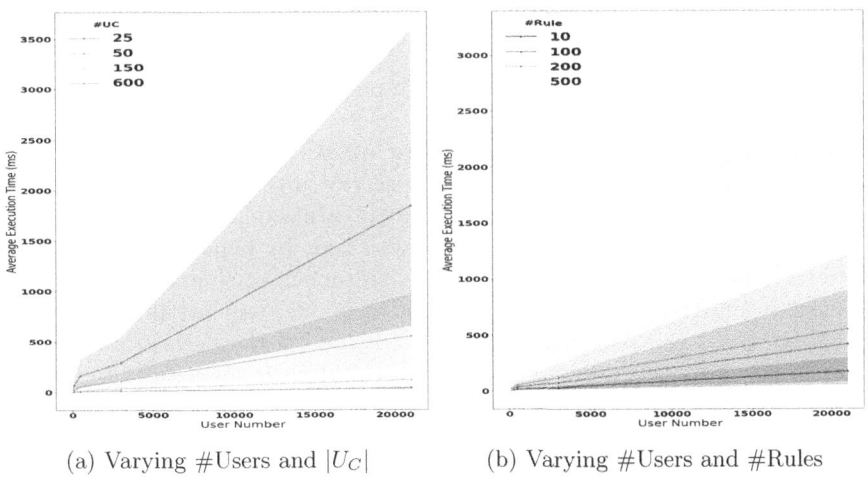

(a) Varying #Users and $|U_C|$

(b) Varying #Users and #Rules

Fig. 4. User-Centric Evaluation with varying the size of rules and attribute conditions

specifying administrative access control policies within the Admin-CBAC framework, an administrative model for Category-Based Access Control. While this is not catered to ABAC, Admin-CBAC's generic nature means these methodologies can be applied across different access control models, by providing specific applications for RBAC and ABAC. Rizvi et al. [16] present an AReBAC model for Neo4j that enhances fine-grained access control in graph databases by integrating access control policies with database queries. It introduces Nano-Cypher, a declarative policy language, and employs an innovative backtracking scheme, Live-End Backjumpint (LBJ), to optimize the performance of graph-database query evaluations. Similar to that of [8], this work does not consider ABAC. More recently, Zhang et al. [21] utilized graphs to represent access control policies by modeling them as edge-labelled directed graphs (digraphs), where vertices signify entities and edges denote access rights. This graphical approach simplifies the visualization and management of complex access relationships, particularly useful in environments like IoT. By employing graph-theoretic techniques such as strong homomorphism, the model abstracts detailed access matrices into simplified graphs, effectively summarizing essential permissions while reducing overall complexity. Again, this is not catered to ABAC.

Additional work that employs graphs include that of Nyanchama et al. [15], which proposes a role graph model, an approach to visualize and manage RBAC. It enhances role management by defining algorithms that manipulate role graphs to handle conflicts and manage access control dynamically and efficiently. The paper also focuses on conflict of interest within roles, detailing methods to prevent privilege-privilege and role-role conflicts by structuring role graphs to partition conflicting roles and privileges. Additionally, Mohamed et al. [13] explore access control mechanisms for graph databases by proposing the XACML4G framework, which extends the XACML to accommodate the characteristics of graph-structured data. This paper addresses the need for more granular and flexible access control by allowing constraints on vertices and edges, thus enabling more complex authorization policies.

There are several works that consider graph based representation for ABAC and some among them implement the graph in Neo4j. Nabil et al. [14] introduced a conceptual graph model for implementing ABAC in composite web services. The model provides a visual tool for web service providers to manage access control policies more effectively. For instance, [5] utilized graph-based structures to verify the completeness of NGAC [9] extractions by populating elements as nodes and relations as edges in Neo4j. In their model, each attribute condition is distinctly represented as a node, and for each rule, user attribute condition and object attribute condition are connected through the permission, forming a sequential chain. They verify the correctness of their modeling by checking the presence of at least one valid path in Neo4j. This graph modeling does not consider more than one attribute conditions in this chain. If it were to be accommodated as multiple paths, it would result in "OR" among the attribute conditions rather "AND", since the checking is only to find a valid path. Ahmadi et al. [6] introduce a graph-based model for evaluating ABAC policies, utilizing the Neo4j Database. Their modeling also is similar to that in [5], and therefore the same concerns are valid in this case. All of these attempts do not utilize Neo4j neither for access control evaluation nor offer any analysis.

8   Conclusions

In this paper, we have proposed a graph based framework that succinctly and accurately represents the ABAC policy as a flow graph. In particular, our graph model employs the directional flow capacities to assess access control, so that access request evaluation and other analyses can be carried out by traversing the potential pathways between a subject and an object. In addition, we demonstrate, using a graph database system called Neo4j, how access request evaluation and user and object centric analyses can be carried out simply through the execution of graph queries. Our framework thus enables any ABAC system to be readily deployed using a commercially available graph databases. Our experimental evaluation shows promising performance results our future work includes comparing it with that of Harmonia [2] and other implementations of ABAC using XACML [1] and Java. We plan to extend our graph based framework to model negative ABAC permissions, and explore other analyses including safety analysis [20], and identification of redundant [19] and conflicting policies [12].

Acknowledgments. This research was supported in part by the National Science Foundation award CNS-1747728, the National Institutes of Health award R35GM134927 and a grant from CISCO Research. The content is solely the responsibility of the authors and does not necessarily represent the official views of the agencies funding the research.

A   Python and Neo4j Code

A.1   Code for Access Request Evaluation

```
def access_check(driver, user_name, object_name, relationship_type):
    sanitized_user_label = sanitize_label(user_name)
    sanitized_object_label = sanitize_label(object_name)
    sanitized_relationship_type = sanitize_label(relationship_type)
    # Initialize the deny flag
    access_granted = False
    with driver.session(database=database_name) as session:
        # Query to find ANDUs connected to the user
        cypher_query_andu = (
            f"MATCH (u:{sanitized_user_label})-[r1]->(uc)-[r2]-(ANDU) "
            "WITH ANDU, count(r2) as inflowofANDU "
            "CALL apoc.neighbors.tohop.count(ANDU, 'AND_Relationship', 1) "
            "YIELD value "
            "WHERE inflowofANDU = value "
            "RETURN ANDU"
        )
        # Run the first query
        result_andu = session.run(cypher_query_andu).data()
        #print(result_andu)
        if not result_andu:
            return "Deny"
```

```
# Process each ANDU result
for andu in result_andu:
    andu_id = andu['ANDU']['id']
    #print("this is andu_id:", andu_id)
    #print("this is the andu in iteration:", andu)
    # Query to find ANDOs connected to ANDUs
    cypher_query_ando = (
        f"MATCH (ANDU)-[r:{sanitized_relationship_type}]-(ANDO) "
        "WHERE ANDU.id = $andu_id "
        "RETURN ANDO"
    )
    # Run the second query
    result_ando = session.run(cypher_query_ando, andu_id=andu_id).data()
    #print("this is result_ando:",result_ando)
    # Process each ANDO result
    for ando in result_ando:
        ando_id = ando['ANDO']['id']
        # Query to check access to the object
        cypher_query_o = (
            f"MATCH (ANDO)<-[r4]-(oc)<-[r3]-(o:{sanitized_object_label}) "
            "WHERE ANDO.id = $ando_id "
            "WITH o, ANDO, COUNT(r4) AS inflowCount "
            "CALL apoc.neighbors.tohop.count(ANDO, 'AND_Relationship', 1) YIELD value "
            "WHERE inflowCount = value "
            "RETURN COUNT(DISTINCT o)> 0 AS CanAccess"
        )
        # Run the third query
        connected_o = session.run(cypher_query_o,ando_id=ando_id,object_name=
        object_name).data()
        #print("this is connected_o",connected_o)
        # Check if CanAccess is True for any o
        if connected_o and connected_o[0]['CanAccess']:
            access_granted = True
            break  # No need to check other ANDOs if access is already granted
    if access_granted:
        break  # No need to check other ANDUs if access is already granted
return "Grant" if access_granted else "Deny"
```

A.2   Code for User-Centric Analysis

```
def find_accessible_objects(driver, user_name):
    sanitized_user_label = sanitize_label(user_name)
    with driver.session(database=database_name) as session:
        # Query to find ANDUs connected to the user
        cypher_query_andu = (
            f"MATCH (u:{sanitized_user_label})-[r1]->(uc)-[r2]-(ANDU) "
            "WITH ANDU, count(r2) as inflowofANDU "
            "CALL apoc.neighbors.tohop.count(ANDU, 'AND_Relationship', 1) "
            "YIELD value "
            "WHERE inflowofANDU = value "
            "RETURN ANDU"
        )
        result_andu = session.run(cypher_query_andu).data()
        #print("this is result_andu:", result_andu)
        if not result_andu:
            return 'No Object Can ' + user_name + ' Access'
        # Initialize a set to hold accessible objects
        accessible_objects = set()
        for andu in result_andu:
            andu_id = andu['ANDU']['id']
            # For each ANDU, find related ANDOs via the specified relationship type
            cypher_query_ando = (
                f"MATCH (ANDU)-[r]-(ANDO) "
                "WHERE ANDU.id = $andu_id "
                "RETURN ANDO"
            )
            result_ando = session.run(cypher_query_ando, andu_id=andu_id).data()
            # For each ANDO, check for accessible Objects
            for ando in result_ando:
                ando_id = ando['ANDO']['id']
                cypher_query_o = (
```

```
            f"MATCH␣p␣=␣(ANDU)-[r5]->(ANDO:ANDO)<-[r4]-(oc)<-[r3]-(o)␣"
            "WHERE␣ANDO.id␣=␣$ando_id␣"
            "WITH␣o,␣r5,␣ANDO␣,␣COUNT(r4)␣AS␣inflowCount␣"
            "CALL␣apoc.neighbors.tohop.count(ANDO,␣'AND_Relationship',␣1)␣YIELD␣value␣"
            "WHERE␣inflowCount␣=␣value␣"
            "RETURN␣labels(o),␣type(r5)"
        )
        # Run the third query
        connected_o = session.run(cypher_query_o, ando_id=ando_id).data()
        #print(connected_o)
        for obj in connected_o:
            object_names = obj['labels(o)'][0]
            permission_names = obj['type(r5)'][0]
            accessible_objects.add((object_names, permission_names))
    return accessible_objects
```

A.3   Code for Object-Centric Analysis

```
def find_accessible_users(driver, object_name):
    sanitized_object_label = sanitize_label(object_name)
    with driver.session(database=database_name) as session:
        # Query to find ANDUs connected to the user
        cypher_query_ando = (
            f"MATCH␣(o:{sanitized_object_label})-[r3]->(oc)-[r4]-(ANDO)␣"
            "WITH␣ANDO,␣count(r4)␣as␣inflowofANDO␣"
            "CALL␣apoc.neighbors.tohop.count(ANDO,␣'AND_Relationship',␣1)␣"
            "YIELD␣value␣"
            "WHERE␣inflowofANDO␣=␣value␣"
            "RETURN␣ANDO"
        )
        result_ando = session.run(cypher_query_ando).data()
        if not result_ando:
            return 'No␣Subject␣Can␣' + object_name + '␣Access'
        # Initialize a set to hold accessible objects
        accessible_users = set()
        for ando in result_ando:
            ando_id = ando['ANDO']['id']
            # For each ANDU, find related ANDOs via the specified relationship type
            cypher_query_andu = (
                f"MATCH␣(ANDU)-[r]-(ANDO)␣"
                "WHERE␣ANDO.id␣=␣$ando_id␣"
                "RETURN␣ANDU"
            )
            result_andu = session.run(cypher_query_andu, ando_id=ando_id).data()
            for andu in result_andu:
                andu_id = andu['ANDU']['id']
                cypher_query_u = (
                    "MATCH␣p␣=␣(ANDO)<-[r5]-(ANDU:ANDU)<-[r2]-(uc)<-[r1]-(u)␣"
                    "WHERE␣ANDU.id␣=␣$andu_id␣"
                    "WITH␣u,␣r5,␣ANDU␣,␣COUNT(r2)␣AS␣inflowCount␣"
                    "CALL␣apoc.neighbors.tohop.count(ANDU,␣'AND_Relationship',␣1)␣YIELD␣value␣"
                    "WHERE␣inflowCount␣=␣value␣"
                    "RETURN␣labels(u),type(r5)"
                )
                # Run the third query
                connected_u = session.run(cypher_query_u, andu_id=andu_id).data()
                for user in connected_u:
                    user_names = user['labels(u)'][0]
                    permission_names = user['type(r5)'][0]
                    accessible_users.add((user_names,permission_names))
        return accessible_users
```

References

1. The eXtensible Access Control Markup Language (XACML), Version 3.0, OASIS Standard, 22 January 2013. http://docs.oasis-open.org/xacml/3.0/xacml-3.0-core-spec-os-en.pdf
2. The Harmonia Open Source Software from NIST. https://github.com/PM-Master/Harmonia-1.6/releases
3. Top Ten Reasons for Choosing Neo4j. https://neo4j.com/top-ten-reasons/
4. Welcome to Neo4j. https://neo4j.com/docs/getting-started/
5. Abdelgawad, M., Ray, I., Alqurashi, S., Venkatesha, V., Shirazi, H.: Synthesizing and analyzing attribute-based access control model generated from natural language policy statements. In: Proceedings of the 28th ACM Symposium on Access Control Models and Technologies, pp. 91–98 (2023)
6. Ahmadi, H., Small, D.: Graph model implementation of attribute-based access control policies. CoRR abs/1909.09904 (2019)
7. Alves, S., Fernández, M.: A graph-based framework for the analysis of access control policies. Theoret. Comput. Sci. **685**, 3–22 (2017)
8. Bertolissi, C., Fernandez, M., Thuraisingham, B.: Graph-based specification of admin-CBAC policies. In: Proceedings of the Eleventh ACM Conference on Data and Application Security and Privacy, pp. 173–184 (2021)
9. Ferraiolo, D., Atluri, V., Gavrila, S.: The policy machine: a novel architecture and framework for access control policy specification and enforcement. J. Syst. Architect. **57**(4), 412–424 (2011)
10. Ferraiolo, D., Sandhu, R., Gavrila, S., Kuhn, D.R., Chandramouli, R.: Proposed NIST standard for role-based access control. ACM Trans. Inf. Syst. Secur. **4**(3), 224–274 (2001)
11. Hu, V., et al.: Guide to attribute based access control (ABAC) definition and considerations (2019)
12. Koch, M., Mancini, L.V., Parisi-Presicce, F.: Conflict detection and resolution in access control policy specifications. In: Foundations of Software Science and Computation Structures, pp. 223–238 (2002)
13. Mohamed, A., Auer, D., Hofer, D., Küng, J.: Extended authorization policy for graph-structured data. SN Comput. Sci. **2**, 351 (2021)
14. Nabil, D., Slimani, H., Nacer, H., Aissani, D., Bey, K.B.: ABAC conceptual graph model for composite web services. In: 2018 IEEE 5th International Congress on Information Science and Technology (CiSt), pp. 36–41 (2018)
15. Nyanchama, M., Osborn, S.: The role graph model and conflict of interest. ACM Trans. Inf. Syst. Secur. **2**(1), 3–33 (1999)
16. Rizvi, S.Z.R., Fong, P.W.L.: Efficient authorization of graph-database queries in an attribute-supporting ReBAC model. ACM Trans. Priv. Secur. **23**(4), 1–33 (2020)
17. Servos, D., Osborn, S.L.: Current research and open problems in attribute-based access control. ACM Comput. Surv. **49**(4), 1–45 (2017)
18. Talegaon, S., Batra, G., Atluri, V., Sural, S., Vaidya, J.: Contemporaneous update and enforcement of ABAC policies. In: Proceedings of the 27th ACM on Symposium on Access Control Models and Technologies (SACMAT), pp. 31–42 (2022)
19. Talukdar, T., Batra, G., Vaidya, J., Atluri, V., Sural, S.: Efficient bottom-up mining of attribute based access control policies. In: IEEE International Conference on Collaboration and Internet Computing, pp. 339–348 (2017)

20. Uzun, E., et al.: Analyzing temporal role based access control models. In: Proceedings of the 17th ACM Symposium on Access Control Models and Technologies, SACMAT 2012, pp. 177–186 (2012)
21. Zhang, S., Fong, P.W.L.: Mining domain-based policies. In: Proceedings of the 14th ACM Conference on Data and Application Security and Privacy (CODASPY) (2024)

Human Digital Twins: Efficient Privacy-Preserving Access Control Through Views Pre-materialisation

Giorgia Sirigu$^{(\boxtimes)}$ ⓘ, Barbara Carminati ⓘ, and Elena Ferrari ⓘ

University of Insubria, Varese, Italy
{gsirigu,barbara.carminati,elena.ferrari}@uninsubria.it

Abstract. Digital Twins (DTs) are virtual copies of physical entities, processes, or systems used for various tasks, such as controlling, monitoring, and analysing the status of the real entity. The DT sector is expected to surpass six billion U.S. dollars by 2025, with the Human Digital Twin (HDT) being a prime example. HDTs are being used in various applications, such as personalised medicine, healthcare, and education. However, the materialisation of HDTs can be costly and lead to delays in HDT-based services. To overcome this, we propose a strategy, HDT-ViewMat, to identify the portions of an HDT that should be pre-materialised, considering the trade-off between potential delays and resource waste. The proposed strategy analyses the process/workflow that requires HDT data to estimate the probability of its tasks being executed. Furthermore, due to the sensitivity of the data maintained by the HDTs, access to them must be limited to guarantee the users' privacy. This strategy also considers the compliance of privacy policies with users' preferences. HDT-ViewMat assesses the user's chance of executing a task in the workflow based on the probability of the task's invocation and the probability of the user accepting the policies of the corresponding service provider.

Keywords: Human Digital Twins · Privacy · Privacy Preferences · Policy Compliance

1 Introduction

A Digital Twin (DT) is a virtual copy of a physical entity, industrial process or system that can be used for different tasks, such as controlling, monitoring, and analysing the status of the real twin, as well as testing it before deploying updates and new patches. Thanks to these features, the DT sector's market size is expected to surpass six billion U.S. dollars by 2025.[1]

The evolution of DTs has extended their scope to humans, leading to the development of the Human Digital Twin (HDT) [18], which mimics the psychophysical aspects of an individual. Research projects are exploring the use of

---

[1] https://www.statista.com/statistics/1296187/global-digital-twin-market-by-industry/.

© IFIP International Federation for Information Processing 2024
Published by Springer Nature Switzerland AG 2024
A. L. Ferrara and R. Krishnan (Eds.): DBSec 2024, LNCS 14901, pp. 24–43, 2024.
https://doi.org/10.1007/978-3-031-65172-4_2

digital twin technology in healthcare to monitor disease status (e.g., oncologic patients [14]), develop personalised therapies (e.g., [2] and [20]), and treat and prevent illnesses (e.g., cardiovascular disease [6]).

Further applications adopt HDTs to optimise training strategies (e.g., [4] and [12]), enhance user experiences, create immersive environments, and improve virtual interactions in the metaverse (e.g., [3]). They are also used to personalise learning experiences and optimise teaching methods for better student outcomes, such as in sports management and healthcare (e.g., [21]).

One of the challenges of using HDT is its materialisation [15]. These operations require time as they may involve heterogeneous data acquisition, parsing, and processing to make the data functional for the data consumer and, if needed, creating a 3D view (e.g., the generation of a Cardiac Digital Twin might take four hours [10]). Generally, data materialisation leads to delays in HDT-based services. Considering the healthcare domain, this delay might prevent timely interventions in the case of a need for prompt response. A naive solution to this issue is to pre-materialise the whole HDT in advance, that is, before launching the HDT-based service. However, this has the drawback of potentially wasting resources. In this paper, we aim to define a strategy for identifying the portions of an HDT, *views*, that should be pre-materialised, taking into consideration the trade-off between potential delays and resource waste.

The strategy to identify pre-materialised views involves analysing the process requiring HDT data, which may include multiple tasks organised in a complex workflow. Each task might require a distinct view of the HDT, and determining which tasks are more likely to be executed is crucial.

Furthermore, sensitive data, like the one in HDT that contains extremely private information about individuals, must be protected in accordance with current laws and standards (e.g., GDPR, HIPAA, and CCPA). A fundamental principle is that service providers must declare their privacy policies to inform users of how they will collect, manage, and disclose their personal data. At the same time, users can decide to accept or refuse the policy based on their preferences and thus obtain or not the service. In identifying the views to be pre-materialised, we also consider the compliance of privacy policies with users' preferences to avoid pre-materialising views that users do not want to release. This scenario is further complicated if we consider that a complex process typically involves multiple providers (e.g., for each different task) with different privacy policies.

To cope with these challenges, in this paper, we propose a strategy, called *HDT-ViewMat*, to determine, given a workflow, which views have to be pre-materialised. To this end, HDT-ViewMat assesses the user's chance of executing a task in the workflow, named *Execution Chance (EC)*: the higher the value is, the more likely the task will be executed. The EC of a task is defined by considering the probability of the task's invocation based on the position within the workflow and the probability of the user accepting the policies of the corresponding service provider.

Recently, controlled HDT data sharing has been investigated. Among the solutions, [5] proposes a Tag-Based Access Control mechanism, where accesses are controlled by tags attached to DT data. Moreover, different projects adopt blockchain to provide secure data sharing for DTs (e.g., [8] and [16]). Notable, [1] proposes a blockchain-based solution for HDTs to model individual patients' conditions and generate instances accessible to care providers. However, in all the above-mentioned proposals, user's privacy preferences are not considered to drive controlled data sharing.

This paper is organised as follows: Sect. 2 illustrates the proposed architecture; Sect. 3 defines the concepts of Execution Chance; and Sect. 4 illustrates the functionalities of HDT-ViewMat. Section 5 shows the experiments we run to test the proposed solution. Finally, Sect. 6 concludes the paper.

2 HDT-ViewMat Overview

Before introducing the strategy to pre-materialise HDT views, let us briefly introduce this technology. HDTs are used to represent the physical person in virtual environments. More specifically, an HDT models human physiological and psychological by gathering data from different sources: sensing devices (e.g., wearable and biomedical devices), optical technologies (e.g., cameras), implantable biomedical sensors, electronic health records (e.g., computed tomography scans, X-rays, medical histories, and diagnoses), as well as social networks (e.g., posts on feelings and emotions). The virtual replica simulates the physical counterpart in real-time through constant data synchronisation. It analyses the physical twin data to validate, optimise, evaluate, diagnose, give suggestions, and make predictions and decisions [18]. Hence, the HDT includes technologies to model and virtualise the digital model (e.g., digitisation technologies and physical geometries), store (e.g., databases and data management frameworks), compute data (e.g., edge-cloud computing), analyse the information, and make decisions (e.g., AI techniques).

In this paper, we consider a scenario where a user undergoes processes that require consuming portions of their HDT's data to provide personalised services. More precisely, we assume the process is modelled through a workflow consisting of different tasks, where each task corresponds to a specific decision-making step that is executed based on the available data. Moreover, each task in the workflow could be implemented by a different service provider. To be compliant with privacy laws, we assume that users define their preferences on personal data management via *privacy preferences*, as well as that the service providers encode their data management procedures via *privacy policies* (see Sect. 3.1 for their definitions). We further suppose that the workflow is encoded via standard process language (e.g., BPEL[2]) to be executed by a workflow engine. As depicted in Fig. 1, this engine receives a request for workflow execution from the user, complemented by the privacy preferences. Before the engine initiates the execution of the workflow, the proposed architecture pre-materialises views of HDTs of the requester user. To achieve this aim, we introduce the *Execution Chance (EC)* measure to estimate the likelihood that a task will be executed.

---

[2] https://docs.oasis-open.org/wsbpel/2.0/wsbpel-v2.0.pdf.

As described in Sect. 3, this estimation takes into account both the position of the task inside the workflow as well as the likelihood that the provider's privacy policy is compliant with the user's preferences. The computation of EC values is delegated to the EC evaluator component (see Fig. 1, steps 2–3). Based on the returned values, the workflow engine decides which views have to be pre-materialised, that is, those associated with tasks whose EC is greater than a given threshold.

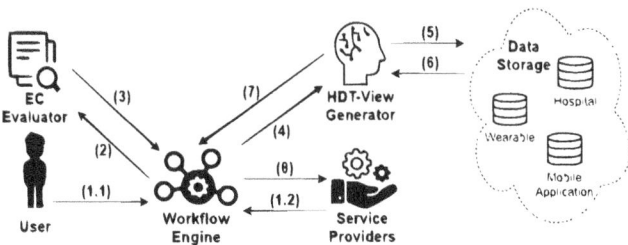

Fig. 1. Overall architecture

A reference to these tasks is then passed to the HDT-View generator (cfr. Fig. 1 steps 4–7). During the workflow execution, the engine requires updates to the EC measures to dynamically decide which views further have to be pre-materialised based on their updated probabilities of being used. In the rest of the paper, we will focus on the workflow engine; before that, we introduce the Execution Chance measure as a novel metric for the selection of views to pre-materialise.

To support the explanation of HDT-ViewMat, in the following, we refer to the sanitary protocol defined by the European Union Standards for Tuberculosis Care (ESTC)[3] as a running example.

Figure 2 shows the graphical representation of this case study, where nodes represent tasks, and edges represent the passages between tasks. For instance, the first task is *First Check (FC)* followed by an initial *Microscopic and Molecular Examination (MME)* to identify tuberculosis and drug resistance. Nodes with a bold border are final - i.e., tasks where the protocol execution could end. For the purpose of this paper, we do not consider loops, assuming that if users agree to share their data with a provider for a task once, they also agree to share them at the next execution. We further assume that workflow is complemented by the probabilities of passing from one task to another, which are represented as labels on edges. In realistic scenarios, probabilities can be obtained based on heuristics (e.g., symptom history, examination results, and therapy histories of previous patients).

---

[3] https://www.ecdc.europa.eu/en/all-topics-z/tuberculosis/prevention-and-control/european-union-standards-tuberculosis-care.

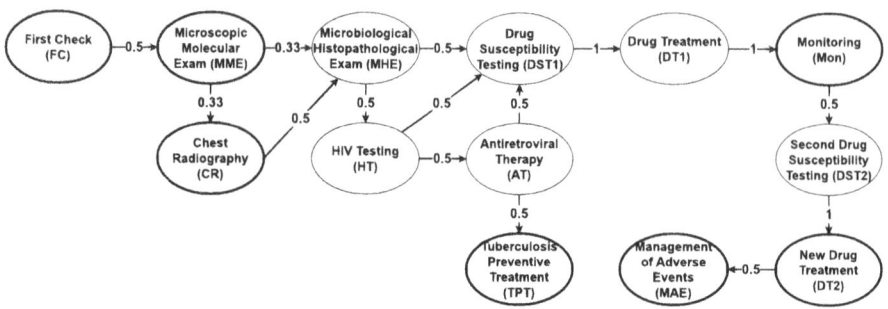

Fig. 2. Tuberculosis process.

3   Execution Chance Assessment

To effectively pre-materialise views, HDT-ViewMat utilises the Execution Chance metric, which provides an estimation of the likelihood of executing a task inside the workflow. This measure is specifically developed to consider the likelihood that the user making the request accepts the privacy policy connected with the task (see Sect. 3.1) and the task's position in the workflow (see Sect. 3.2).

3.1   Privacy Level Agreement

According to current laws and norms, users have to be notified and provide informed consent before the collection and use of their personal data. This consent can be automatically checked if the preferences of the user regarding their personal data management, as well as the privacy policy of a provider, are encoded in machine-readable format. That is, if users state a set of conditions under which the usage of their data is allowed (e.g., the purpose of data collection, third-party sharing), it is possible to automatically verify if the provider's privacy policy is compliant with them. Given a task, determining if the associated policy is compliant with the user's preference is relevant for view materialisation, aka EC estimation, as we can assume that the user will consent to the task execution (e.g., the data collection for task execution) if the policy is compliant. Nevertheless, in practical situations, users may deviate slightly from their initial preferences depending on the objective of the task or workflow they are carrying out. Therefore, the presence of a non-compliant policy does not automatically indicate that the user will not consent to using their data. However, we can observe that the more the policy is far from satisfying the user's preferences, the smaller the likelihood that the user will give consent to data collection. Thus, to define EC taking into account the likelihood that the user accepts the privacy policy connected with the task, we introduce a *measure of compliance* between the privacy policies associated with a task and the user's privacy preferences, called *Privacy Level Agreement (PLA)*. Before introducing PLA, we formally define the privacy policy and preferences. We adopt a definition based on data practice [22], where a privacy policy can be defined as a tuple

$pol = \langle dt, prp, tp, rt \rangle$, stating that the provider will collect the data dt, for the purposes prp, that it could share the collected data with a set of third parties, tp, and that the time the collected data will be stored at provider servers is at maximum rt days. Privacy preferences have a similar structure as they define constraints/preferences regarding providers' data collection procedures. Specifically, a preference $pp = \langle dt, prp, tp, rt \rangle$ indicates that the user allows collecting data dt, for certain purposes prp, for a limited period of time rt, and that the data could be shared with certain third parties tp. We assume that data and purpose fields are organised in taxonomies, representing their semantic relationships and hierarchies (see Figs. 3a–3b as examples). In contrast, the third-party component is a set, and the retention component is an integer expressing the number of days.

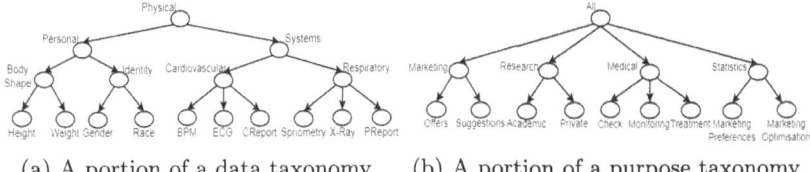

(a) A portion of a data taxonomy   (b) A portion of a purpose taxonomy

Fig. 3. Samples of data and purposes taxonomies

Example 1. Let us consider the example in Fig. 2. Suppose that the specialist S performing FC task requires to collect cardiovascular reports for private research purposes, which will be kept at maximum for one year, and that the collected data could be shared with external hospitals and private clinics. The privacy policy modelling this data management is defined as $pol_S = \langle CReport, Private, \{Hospitals, PrivClinics\}, 365 \rangle$. In contrast, suppose that a patient, say P, agrees to share the cardiovascular reports only for academic research, with a retention period of at most nine months and prefers sharing this data only with external hospitals. These requirements can be modelled as $pp_P = \langle CReport, Academic, \{Hospitals\}, 270 \rangle$.

Given a privacy policy and preference, the PLA between them is defined based on a set of similarity functions, each one comparing the corresponding components (i.e., dt, prp, tp, and rt), as described in the following:

– dt, prp - since data and purpose components are modelled as taxonomies, we can apply the Wu and Palmer Similarity Measure [23]. This similarity measure computes the distance between two concepts (i.e., nodes) in a taxonomy based on the number of arcs separating them. More precisely, given two nodes n_1 and n_2 in a taxonomy, $WPSim(n_1, n_2) = \frac{2*ra}{nr_1 + nr_2}$, where ra is the number of nodes between the root and the closest common ancestor of n_1 and n_2; and nr_1 (nr_2 resp.) is the number of nodes between n_1 (n_2 resp.) and the root;

- tp - since these are defined as sets, we use the Jaccard Similarity that, given two sets tp_1 and tp_2, evaluates their similarity as $JacSim(tp_1, tp_2) = \frac{|tp_1 \cap tp_2|}{|tp_1 \cup tp_2|}$;
- rt - the similarity between these components is computed through the normalised Euclidean Distance. Thus, given two retention times rt_1 and rt_2, $EucSim(rt_1, rt_2) = 1 - \frac{|rt_1 - rt_2|}{max(rt_1, rt_2)}$.

We opted for these similarity functions as they are easy to implement and are commonly used. However, alternatives could be adopted without compromising the overall logic of the system. The Privacy Level Agreement between a privacy policy and a privacy preference is formally defined as follows.

Definition 1 (Privacy Level Agreement - PLA). *Given a privacy policy pol and a privacy preference pp, we define their Privacy Level Agreement as a weighted mean:*[4]

$$PLA(pol, pp) = WPSim(pol.dt, pp.dt) * \omega_{dt} + WPSim(pol.prp, pp.prp) * \omega_{prp}$$
$$+ JacSim(pol.tp, pp.tp) * \omega_{tp} + EucSim(pol.rt, pp.rt) * \omega_{rt}$$

The user-defined weights ω_{dt}, ω_{prp}, ω_{tp}, and ω_{rt} represent the relative importance of each privacy component in accordance with users perspectives. Weights assume a value in the $[0, 1]$ range, and their sum must equal 1.

Example 2. Let us consider Example 1. Both policy and privacy preferences specify the same data type $pol_S.dt = pp_P.dt = CReport$ and their similarity is $WPSim(pol_S.dt, pp_P.dt) = 1$. Then, having $pol_S.prp = Private$ and $pp_P.prp = Academic$, $WPSim(pol_S.prp, pp_P.prp) = \frac{2*1}{2+2} = 0.5$ (cfr. Fig. 3b). The third-party component of the policy is $\{Hospitals, PrivClinics\}$, whereas the one of the privacy preferences is $\{Hospitals\}$. Therefore, $JacSim(pol_S.tp, pp_P.tp) = \frac{1}{2+1-1} = 0.5$. Finally, $EucSim(pol_S.rt, pp_P.rt) = 1 - \frac{95}{365} = 0.74$. Suppose that the user assigned a higher weight to the data and purpose component (e.g., 0.3) and a lower weight to the third party and retention period ones (e.g., 0.2). Therefore, $PLA(pol_S, pp_P) = 1 * 0.3 + 0.5 * 0.3 + 0.5 * 0.2 + 0.74 * 0.2 = 0.70$

PLA measures the compliance between one policy and one preference. However, in our scenario, we are interested in measuring PLA between the set of policies and preferences applied to the data included in the view to be materialised. In particular, a task t_k executed by a provider S could require a set of data types, denoted $view(t_k)$. We assume that for each $dt \in view(t_k)$, the provider S has specified a policy on how the data type will be handled. On the other hand, the user might not have defined a preference for each $dt \in view(t_k)$. In case there is no preference for a dt, for the PLA computation, we consider the preference defined for the data type most closely related to dt, that is, the preference defined by the user for a data type having the minimum Wu and Palmer distance with dt. Let us consider, as an example, a $view(t_k)$ containing the data type *Identity* in Fig. 3a, and assume that the user has defined privacy

---

[4] We refer to single components of a privacy policy/preference via the dot notation.

preferences for data *Body Shape* and *Gender*. For PLA computation, we consider the *Gender* data type as its Wu and Palmer similarity (i.e., 0.8) is greater than *Body Shape*'s distance (i.e., 0.5).

Thus, given a task, we compute a set of PLA values, one for each data type $dt \in view(t_k)$. The likelihood that a user will accept the policies linked to t_k can be approximated by calculating the average of their PLA values, obtaining a single value, the *Overall PLA (OPLA)* of t_k.

3.2 Probability of Task Invocation

To estimate the probability of invoking a task, we leverage the Discrete-Time Markov Chain Probabilistic model (DTMC) [13]. A DTMC is a model that describes the behaviour of a system that transits from one state to another with a specific probability. In particular, DTMC considers a discrete set of states $S = \{s_1, ..., s_k\}$, where the probability p_{ij} of transiting from a state s_i to state s_j does not depend on the previous states in the system. With reference to our scenario, the workflow's tasks represent DTMC states, whereas their probabilities (i.e., labels on workflow edges) indicate the transition probabilities between two consequent states. For instance, considering Fig. 2, the transition probability from *MME* to *MHE* is $p_{MME,MHE} = 0.33$.

In general, to estimate the probability of invoking task t_k, we need to estimate the probability of multiple transitions/steps corresponding to the path to reach t_k. In DTMC, we define the probability of transitioning from state s_i to s_j in n steps as $p_{i,j}^{(n)} = Pr(X_{m+n} = s_j | X_m = s_i)$, where X_i is a random variable representing a state in S at time step i, and m represents the time step at which the transition began. To apply the n-steps transitioning probability in our scenario, we have to consider that a task t_k could be reached through different paths. Indeed, on the basis of input data, a workflow may generate different *execution paths*, that is, different sequences of tasks that are executed sequentially during the workflow completion. Thus, given a task t_k, we denote with $Paths(t_k)$ all the execution paths of the workflow that include t_k, that is, paths that during workflow execution require to invoke t_k. Moreover, we denote with $sbPths(t_k)$ the portion of execution paths, named subpaths, starting from the initial task t_1 and ending with the task preceding t_k. As an example, considering the task *MHE* in Fig. 2, the $sbPths(MHE)$ contains $path1 = FC \rightarrow MME$ and $path2 = FC \rightarrow MME \rightarrow CR$.

Finally, we compute the DTMC probability of transitioning between states over multiple steps by applying the Chapman-Kolmogorov equation and considering all subpaths in $sbPths(t_k)$. This equation, which is widely used in research (e.g., process management in cloud computing [7]), computes the probability as follows: $p_{i,j}^{(m+n)} = \sum_{k \in S} p_{i,k}^{(m)} * p_{k,j}^{(n)}$, where m is the number of steps already taken and n is the number of future steps from state s_i to s_j. Thus, given a task t_k, we estimate the probability of its invocation, denoted as $prob_{t_k}$, by computing the Chapman-Kolmogorov equation, summarising the probabilities of invoking t_k in all subpaths. Let us consider Fig. 2 again, the Chapman-Kolmogorov formula for probability of invoking MHE from FC computes $p_{FC,MHE}^{(2)} = 0.5 * 0.33 = 0.165$ and $p_{FC,MHE}^{(3)} = 0.5 * 0.33 * 0.5 = 0.0825$ for $path1$ and $path2$, respectively. Hence, $p_{MHE} = 0.2475$.

3.3   Execution Chance

The Execution Chance of a task t_k is determined by two primary factors: (1) the likelihood of the task being executed ($prob_{t_k}$, cfr. Sect. 3.2); and (2) the likelihood of the user accepting the policies of the service provider associated with the task t_k ($OPLA(t_k)$, cfr. Sect. 3.1).

Regarding the first component, we have to consider that the event "t_k execution" is possible only if the user has consented to the privacy policies of each previous task in the execution path. Therefore, in addition to the likelihood of t_k invocation, $prob_{t_k}$, we have to consider a further element (3): the probability that the user accepts policies for all preceding tasks. This can be estimated similarly to (2), that is, by employing OPLA values assigned to previous tasks. More precisely, given a subpath $sbP \in sbPths(t_k)$ connecting the initial task t_1 to t_k, the higher the OPLAs values assigned to each task in sbP are, the higher the probability that the user will accept policies for all preceding tasks. As such, the OPLA assigned to a single subpath sbP can be computed as the average of OPLA values of each task in sbP. More in general, the OPLA value assigned to a set of subpaths $sbPths(t_k)$, denoted as $prevOPLA(t_k)$, is computed as the average of OPLA assigned to each single subpath $sbP \in sbPths(t_k)$.

$$prevOPLA(t_k) = \frac{1}{|sbPths(t_k)|} \sum_{sbP \in sbPths(t_k)} \left(\frac{1}{|sbP|} \sum_{t_x \in sbP} OPLA(t_x) \right)$$

Before presenting the Execution Chance definition, we introduce a further element that affects the second component (2), that is, the likelihood of the user accepting the policies of the service provider associated with the task t_k, $OPLA(t_k)$. Indeed, it is crucial to acknowledge that users may choose to deviate from their usual privacy settings in specific situations. In particular, this situation may arise when they are approaching the end of a process, and the violation of a privacy preference relates to the last tasks of that workflow. This could potentially have an impact on the Execution Chance of t_k. To this end, we introduce a weight α to adjust the $OPLA(t_k)$ value. In particular, if task t_k is near the end of the process, the user could be more likely to relax their preferences and accept the policies because they have already executed most of the tasks. In this case, the weight should decrease the value of $OPLA(t_k)$. As such, we defined α based on the distance of t_k to the end of an execution path in $Paths(t_k)$. In particular, given a $path$ in $Paths(t_k)$, α is defined as relative position of t_k with respect to $path$, that is, $\frac{pos(path, t_k)}{|path|+1}$, where $pos(path, t_k)$ returns the position of t_k in $path$. Considering that several execution paths, i.e., $Paths(t_k)$, could pass via t_k, we can compute distinct α value for each path in $Paths(t_k)$ and then compute their average as follows.

$$\alpha_{t_k} = \frac{1}{|Paths(t_k)|} \sum_{path \in Paths(t_k)} \frac{pos(path, t_k)}{|path| + 1}$$

We can define the concept of Execution Chance.

Definition 2. (Execution Chance - EC). *Given a task t_k in a workflow, its Execution Chance (EC) is defined as follows:*

$$EC(t_k) = prevOPLA(t_k) * \left(OPLA(t_k) + \alpha_{t_k} * \left(1 - OPLA(t_k)\right)\right) * prob_{t_k}$$

Example 3. Considering the example in Fig. 2, let us assume to evaluate the EC for TPT task, supposing $OPLA = 0.9$ for each task. According to Fig. 2,

$$Path(TPT) = \{FC \rightarrow MME \rightarrow MHE \rightarrow HT \rightarrow AT,$$
$$FC \rightarrow MME \rightarrow CR \rightarrow MHE \rightarrow HT \rightarrow AT\}$$

$$\alpha_{TPT} = \frac{1}{2} * \left(\frac{6}{7} + \frac{7}{8}\right) = 0.87$$

$$prevOPLA(TPT) = \frac{1}{2}\left[\frac{1}{5}\left(OPLA(FC) + OPLA(MME) + OPLA(MHE)\right.\right.$$
$$+ OPLA(HT) + OPLA(AT)) + \frac{1}{6}\left(OPLA(FC) + OPLA(MME)\right.$$
$$\left.\left. + OPLA(CR) + OPLA(MHE) + OPLA(HT) + OPLA(AT)\right)\right]$$

$$prob_{TPT} = (0.5 * 0.33 * 0.5 * 0.5 * 0.5) + (0.5 * 0.33 * 0.5 * 0.5 * 0.5 * 0.5) = 0.03$$

$$EC(TPT) = prevOPLA(TPT)\left[OPLA(TPT) + \alpha_{TPT}\left(1 - OPLA(TPT)\right)\right]prob_{TPT}$$
$$= 0.9 * \left(0.9 + 0.87 * (1 - 0.9)\right) * 0.03 = 0.03$$

4   Workflow Engine

In general, a workflow engine aims at properly invoking tasks of the workflow based on the execution order specified in the workflow definition. In HDT-Views, the workflow engine has been modified to add a preliminary computation of EC values, as well as the continuous update of these values, to determine the views to be pre-materialised, that is, those associated with a task with EC greater than a threshold.

ï¿£

The core functionalities of the engine are represented in Algorithm 1, which receives as input the BPEL representing the workflow (*bpel*), the threshold value (*th*), the policies (*Pols*) specified by involved service providers and the privacy preferences of the user requesting the workflow execution (*Pps*).

The algorithm starts by performing an initial EC evaluation. This is done via the *ECEvaluation*() function run by the EC evaluator component (see Fig. 1). The function takes as input *Pols*, *Pps*, *bpel*, and t_k, a reference to a task in *bpel* (e.g., task ID). For the initial execution of *ECEvaluation*(), t_k corresponds to

---

Algorithm 1 EngineExecution

Require: *Pols*, privacy policies of involved providers; *Pps*, privacy preferences of requesting user; *th*, a threshold; *bpel*, the BPEL document encoding the workflow.

1: $enbpel \leftarrow ECEvaluation(Pols, Pps, bpel, t_1)$
2: $toMaterialise = \{t \in enbpel | EC(t) \geq th\}$
3: $HDT\text{-}VGenerate(toMaterialise)$
4: $materialised = toMaterialise$
5: Wait invocation t_1
6: Invoke t_1
7: $currentTask = t_1$
8: **while** processIsRunning **do**
9:    **if** $currentTask \neq t_1$ **then**
10:        $enbpel \leftarrow ECEvaluation(Pols, Pps, bpel, currentTask)$
11:        $UpdateMaterialise = \{t \in enbpel | EC(t) \geq th\}$
12:        $toMaterialise = UpdateMaterialise - materialised$
13:        $HDT\text{-}VGenerate(toMaterialise)$
14:        $materialised = materialised \bigcup toMaterialise$
15:    **end if**
16:    Wait termination $currentTask$
17:    Let $newTask$ be the task to be invoked after $currentTask$ in $bpel$
18:    Wait invocation $newTask$
19:    $currentTask = newTask$
20: **end while**

---

the starting task t_1 of the workflow. The function returns an enhanced BPEL version, *enbpel*, where all tasks reachable from t_k, t_k included, are complemented with the corresponding EC and OPLA values (Line 1). We defined a task t_r reachable from t_k if it belongs to at least one path in $Paths(t_k)$ and it is after t_k. Note that the execution of $ECEvaluation()$ with t_1 implies the computation of EC and OPLA values for each task in *enbpel*.

From *enbpel*, the algorithm extracts the set of tasks (*toMaterialise*) with an EC greater than (or equal to) threshold *th* (Line 2). The views corresponding to these tasks have to be pre-materialised and thus are passed to the HDT-View generator (Line 3). The variable *materialised* stores the references to tasks whose views have been materialised (Line 4). The engine then waits for the request to invoke the first task, t_1, and invokes it as a consequence (Lines 5 and 6). Until the engine reaches the end of the workflow, it continuously waits for the request for the invocation of a new task, which arrives upon the completion of the current task (While loop in Line 8). Once it arrives, the algorithm updates the EC and OPLA values by re-executing the $ECEvaluation()$ function with reference to the current task (Line 10). It determines the set of views to be materialised, that is, those associated with tasks with EC greater (or equal) than the threshold and that have not yet been materialised (Lines 11 and 12). These are passed to HDT-View Generator (Line 13), and the variable *materialised* is updated accordingly (Line 14).

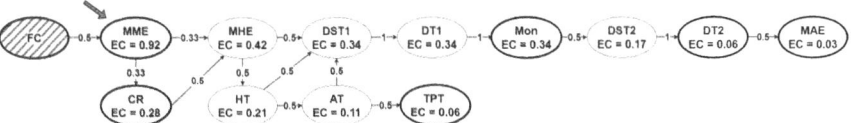

(a) Initial evaluation

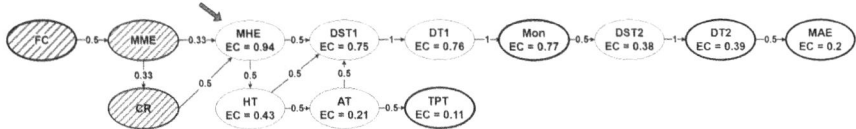

(b) Dynamic evaluation when the user is executing task MME

(c) Dynamic evaluation when the user is executing task MHE

Fig. 4. Graphical representation of EC evaluation of Tuberculosis example.

Example 4. Consider the workflow represented in Fig. 2, and assume that the OPLA of each task is 0.9. The results of the initial EC computation performed by *ECEvaluation*() (Line 1 of Algorithm 1) are depicted in Fig. 4a. For simplicity, we replaced the names of tasks with their acronyms and reported EC values inside nodes. For instance, task MME has $EC(MME) = 0.42$. We can observe that this first EC evaluation generates smaller values for tasks at the end of the path (e.g., MAE and TPT).

Let us assume that task MME is invoked. According to Algorithm 1, EC values are updated. Figure 4b shows the ECs after their new assessment. We can observe for tasks following MME greater EC values than those generated by the initial evaluation (Fig. 4a). For instance, initially, HT has $EC(HT) = 0.11$, while the new computed value is $EC(HT) = 0.21$. Suppose now task MHE is invoked. The Algorithm 1 updates EC values (Fig. 4c), where values of tasks following MHE further increase again compared to the previous evaluation. For instance, task HT passes from $EC(HT) = 0.21$ to $EC(HT) = 0.43$.

Consider the EC evaluation depicted in Fig. 4 and a threshold $th = 0.2$. After executing the initial evaluation, only the HDT-Views of three tasks are pre-materialised: FC, MME and MHE. When MME is invoked, according to the new EC values also the views of CR, HT, $DST1$, $DT1$ and Mon tasks are pre-materialised. Then, once MHE is invoked, further EC values reach th: AT, $DST2$, $DT2$ and MAE. This ensures that once the user reaches these tasks, they do not have to wait for the pre-materialisation.

5   Experimental Evaluation

This section presents a series of tests to assess the efficacy of HDT-ViewMat. The evaluation is conducted using two datasets: real-world processes generated considering the healthcare scenario and a benchmark of BPEL processes. In the following, we present results on the BPEL benchmark, whereas results on the healthcare scenario are presented in Appendix A.2.

5.1   BPEL Benchmark Dataset

We considered the real-world BPEL processes benchmark from [19] by filtering out those processes that represent a unique sequence of task invocations since they are not relevant for testing the pre-materialisation strategy as all tasks will be most likely executed. The final set consists of 27 BPEL processes that we then organised in groups based on the number of execution paths and, thus, the number of possible views (aka tasks in the path) to be materialised. Features of resulting clusters are represented in Table 1, where for each *Cluster*, we present the number of its *#Processes*, the number of execution *#Paths* of these processes (i.e., the minimum and maximum number of paths among all processes), and the number of *#Tasks* in paths (i.e., the minimum and maximum number of tasks among all paths). For instance, cluster C7 contains three processes, each with a number of paths in the interval 5–6 and tasks between 6–13.

Table 1. Clusters description of the benchmark dataset

Cluster	#Paths	#Tasks	#Processes	Cluster	#Paths	#Tasks	#Processes
C1	2	3-4	4	C6	4	6-13	3
C2	2	5-10	3	C7	5-6	6-13	3
C3	3	3-5	3	C8	10-12	6-12	3
C4	3	7	2	C9	44	110	1
C5	4	2-5	4	C10	45	44	1

Each process in the dataset is further complemented with privacy preferences and policies to compute OPLA for each task. In doing that, we acknowledge that different users might have different attitudes with respect to privacy and thus generate privacy preferences that are more or less stringent. This impacts the OPLA values. Indeed, given a task and corresponding privacy policy, a more privacy-conscious user might have preferences with more stringent constraints and, thus, an OPLA value smaller than those obtained when comparing the privacy preferences of users with fewer constraints. In particular, we used [17] to model four different types of users according to their privacy concerns: Unconcerned (*Type1*), Circumspect (*Type2*), Wary (*Type3*) and Alarmed (*Type4*). We assume that each user type corresponds to a different range of OPLA values.

As described [17], users of Type1 register to any website, and, most of the time, they share accurate information (i.e., $OPLA = [0.83, 1]$); Type2 users pay more attention but are still permissive (i.e., $OPLA = [0.50, 0.82]$); Type3 users pay attention to registering on any website and avoid sharing accurate personal information (i.e., $OPLA = [0.18, 0.49]$); Type4 users register to a website only if it is strictly necessary (i.e., $OPLA = [0, 0.17]$).

For testing purposes, we simulate that each process of the benchmark is executed four times, one per user type. At each simulation, all tasks are assigned an OPLA value randomly taken from the range corresponding to the considered user type.

5.2   Experiment Settings and Results

To evaluate the benefits of the proposed strategy for each process p in the dataset, aka its workflow, we compute its execution paths, $ExPaths(p)$, and simulate their executions. More precisely, given a process p and an execution path $ep \in ExPaths(p)$, its execution is simulated using Algorithm 1. Firstly, EC values are calculated for each task in p, and a first set of views is pre-materialised. Next, the path is traversed, which means that for each task in ep, EC values are updated, and new views are pre-materialised. As described in Algorithm 1, the simulation requires a threshold th. To use a value that is not too stringent or too relaxed, for each process p, we set a distinct th value. This is defined as the mean of EC values of p's tasks, where the task's $OPLA$ is set to 1.

To estimate the effectiveness of the strategy, we are interested in measuring the percentage of views that have been pre-materialised and actually used within ep execution (i.e., views associated with tasks in ep) as well as the percentage of views that have been pre-materialised but not used, aka wasted, during ep execution (i.e., views associated with tasks not in ep). Let us assume an execution path $ep \in ExPaths(p)$, where $|ep|$ is the number of tasks in ep, and $premat(ep)$ is the number of pre-materialised views by Algorithm 1 for tasks in ep. We compute the percentage of the pre-materialised views actually used in ep as $\%prematEP(ep) = \frac{premat(ep)}{|ep|}$. Similarly, we define the percentage of the pre-materialised views wasted in ep as $\%wastedEP(ep) = \frac{wasted(ep)}{|ep|}$, where $wasted(ep)$ is the number of views pre-materialised by Algorithm 1 and not used by tasks in ep. More in general, once all paths in $ExPaths(p)$ are simulated, we define the *average of the employed views*, *Emp*, as the average of $\%prematEP(ep)$, for each $ep \in ExPaths(p)$. Similarly, we define the *average of the wasted views*, *Wst*, as the average of $\%wastedEP(ep)$, for each $ep \in ExPaths(p)$.

The simulation platform is deployed in Python 3.10+ and is available on GitHub.[5]

Figure 5 presents the results of evaluation on the BPEL benchmark dataset by varying the user types and showing each BPEL process's percentage of used (Emp) and wasted (Wst) pre-materialised views.

---

[5] https://github.com/GiorgiaS/hdt-viewmat.git.

Overall, as reported in Fig. 5, the percentage of pre-materialised views is affected by OPLA values (aka user types): both Emp and Wst decline when OPLA decreases (i.e., Type4). Emp and Wst do not show linear results between clusters. For instance, clusters C1 and C2 do not have any wasted views. In cluster C3, Wst is 30% and 28% for Type1 and Type2 users. Then, this value returns null in C4. Also, Emp does not have linear values from C1 to C10. These anomalies are due to the BPEL process structures and not only to the number of paths and tasks. Indeed, by manually inspecting the results, we noticed that BPEL processes where tasks appear over multiple paths have a higher Emp than those with numerous paths with few tasks in common. For instance, C9 is the cluster with the highest Wst rate. This is due to the structure of its unique process, which is characterised by numerous short paths with few tasks in common. Despite cluster C9, the other results demonstrate that the rate of Emp is always higher than Wst. Also, the difference is usually high - e.g., 94% Emp and 30% Wst for C3. In most cases, the Wst even equals 0% for all user types, as for clusters C1, C2 and C4.

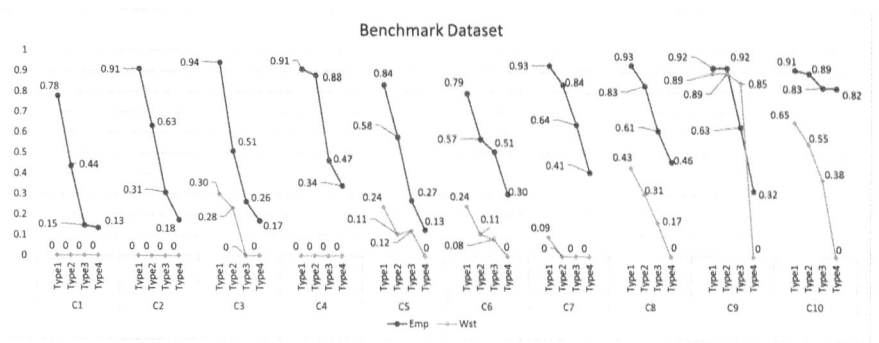

Fig. 5. *Emp* and *Wst* by varying users' types on the BPEL benchmark dataset.

To additionally prove the benefits produced by HDT-ViewMat, we assess how much the pre-materialisation reduces the waiting time for a user running a process. For further details, refer to Appendix A.3, which explains the evaluation measures and includes results tables for both healthcare and benchmark datasets.

Overall, the results of those tests demonstrate that HDT-ViewMat allows a user to use tasks within a business process by reducing the waiting time for generating the required HDT-Views.

6 Conclusion

In the paper, we discussed the growing significance of Human Digital Twins (HDTs) and the challenges of adopting them, such as the time-consuming materialisation process and privacy concerns. To tackle these challenges, we proposed to identify views of HDT to be pre-materialised based on *Execution Chance* (EC) metric, which indicates the user's probability of invoking a given task in the workflow. As a future work, we plan to have a more in-depth evaluation to further explore the impact of different factors on the EC metric and its effectiveness in predicting task execution. In doing this, we plan to consider realistic data to estimate the saved time. In fact, HDT-View materialisation can require a few minutes in the case of small amounts of data or different hours if the data to be processed is large and/or the view consists of complex data representation (e.g., a high-quality virtual model). A more in-depth evaluation will also allow us to evaluate alternative similarity metrics to detect their impacts on predicting task execution. Moreover, we plan to explore the possibility of incorporating user feedback to refine the calculation of the EC metric and enhance the accuracy of predicting task execution. This could potentially lead to a more personalised and efficient workflow management system for users.

Another future work is the investigation of more complex scenarios in terms of supported workflows (e.g., including parallel executions) and data types. This latter is of particular interest as supporting, for example, data with frequent updates, like continuously generated data streams (e.g., heartbeats), might require revising the strategy to pre-materialise the view. Another interesting direction is to analyse the security of the proposed framework. In particular, to avoid concerns similar to those affecting speculative execution, where attackers exploit the pre-materialised view (e.g., access a system's memory) to steal personal data.

Acknowledgments. The work was partially supported by project SERICS (PE00000014) under the NRRP MUR program funded by the EU - NGEU.

A   Healthcare Scenario Experimental Evaluation

A.1   Healthcare Dataset

The healthcare scenario was further considered to generate three additional BPEL processes representing real case studies.

Advanced Practitioner Physiotherapist (APP), Fig. 6a. This process is taken from [9], a survey exploring patients' clinical journeys attending APP services in two different hospitals.

Gestational Diabetes (GD), Fig. 6b. It is based on the International Federation of Gynecology and Obstetrics guidelines for diagnosis, management and care of gestational diabetes mellitus [11].

Coronary Artery Disease (CAD), Fig. 6c. This process is based on guidelines issued by the National Library of Medicine.[6] It is defined as a linear process of checks, cardiac catheterisation, medications, and drug therapy for patients.

---

[6] https://www.ncbi.nlm.nih.gov/books/NBK564304/.

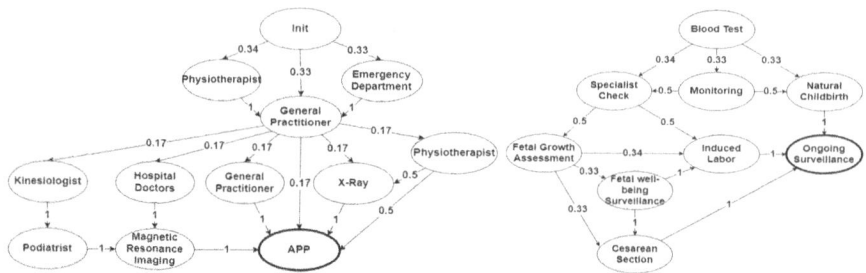

(a) Advanced Practitioner Physiotherapist pro- (b) Gestational Diabetes process
cess

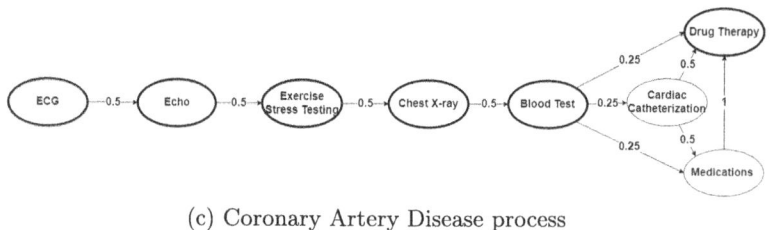

(c) Coronary Artery Disease process

Fig. 6. Graphical representation of the case studies in healthcare.

A.2   Healthcare Simulation Experiment Results

The percentages of employed (Emp) and wasted (Wst) pre-materialised views
for the different user types in healthcare processes are shown in Fig. 7. Generally,
the difference between Emp and Wst is high, and Emp sometimes maintains a
high value while Wst decreases. An instance is the *Tub* process for Type1, Type2
and Type3 users, where Emp always equals 100%, and Wst falls from 31% to 2%.
This variation occurs because HDT-ViewMat pre-materialises fewer HDT-Views
in the whole business process for Type2 and Type3 users, but those views are
included in the execution path. The rate differences between processes are due
to their structure and how many tasks the execution paths share. For instance,
CAD has higher Emp than APP; indeed, tasks in the latter are traversed by
fewer paths than those in the former (cfr. Fig. 6). Running HDT-ViewMat over
those realistic processes highlights, even more than the benchmark dataset, the
positive effects of this solution against wasted resources.

A.3   Waiting Time Assessment

Aiming at demonstrating the efficacy of HDT-ViewMat, we also evaluate the
benefit for users in terms of decreased waiting time. To estimate this gain, we
assumed that each task requires a fixed amount of time δ_t to compute a view (e.g.,
$delta_t = 100$ units of time) and calculated the total time saved by comparing the
time of execution of a path with the pre-materialisation strategy and without. In
particular, without the pre-materialisation strategy, views are materialised when

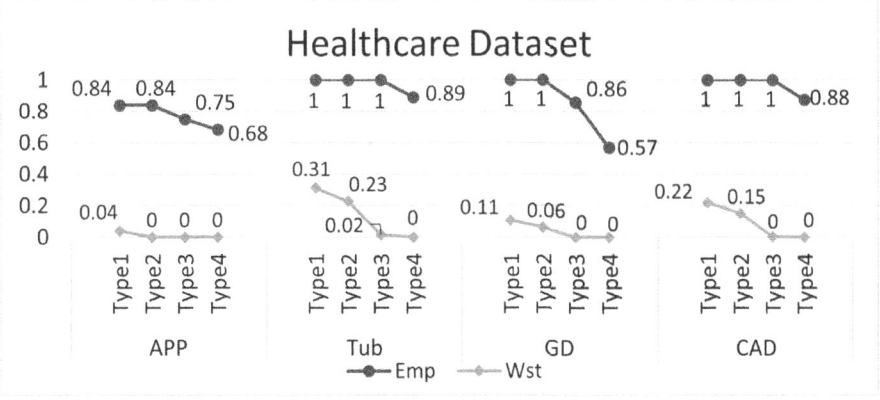

Fig. 7. Percentage of pre-materialised HDT-Views according to the users' types on the healthcare case study dataset.

the tasks are invoked, thus the total waiting time of an execution path is the number of tasks multiplied by δ_t. Given a process p, we compute its average Total Time (aTT) by averaging the total waiting times of each path in $ExPaths(p)$. On the other hand, with a pre-materialisation strategy, the delay in an execution path results from the time required to materialise those views that Algorithm 1 has not yet materialised (i.e., #views not materialised multiplied by δ_t). Thus, given a process p, the average Waiting Time (aWT) is defined as the average of the waiting times of each path in $ExPaths(p)$.

Table 2. Saved Time for Healthcare Dataset

	APP	Tub	GD	CAD
aTT	411	836	533	477
aWT	66	0	0	0

Tables 2 and 3 show the comparison of time without (i.e., aTT) and with (i.e., aWT) the pre-materialisation strategy on the processes within the healthcare and benchmark datasets. Results in the tables refer to process execution where $OPLA = 1$ for all tasks. The benefits brought by HDT-ViewMat are evident from these tables, which demonstrate the advantages of pre-materialisation. Specifically, Table 2, for the healthcare dataset, highlights that the waiting time is drastically reduced for all processes. Indeed, the average time required for a user to traverse an execution path without pre-materialisation in APP is almost 441. In contrast, with HDT-ViewMat, this time decreases to 66. Meanwhile, the waiting time for Tub, GD, and CAD processes is null. Table 3, for the BPEL benchmark dataset, shows that our system allows users to traverse most of the execution paths with an insignificant waiting time, except C5 and C6. For processes in cluster C5, the $\overline{aWT}$ equals 29 compared to an $\overline{aTT}$ of 233, while in C6, $\overline{aWT} = 43$ and $\overline{aTT} = 510$.

Table 3. Saved Time for Benchmark Dataset

	C1	C2	C3	C4	C5	C6	C7	C8	C9	C10
$\overline{aTT}$	275	500	99	500	233	510	521	330	300	978
$\overline{aWT}$	0	0	0	0	29	43	0	0	0	0

References

1. Amofa, S., et al.: Blockchain-secure patient digital twin in healthcare using smart contracts. PLoS ONE **19**(2) (2024)
2. Bagaria, N., Laamarti, F., Badawi, H.F., Albraikan, A., Martinez Velazquez, R.A., El Saddik, A.: Health 4.0: digital twins for health and well-being. Connect. Health Smart Cities (1), 143–152 (2020)
3. Banaeian Far, S., Imani Rad, A.: Applying digital twins in metaverse: user interface, security and privacy challenges. J. Metaverse **2**(1), 8–15 (2022)
4. Barricelli, B.R., Casiraghi, E., Gliozzo, J., Petrini, A., Valtolina, S.: Human digital twin for fitness management. IEEE Access **8**, 26637–26664 (2020)
5. Cathey, G., Benson, J., Gupta, M., Sandhu, R.: Edge centric secure data sharing with digital twins in smart ecosystems. In: 2021 Third IEEE International Conference on Trust. Privacy and Security in Intelligent Systems and Applications (TPS-ISA), pp. 70–79. IEEE, USA (2021)
6. Corral-Acero, J., et al.: The 'digital twin' to enable the vision of precision cardiology. Eur. Heart J. **41**(48), 4556–4564 (2020)
7. Dai, Y.S., Yang, B., Dongarra, J., Zhang, G.: Cloud service reliability: modeling and analysis. In: 15th IEEE Pacific Rim International Symposium on Dependable Computing, pp. 1–17. Citeseer (2009)
8. Dietz, M., Putz, B., Pernul, G.: A distributed ledger approach to digital twin secure data sharing. In: Foley, S.N. (ed.) DBSec 2019. LNCS, vol. 11559, pp. 281–300. Springer, Cham (2019). https://doi.org/10.1007/978-3-030-22479-0_15
9. Fennelly, O., et al.: Advanced musculoskeletal physiotherapy practice: the patient journey and experience. Musculoskelet. Sci. Pract. **45**, 102077 (2020)
10. Gillette, K., et al.: A framework for the generation of digital twins of cardiac electrophysiology from clinical 12-leads ECGs. Med. Image Anal. **71**, 102080 (2021)
11. Hod, M., et al.: The international federation of gynecology and obstetrics (FIGO) initiative on gestational diabetes mellitus: a pragmatic guide for diagnosis, management, and care. Int. J. Gynecol. Obstet. **131**(S3), S173–S211 (2015)
12. Lukač, L., Fister, I., Fister, I.: Digital twin in sport: from an idea to realization. Appl. Sci. **12**(24), 12741 (2022)
13. Miller, S.L., Childers, D.: Markov processes. In: Miller, S.L., Childers, D. (eds.) Probability and Random Processes, 2nd edn, pp. 383–428. Academic Press, Boston (2012)

14. Mourtzis, D., Angelopoulos, J., Panopoulos, N., Kardamakis, D.: A smart IoT platform for oncology patient diagnosis based on AI: towards the human digital twin. Procedia CIRP **104**, 1686–1691 (2021)
15. Ou, H., et al.: Development of a low-cost and user-friendly system to create personalized human digital twin. In: 2023 45th Annual International Conference of the IEEE Engineering in Medicine & Biology Society (EMBC), pp. 1–4 (2023)
16. Putz, B., Dietz, M., Empl, P., Pernul, G.: Ethertwin: blockchain-based secure digital twin information management. Inf. Process. Manag. **58**(1), 102425 (2021)
17. Sheehan, K.B.: Toward a typology of internet users and online privacy concerns. Inf. Soc. **18**(1), 21–32 (2002)
18. Shengli, W.: Is human digital twin possible? Comput. Methods Progr. Biomed. Update **1**, 100014 (2021)
19. Song, W., Zhang, C., Jacobsen, H.A.: An empirical study on data flow bugs in business processes. IEEE Trans. Cloud Comput. **9**(1), 88–101 (2021)
20. Subramanian, K.: Digital twin for drug discovery and development-the virtual liver. J. Indian Inst. Sci. **100**(4), 653–662 (2020)
21. Sun, T., He, X., Li, Z.: Digital twin in healthcare: recent updates and challenges. Digit. Health **9** (2023)
22. Wilson, S., et al.: The creation and analysis of a website privacy policy corpus. In: Proceedings of the 54th Annual Meeting of the Association for Computational Linguistics (Volume 1: Long Papers), pp. 1330–1340 (2016)
23. Wu, Z., Palmer, M.: Verbs semantics and lexical selection. In: Proceedings of the 32nd Annual Meeting on Association for Computational Linguistics, ACL 1994, pp. 133–138. Association for Computational Linguistics, USA (1994)

IAM Meets CTI: Make Identity and Access Management Ready for Cyber Threat Intelligence

Alexander Puchta[1]([✉])(iD), Thomas Baumer[1]([✉])(iD), Mathis Müller[2]([✉])(iD), and Günther Pernul[2]([✉])(iD)

[1] Nexis GmbH, Rudolf-Vogt-Straße 6, Regensburg 93053, Germany
{alexander.puchta,thomas.baumer}@nexis-secure.com
[2] University of Regensburg, Universitätsstraße 31, Regensburg 93053, Germany
{mathis.muller,gunther.pernul}@informatik.uni-regensburg.de
https://nexis-secure.com/en/ ,
https://www.uni-regensburg.de/informatik-data-science/wi-pernul/
startseite/index.html

Abstract. Enterprises rely on Identity and Access Management (IAM) systems as their primary solution for digital identity management and access control. While regulatory compliance is often a driving factor for such systems, they also serve as an essential security gate fortifying the defense against cyber attacks. However, when analyzing suspected or actual attacks, Security Information and Event Management (SIEM) systems and Cyber Threat Intelligence (CTI) are commonly employed but under-utilize valuable IAM data. IAM analysts can overcome such challenges by designing and implementing suitable mechanisms for a swift, easy-to-use, and faultless data transfer from IAM to SIEM. We contribute with a survey to identify ten central IAM findings relevant to CTI. We also evaluate their real-world feasibility by applying them within an anonymized data set of *TrustCorp* and make our tools open-source.

Keywords: IAM · CTI · SIEM · IDS

1 Introduction

The 2021 OWASP Top Ten lists *Broken Access Control* as first place, indicating its severity for 94% of the tested applications. Regulative authorities acknowledge this risk and demand effective IAM controls. Notable regulations include the GDPR, Basel III, or SOX. Such IAM systems allow observing access control data: For example, a critical permission assignment to a recently created identity can indicate malicious actions. IAM can capture these patterns and can communicate them via CTI to SIEM systems [35]. However it remains unclear which IAM findings are relevant to CTI and SIEM. Thus, this paper provides the following central contributions:

The research leading to these results was supported by the German Federal Ministry of Education and Research as part of the DEVISE project (https://devise.ur.de).

A. L. Ferrara and R. Krishnan (Eds.): DBSec 2024, LNCS 14901, pp. 44–52, 2024.
https://doi.org/10.1007/978-3-031-65172-4_3

- We synthesize ten distinct IAM findings from theory and practice.
- We show real-world practicability by a case study for the role model of an anonymized enterprise *TrustCorp* (maintaining ca. 5.5k digital identities).
- We make our tools open-source to ease a vendor-independent impact.

To realize these contributions, the remainder of this work is outlined as follows: Sect. 2 covers related work and Sect. 3 details our method. In Sect. 4, we review various sources to synthesize a set of ten relevant IAM findings. Afterward, in Sect. 5, we show the feasibility by evaluating a selection of the IAM findings within a case study for the anonymized enterprise *TrustCorp*. Section 6 concludes our work and highlights future research.

2   Related Work

Research on CTI or SIEM has recently received much interest, with many published overviews and frameworks (e.g., [40,47]). However, integrating IAM data to leverage insights for improved detection of cyber attacks has yet to be noticed. On the contrary, there is already work published to identify individual anomalies or findings within access control or IAM (e.g., [7,28,37] to name only a few).

When looking into practice, one can find various frameworks like the Lockheed Martin Cyber Kill Chain [20] or the MITRE ATT&CK framework [44]. However, we have consulted more than 70 enterprises regarding IAM, and less than five of them integrate IAM information into their CTI processes.

Summarizing the research and practice experience above, we observe a research gap for applying IAM or access control data to CTI. An important enabler for this is also an overview of proposed detectable IAM anomalies. To the best of our knowledge, there is no comprehensive work considering IAM to include within CTI to enhance analytical results. Therefore, we close this research gap by defining IAM findings relevant to CTI.

3   Method

We research literature and practitioner's sources to identify our IAM findings on a comprehensive basis. In Sect. 4, we summarize these IAM findings. While we are aware that there are more findings within an IAM landscape, we chose the most relevant ones for this work. For these findings, we show their applicability for the real-world data set of *TrustCorp*. We thus develop and apply scripts to transform the IAM data into CTI-readable information.

Literature Survey. We employ a structured approach by defining suitable keywords to identify relevant IAM literature based on Levy and Ellis [26]. After defining suitable search terms[1,2], we apply them to academic databases (ACM, IEEE XPlore, DBLP, Google Scholar) and integrate research from the related topic "*access control*". Applying our search terms yields a feasible number of

---

[1] Identity and access management.
[2] Identity|access management analysis|anomaly|signature|pattern.

relevant results. We decreased the amount by manually filtering the results based on title, abstract, and content. This results in 23 relevant scientific entries (c.f. Table 1) mentioning or defining relevant IAM findings. We group the results, which leads to ten IAM findings (see Sect. 4).

Practitioners' View. Using the initial IAM findings as input, we further improve the results. We incorporate various sources for our business analysis by relying, among others, on analysts having specific expertise in IAM like Kuppinger Cole[3]. They have built up profound IAM knowledge and serve as a further source of information. Additionally, more general analysts like Gartner[4] and Forrester[5] also highlight IAM topics relevant to our work. We also include information security or IAM standards as they often have specific regulations (e.g., ISO/IEC 27001, SOX, BASEL). Violations of such lead to IAM findings. Hence, these sources offer further results for our analysis.

4   IAM Findings

Table 1 provides an overview of IAM findings based on theory and practice. We exclude valid findings which are only of small value for CTI processes as they do not pose any threat. Next, we detail each finding. We resort to some elements from Structured Threat Information eXpression (STIX) to provide additional, structured information, namely *Vulnerability* and *Course of Action* [45].

Table 1. IAM findings.

Name	Literature	Practitioner
F1 - Orphan accounts	[6]	[30]
F2 - Excessive permission assignments	[10,38]	[21,29,30,44]
F3 - Assignment rule errors	[15,23,46]	
F4 - Attribute data quality	[3–5,13,24,25]	[36,42]
F5 - Privacy leak of identity data	[1,33,34,41]	[8,27,29,44]
F6 - Process errors	[18]	[21]
F7 - Policy violations	[15–17,19,32]	[2,21,43]
F8 - Dead access control policies	[10,32]	[30]
F9 - Liveness	[32,41,44]	[30]
F10 - Missing recertification	[12]	[21,36]

F1 - Orphan Accounts. An orphan account is an account without a connection to a valid identity [6,9,11,34]. An orphan account may prove a higher threat than others because nobody is responsible. A *vulnerability* arises as an orphan

---

[3] https://www.kuppingercole.com/.
[4] https://www.gartner.com.
[5] https://www.forrester.com.

account is still usable and has irregularly examined permissions. Especially having administrative permissions assigned can lead to a privilege escalation attack. Suitable *courses of action* within IAM would be regular recertifications of all account types and the definition of sufficient identity life cycle processes [36].

F2 - Excessive Permission Assignments. Excessive permissions are a basic reason for IAM. Identities have unnecessary permissions directly assigned or inherited via roles, leading to *vulnerabilities*. Identification methods rely on visualization, access reviews, or automated algorithms for analysis [7,28]. Regarding *courses of actions*, automated account cleansing or automation of role assignments, or attribute-based access control are suitable options [15,39].

F3 - Assignment Rule Errors. Assignment rules typically encompass a condition set. All identities fulfilling the condition are automatically assigned to permissions (e.g., all internal identities get permission to the intranet). Errors (like a specific attribute is incorrectly set within the IAM) can lead to a wrong membership calculation, thus resulting in the *vulnerability* since identities get assigned the wrong permissions. Processes for assignment rule changes or automated validity checks before error occurrence are suitable *courses of action*.

F4 - Attribute Data Quality. Data quality is extensively covered by various authors with quality metrics and frameworks [3,13]. Lacking attribute data quality can impair the reasoning of permission assignment because of missing descriptions. Other data quality errors are missing ownership or criticality flags. Such cases lead to the *vulnerability* that poor data quality enables attackers to obtain permissions because of a poor criticality assessment. Organizations can employ *courses of action* like the regularly checked attribute quality rules.

F5 - Privacy Leak of Identity Data. IAM systems typically cache privacy-related information (e.g., name, address, contact information) due to their connection to HR systems [6]. An information leakage bears the *vulnerability* that attackers may use privacy-related information retrieved from IAM systems (e.g., spear phishing) [14]. A suitable *course of action* is the encryption or anonymization of privacy-related information within IAM systems and processes [35].

F6 - Process Errors. Organizations may lack automation, especially regarding leaver or mover processes [25]. Attackers may want to bypass defined processes to gain elevated permissions, resulting in a *vulnerability*. A suitable *course of action* is the automation of processes and integration into IAM tools.

F7 - Policy Violations. Organizations often constrain their access control with policies like Separation of Duty (SoD). *Vulnerabilities* arise when attackers can directly exploit toxic permission combinations. As *course of action*, organizations can use automated SoD checks and regular maintenance of SoD rules.

F8 - Dead Access Control Policies. Dead Access Control Policies (ACPs) refer to unused permissions or roles [32]. As a *vulnerability*, attackers might assign dead ACP to themselves by circumventing processes without further approval. Fortunately, *courses of action* are easy to deploy [22]. Organizations can automatically check for dead ACPs and process permanent deletion or reassignment.

F9 - Liveness. Liveness comprises ACPs assigned to exactly one identity. Liveness indicates that only a single person conducts a specific task. A *vulnerability* is that these ACPs become dead ACPs when this person leaves the organization or has the assignment removed. To avoid such cases, organizations can include such ACPs in a comprehensive role model as a *course of action* [39].

F10 - Missing Recertification. A recertification (or access review) is a regular inspection of identities, accounts, permissions, or roles to maintain an organization's ACPs [31]. Uncertified identities are a *vulnerability* as these might have excessive permissions. Therefore, attackers might misuse these identities to gain access. A *course of action* is an automated reporting of uncertified entities.

5   Evaluation

This Section evaluates the practical feasibility of our IAM findings. Hence, we showcase generating a CTI threat report using an anonymized data set of the real-world enterprise *TrustCorp* (pseudonym). We make our implementations open-source: https://github.com/IAMmeetsCTI/IAM-meets-CTI

TrustCorp is a German enterprise in the finance and insurance sector. The enterprise has an IAM system with eight connected applications based on Role-Based Access Control (RBAC) [39]. We extracted several of our previously defined IAM findings during the analysis and integrated them in CTI.

TrustCrop also narrowed down the goals and wished an exclusive analysis of the existing role model and selected IAM findings, including *F3*, *F8*, *F9*, and *F10*. With this generously provided *TrustCorp* data and scope, we generated a threat report with our scripts by querying the dataset for our IAM findings.

Table 2 summarizes the analysis results. We apply further filter criteria only to include the most critical findings if analyzing and discussing the resulting number of defect roles with *TrustCorp* would become too cumbersome. Thus, our analysis covers 221 findings, comprising 3 of the 4 investigated IAM findings.

Table 2. Analysis results of TrustCorp.

Analyzed finding	#Results	Filter criteria
F3 - Assignment rule errors	0	No filter applied
F8 - Dead access control polices	49 roles	Criticality = 'high'
F9 - Liveness	162 roles	Criticality = 'high'
F10 - Missing recertification	10 roles	No filter applied

TrustCorp returned positive feedback. The swift and easy-to-use scripts allowed quick integration within their IAM system without tedious change requests. The reports contain interactive visualizations[6] and ease the awareness

---

[6] https://oasis-open.github.io/cti-stix-visualization/?url=https://raw.githubusercontent. com/IAMmeetsCTI/IAM-meets-CTI/main/example-trustcorp.json.

of solvable findings. *TrustCorp* further discussed a deeper integration to their SIEM infrastructure and also reporting the other IAM findings periodically.

6   Conclusion

We derived ten IAM findings (non-exhaustive) that would enhance the SIEM effectiveness from theory and practice. We focused on the most essential IAM findings relevant to CTI, which we verified throughout the practice-oriented analysis. Due to the scoping of our work, we excluded the authentication side of IAM (e.g., multiple login errors of the same identity at a suspicious time). IAM solutions predominantly focus on authorization, as having critical permissions assigned without suitable justification is often an immediate compliance violation and of particular interest for the domain of CTI. Moreover, we showed practical feasibility with our *TrustCorp* use case by successfully structuring and representing IAM data with STIX. We aim to empower others to leverage IAM data in their CTI efforts by sharing our open-source repository. As future work, we want to research the authentication side of IAM and its implications for CTI.

References

1. Asghar, M.R., Backes, M., Simeonovski, M.: PRIMA: privacy-preserving identity and access management at internet-scale. In: 2018 IEEE International Conference on Communications (ICC), Kansas City, MO, USA, pp. 1–6. IEEE (2018)
2. Basel Comittee on Banking Supervisions: Basel III: International framework for liquidity risk measurement, standards and monitoring (2010)
3. Batini, C., Cappiello, C., Francalanci, C., Maurino, A.: Methodologies for data quality assessment and improvement. ACM Comput. Surv. **41**(3), 1–16 (2009)
4. Batini, C., Scannapieco, M.: Data quality: Concepts, Methodologies and Techniques. Springer, New York (2006). https://doi.org/10.1007/3-540-33173-5
5. Batini, C., Scannapieco, M.: Data and Information Quality: Dimensions, Principles and Techniques, 1st edn. Springer, Cham (2016). https://doi.org/10.1007/978-3-319-24106-7
6. Baumer, T., Müller, M., Pernul, G.: System for cross-domain identity management (SCIM): Survey and enhancement with RBAC. IEEE Access **11**, 86872–86894 (2023)
7. Colantonio, A., Di Pietro, R., Ocello, A., Verde, N.: Visual role mining: a picture is worth a thousand roles. IEEE Trans. Knowl. Data Eng. **24**(6), 1120–1133 (2012)
8. Diodati, M., Ruddy, M., Rabinovich, P., Mezzera, P.: Gartner - 2020 planning guide for identity and access management (2019)
9. Everett, C.: Identity and access management: the second wave. Comput. Fraud Secur. **2011**(5), 11–13 (2011)
10. Fuchs, L., Kunz, M., Pernul, G.: Role model optimization for secure role-based identity management. In: European Conference on Information Systems (ECIS), Tel Aviv, Israel, pp. 1–15. AIS (2014)
11. Fuchs, L., Pernul, G.: Supporting compliant and secure user handling - a structured approach for in-house identity management. In: The Second International Conference on Availability, Reliability and Security (ARES'07), Vienna, Austria, pp. 374–384. IEEE (2007). https://doi.org/10.1109/ARES.2007.145

12. Groll, S., Kern, S., Fuchs, L., Pernul, G.: Monitoring access reviews by crowd labelling. In: Fischer-Hübner, S., Lambrinoudakis, C., Kotsis, G., Tjoa, A.M., Khalil, I. (eds.) Trust, Privacy and Security in Digital Business, pp. 3–17. Springer, Cham (2021). https://doi.org/10.1007/978-3-030-86586-3_1
13. Heinrich, B., Kaiser, M., Klier, M.: How to measure data quality? a metric-based approach. In: Rivard, S., Webster, J. (eds.) Proceedings of the 28th International Conference on Information Systems (ICIS). Montreal, Queen's University, Montreal, pp. 1–15. AISeL (2007). https://epub.uni-regensburg.de/23633/
14. Ho, G., Sharma, A., Javed, M., Paxson, V., Wagner, D.: Detecting credential spearphishing attacks in enterprise settings. In: Proceedings of the 26th USENIX Conference on Security Symposium. SEC'17, USA, pp. 469–485. USENIX Association (2017)
15. Hu, V., et al.: Guide to attribute based access control (ABAC) definition and considerations. Technical report, NIST (2014)
16. Hu, V., Ferraiolo, D.F., Kuhn, D.R., Kacker, R.N., Lei, Y.: Implementing and managing policy rules in attribute based access control. In: 2015 IEEE International Conference on Information Reuse and Integration, San Francisco, CA, USA, pp. 518–525. IEEE (2015).https://doi.org/10.1109/IRI.2015.98
17. Hu, V., Scarfone, K.: Guidelines for access control system evaluation metrics. Technical report, NIST (2012)
18. Hummer, M., Groll, S., Kunz, M., Fuchs, L., Pernul, G.: Measuring identity and access management performance - an expert survey on possible performance indicators. In: Proceedings of the 4th International Conference on Information Systems Security and Privacy, Funchal, Madeira, Portugal, pp. 233–240. SCITEPRESS - Science and Technology Publications (2018)
19. Hummer, M., Kunz, M., Netter, M., Fuchs, L., Pernul, G.: Adaptive identity and access management–contextual data based policies. J. Inf. Secur. **2016**(1), 19 (2016)
20. Hutchins, E., Cloppert, M., Amin, R.: Intelligence-driven computer network defense informed by analysis of adversary campaigns and intrusion kill chains. Technical report, Lockheed Martin (2011)
21. ISO 27001: Information technology - Security techniques - Information security management systems - Requirements. Standard, International Organization for Standardization (2013)
22. Kern, S., Baumer, T., Fuchs, L., Pernul, G.: Maintain high-quality access control policies: an academic and practice-driven approach. In: Atluri, V., Ferrara, A.L. (eds.) Data and Applications Security and Privacy XXXVII, pp. 223–242. Springer, Cham (2023). https://doi.org/10.1007/978-3-031-37586-6_14
23. Kern, S., Baumer, T., Groll, S., Fuchs, L., Pernul, G.: Optimization of access control policies. J. Inf. Secur. Appl. **70**, 103301 (2022)
24. Kunz, M., Fuchs, L., Hummer, M., Pernul, G.: Introducing dynamic identity and access management in organizations. In: Jajodia, S., Mazumdar, C. (eds.) ICISS 2015. LNCS, vol. 9478, pp. 139–158. Springer, Cham (2015). https://doi.org/10.1007/978-3-319-26961-0_9
25. Kunz, M., Puchta, A., Groll, S., Fuchs, L., Pernul, G.: Attribute quality management for dynamic identity and access management. J. Inf. Secur. Appl. **44**, 64–79 (2019)
26. Levy, Y., Ellis, T.: A systems approach to conduct an effective literature review in support of information systems research. Int. J. Emerg. Transdiscipl. **9** (2006)
27. Maxim, M., Cser, A.: Forrester - Top trends shaping IAM in 2022 (2022)

28. Meier, S., Fuchs, L., Pernul, G.: Managing the access grid - a process view to minimize insider misuse risks. In: 11. Internationale Tagung Wirtschaftsinformatik, Leipzig, Germany, February 27 – March 1, 2013. p. 66. AIS (2013). http://aisel.aisnet.org/wi2013/66

29. MITRE: CAPEC common attack pattern enumeration and classification (2023). https://capec.mitre.org/index.html. Accessed 22 May 2023

30. MITRE: CWE common weakness enumeration (2023). https://cwe.mitre.org/index.html. Accessed 22 May 2023

31. Osmanoglu, E.: Identity and Access Management: Business Performance Through Connected Intelligence. Newnes, Waltham (2013)

32. Parkinson, S., Khan, S.: A survey on empirical security analysis of access-control systems: A real-world perspective. ACM Comput. Surv. **55**(6), 1–28 (2022)

33. Pfitzmann, A., Hansen, M.: A terminology for talking about privacy by data minimization: anonymity, unlinkability, undetectability, unobservability, pseudonymity, and identity management (2010)

34. Puchta, A., Böhm, F., Pernul, G.: Contributing to current challenges in identity and access management with visual analytics. In: Foley, S.N. (ed.) DBSec 2019. LNCS, vol. 11559, pp. 221–239. Springer, Cham (2019). https://doi.org/10.1007/978-3-030-22479-0_12

35. Puchta, A., Groll, S., Pernul., G.: Leveraging dynamic information for identity and access management: an extension of current enterprise iam architecture. In: Proceedings of the 7th International Conference on Information Systems Security and Privacy - ICISSP, pp. 611–618. INSTICC, SciTePress, Online Streaming (2021)

36. Reinwarth, M.: Access reviews done right. Technical report, Kuppingercole Analysts (2019). https://www.kuppingercole.com/report/lb80195

37. Samarati, P., de Vimercati, S.C.: Access control: policies, models, and mechanisms. In: Focardi, R., Gorrieri, R. (eds.) FOSAD 2000. LNCS, vol. 2171, pp. 137–196. Springer, Heidelberg (2001). https://doi.org/10.1007/3-540-45608-2_3

38. Sandhu, R.: The authorization leap from rights to attributes: maturation or chaos? In: Proceedings of the 17th ACM Symposium on Access Control Models and Technologies. SACMAT '12, New York, NY, USA, pp. 69–70. Association for Computing Machinery (2012)

39. Sandhu, R., Coyne, E., Feinstein, H., Youman, C.: Role-based access control models. Computer **29**(2), 38–47 (1996)

40. Schlette, D., Caselli, M., Pernul, G.: A comparative study on cyber threat intelligence: the security incident response perspective. IEEE Commun. Surv. Tutor. **23**(4), 2525–2556 (2021)

41. Servos, D., Osborn, S.: Current research and open problems in attribute-based access control. ACM Comput. Surv. **49**(4), 1–65 (2017)

42. Small, M.: Kuppingercole report - advisory note - big data security, governance, stewardship (2018)

43. SOX: Sarbanes-Oxley Act of 2002, pl 107-204, 116 stat 745 (2002)

44. Strom, B., Applebaum, A., Miller, D., Pennington, A., Thomas, C.: MITRE ATT&CK: Design and Philosophy. Framework, The MITRE Corporation, McLean, USA (2020)

45. Struse, R., Darley, T.: STIX Version 2.1. OASIS Standard, OASIS (2021)

46. Vijayalakshmi, K., Jayalakshmi, V.: Identifying considerable anomalies and conflicts in ABAC security policies. In: 2021 5th International Conference on Intelligent Computing and Control Systems (ICICCS), Madurai, India, pp. 1273–1280. IEEE (2021)
47. Wagner, T., Mahbub, K., Palomar, E., Abdallah, A.: Cyber threat intelligence sharing: survey and research directions. Comput. Secur. **87**, 101589 (2019)

Crypto Application

SmartSSD-Accelerated Cryptographic Shuffling for Enhancing Database Security

Tieming Geng[1]([✉]) and Chin-Tser Huang[2]([✉])

[1] Department of Mathematics and Computer Science, Fayetteville State University,
Fayetteville, NC, USA
tgeng@uncfsu.edu

[2] Department of Computer Science and Engineering, University of South Carolina,
Columbia, SC, USA
huangct@cse.sc.edu

Abstract. Given that databases often house sensitive and valuable data, ensuring data confidentiality and integrity of the databases is imperative. Encryption emerges as one of the predominant techniques employed in bolstering database security. Alongside encryption, shuffling also offers a viable approach to fortify the security of database. However, both encryption and shuffling requires huge amounts of system I/O requests which bring performance burden to the database server. In this paper, we propose a design to further enhance the security of shuffling algorithm and improve its efficiency by employing SmartSSD computational storage device from Samsung and AMD. We conduct experiments to evaluate the overhead of the improved effectiveness.

Keywords: shuffling · relational database · SmartSSD · computational storage device · database security · in storage computation · near storage computation

1 Introduction

In the age of the Internet and big data, database security techniques to protect database data play a particularly crucial role in preventing the leakage of sensitive information. Databases, serving as the repository for sensitive and proprietary data, have become one of the primary targets for cybercriminals seeking valuable information. The escalating frequency of data leaks and the increasing complexity and sophistication of attacks pose significant threats to organizations and communities, resulting in substantial financial losses, legal liabilities, and damage to their reputation [4]. These threats ultimately have repercussions on individuals as well. In August 2023, the aftermath of the MOVEit security incident is still ongoing, with the ransomware group CL0P launching more attacks against government agencies [19], including the Colorado Department of Health Care Policy & Financing and many government contractors. Sensitive data and personal information, including SSNs and health records, of millions of people have been compromised, and more data is still being exposed.

© IFIP International Federation for Information Processing 2024
Published by Springer Nature Switzerland AG 2024
A. L. Ferrara and R. Krishnan (Eds.): DBSec 2024, LNCS 14901, pp. 55–70, 2024.
https://doi.org/10.1007/978-3-031-65172-4_4

Database encryption is an effective and widely adopted method for mitigating data leaks and enhancing database security. It uses cryptographic techniques to encrypt information, thereby providing information protection. In addition to encryption, we propose a cryptographically secure shuffling algorithm [14] that uses random deranged permutation to completely disrupt the original associations of attributes within the relation tuple. This shuffling algorithm can be applied independently in the database or used in conjunction with encryption as an supplementary security layer. While the security benefits of encryption and shuffling often outweigh the performance overhead considerations, they do increase I/O demands on the system.

Looking back at the era when mechanical hard drives were the absolute dominant storage medium, one significant bottleneck for database performance was the read and write speed of mechanical hard drives, especially for random read and write operations. This was due to the internal structure of mechanical hard drives. In comparison to mechanical hard drives, solid state drives (SSDs) are better suited for scenarios with high-speed read and write and high I/O demands. SSDs offer faster access speeds, lower latency, better random read and write capabilities, and lower power consumption [32]. Consequently, the rapid development of SSDs has shifted the system bottleneck from storage medium to the time spent on data transfer [20]. For example, when a significant amount of data pages are requested for tasks such as shuffling and encryption, these data pages must be transferred from storage to memory through interfaces like PCIe, SATA, or SAS. Afterward, the CPU accesses and processes the data, with high-speed cache playing a supporting role when necessary.

The conception of "in storage computing" or "near storage computation" was proposed as early as the 1990s, which was designed to execute most processing tasks to the processor inside the disk [5,25]. However, this conception just began to gain attention and development around the early 2010s [11,18]. By deploying the "in storage computing" devices in database servers, certain tasks like shuffling and encryption can be offloaded to the storage devices. When shuffling or encryption requests are received, the host forwards these requests to the storage. The processor within the storage retrieves the necessary data to its internal DRAM, where it handles the processing requests. Subsequently, the outcomes are written back. This architecture avoids competition on system I/O between data requiring shuffling or encryption and other data. It also eliminates the need for such data to traverse the lengthy path through the system bus, spanning storage, RAM, cache, and processors.

A computational storage drive (CSD) is an "in storage computing" device. It reduces data transfer latency and enhances data processing efficiency by introducing processing capabilities inside the storage. There are several commercial products in the market from companies like NGD Systems, ScaleFlux, Pliops, SNIA, and Xilinx (acquired by AMD). The core of the SmartSSD CSD used in our design is consisted of the AMD FPGA programmable platform and Samsung's Enterprise SSD controller [2]. A dedicated high-speed peer-to-peer (P2P) link connects the SSD controller to the FPGA for low-latency processing.

In this paper, we propose a design that combines SmartSSD CSD and shuffling algorithms to enhance database data security with minimal additional performance overhead. Our main contributions are listed as follows:

- We demonstrate how to use SmartSSD CSD to accelerate the shuffling algorithm, including software and hardware architectures, ensuring that the performance overhead of shuffling algorithms does not affect the execution of other tasks in the database system.
- We propose a novel matrix-based shuffling scheme which effectively enhances the security of the shuffling algorithm.
- We optimize the shuffling algorithm further by utilizing the built-in FPGA accelerator of the SmartSSD CSD to improve efficiency.
- We conduct performance evaluations to show the performance improvements the SmartSSD CSD brings to the shuffling algorithm.

The remainder of the paper is organized as follows. In Sect. 2, we discuss related work on database security and computational storage. In Sect. 3, we provide the background on database shuffling algorithm. In Sect. 4, we introduce the details of our design which can improve the performance of the database shuffling algorithm. In Sect. 5, we describe our experiment and evaluate its performance. We finally conclude the paper in Sect. 6.

2   Related Work

2.1   Computational Storage Drive

Wang et al. [34] explored offloading list intersection tasks, which are crucial in search engine and analytic, onto the Samsung SmartSSD CSD research prototype. They developed an analytical model to understand key performance factors, and their experiments indicated that SmartSSD CSD can enhance list intersection processing while reducing energy consumption. Lee et al. [21] proposed a design that evaluated the performance of offloading data analytic operations into SmartSSD CSD with Spark SQL and the Parquet columnar data format. NASCENT [26] and its successor NASCENT2 [27] are near-storage sort accelerators for data analytics on SmartSSD CSD. SmartSSD CSD are not limited to data analytics; they are also employed to accelerate neural network model training [28,31]. Additionally, Hedam et al. [17] introduced Delilah, the first public description of a CSD supporting eBPF-based code offload. eBPF, an operating system kernel technique, filters applications in network monitoring, packet filtering, performance profiling, and virtualization.

2.2   Database Security

The adoption of database encryption has emerged as a pivotal strategy in the realm of data security, aiming to ensure both data confidentiality and integrity. This is typically achieved through two fundamental approaches: executing queries directly on encrypted data or decrypting data prior to query execution.

CryptDB, a seminal work by Popa et al. [23], introduces a sophisticated scheme that deploys a database proxy situated between the database and the application server. This proxy assumes the role of intercepting SQL queries and subsequently rewriting them to facilitate execution on encrypted data, while preserving the original semantics of the queries. Built upon CryptDB, the MONOMI framework [33] further refines the database proxy concept, dividing the execution process into discrete client and server components. This architectural distinction enables the server to perform computations without necessitating access to the encryption key. The exploration of searching encrypted data has yielded significant advancements, ranging from symmetric to asymmetric ciphers, as documented in seminal works by Song and Wagner [29] and Boneh and Franklin [7], respectively. Subsequent research has delved into executing intricate search queries on encrypted databases, encompassing conjunctive searches [8], rich and range queries [16,35,36], and dynamic queries accommodating data modifications subsequent to encryption [9,30].

The adoption of transparent data encryption (TDE) [10,13] by industries including Microsoft, IBM, and Oracle, constitutes a noteworthy development. TDE offers real-time encryption and decryption capabilities for database data and log files employing symmetric encryption algorithms.

Furthermore, the promising avenue of leveraging auxiliary hardware as secure coprocessing platforms has been a subject of study, offering security assurance for specific operations. Prior investigations [6,24] have effectively demonstrated the practicality of allocating routine operations to untrusted database servers while entrusting sensitive data or operations to dedicated hardware modules, such as FPGA-based coprocessors, and Intel SGX processing units.

3   Background

In this section, we briefly describe the aforementioned shuffling algorithm and column-oriented database design in our previous work, as well as some basic knowledge about the SmartSSD CSD utilized in our design.

3.1   Shuffling Algorithm

We introduced SCORD which stands for Shuffled Column-Oriented Relational Database in our previous publication [15]. It is an approach that integrates a shuffling algorithm with a column-oriented storage scheme to enhance the security and efficiency of relational database. SCORD can accomplish shuffling without physically relocating the data, resulting in significant savings in time and system resources. The individual shuffling in each column would break the original relationship among attributes in one tuple. Table 1 and Table 2 show the example table before and after shuffling. For example in the first tuple of

Table 1. Tuples Before Shuffling   **Table 2.** Tuples After Shuffling

name	credit_card_no	SSN
Alice	4251546446736274	473-34-5897
Bob	4703038231755507	768-12-5834
Cam	5393619347816820	583-35-5681
David	6011220281842771	981-34-6481

name	credit_card_no	SSN
Bob	5393619347816820	981-34-6481
Alice	6011220281842771	473-34-5897
David	4703038231755507	583-35-5681
Cam	4251546446736274	768-12-5834

Table 2, the attribute *name*, *credit_card_no*, and *SSN* were located in different tuples.

The tuples in each column are shuffled based on a random permutation sequence. To enhance shuffling security, we exclusively employ *deranged permutations* instead of considering all possible permutations. This precaution arises from the possibility that shuffling based on non-deranged permutations could leave multiple attributes unaltered in a tuple. In the event of a security breach, such tuples with multiple unaltered attributes may inadvertently disclose sensitive information. Though the likelihood of this occurrence is low, using deranged permutations could eliminates this potential risk.

3.2 Multi-level Shuffling

For the order of performance, the shuffling is performed in a fashion of multi-level. To illustrate how the multi-level shuffling operates, let's consider a single column containing a total of l tuples. l can be organized into equal-sized blocks, except for the last block, which is smaller in size. We denote the number of tuples within each equal-sized block as w, while the last block contains $l \mod w$ tuples. Shuffling occurs independently within each block. In each subsequent level, if the number of blocks at a given level exceeds or equals the block size w (meaning $l/w >= w$), the block itself becomes the fundamental element and builds a larger block, and then conduct the shuffling until the number of blocks falls below w in the level. For instance, $l = 825$ and $w = 20$ in one relation, the multi-level shuffling is conducted as shown in Fig. 1. The relation is divided into 42 blocks including 41 full blocks and one block with only 5 tuples. After the shuffling of the current level, these 42 blocks are regarded as the fundamental elements to construct three blocks for one more level of shuffling.

3.3 Binary Association Index Table

In SCORD, the Binary Association Index Table (BAIT) is a new design on the physical data model. BAIT manages the querying indexes and the unordered data records.

Each BAIT within the database system consists of two distinct columns: the left identifier column and the right data location column. In the left column, unique identifiers for each record in the database are maintained while the right

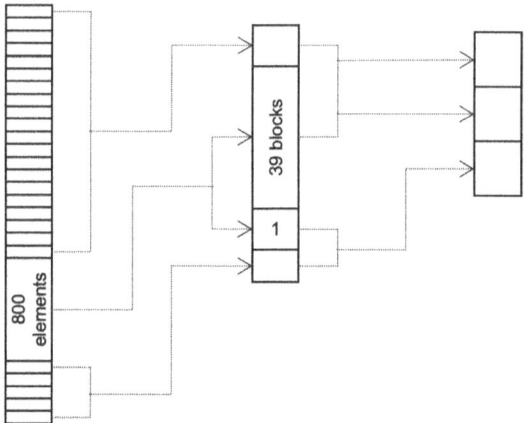

Fig. 1. Example of shuffling the blocks in multiple levels.

column is primarily responsible for the storage of data addresses within the system. Rather than storing the actual data values, this column houses concatenated addresses, expressed as `page_id + slot_id` (slotted page layout is used). The utilization of BAITs facilitates efficient data shuffling operations, performed exclusively within the right columns of the BAITs. This approach obviates the need for shuffling the actual data values in storage, a process that would otherwise necessitate a substantial volume of I/O requests.

3.4 SmartSSD

The SmartSSD CSD possesses the capability to perform computations directly within the storage device itself. As illustrated in Fig. 2, this SmartSSD CSD comprises a high-performance Samsung Enterprise SSD and a specialized Kintex Ultrascale+ FPGA from Xilinx (now part of AMD). These components are interconnected by a fast private P2P link, facilitating seamless data transfer between the SSD controller and the FPGA [2].

Xilinx provides a sophisticated development target shell characterized by high-performance capabilities, thereby facilitating the creation of customized acceleration applications tailored specifically for the SmartSSD CSD platform [3]. Within the architectural framework of the SmartSSD CSD, a pivotal component is the PCIe switch, which exhibits three distinct ports: the upstream port establishes a connection to the host system, the downstream port establishes a link with the SSD controller, and an internal end port is dedicated to interact with the FPGA. This dedicated internal port serves as the conduit for enabling acceleration functions within the SmartSSD CSD.

One of the salient advantages offered by this architectural design is the ability to offload user application code either partially or in its entirety to the FPGA's DRAM. Additionally, data essential for FPGA-based operations, residing within

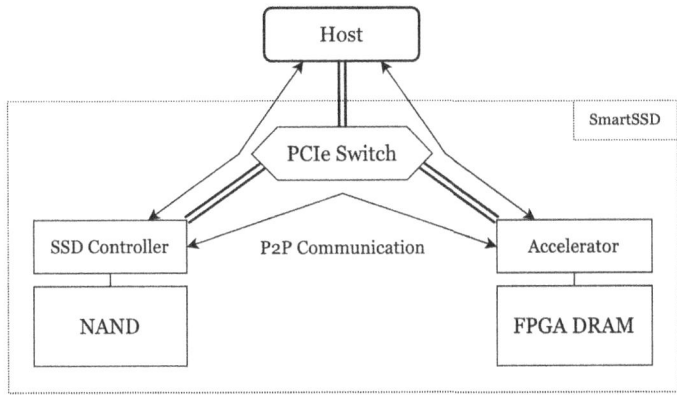

Fig. 2. SmartSSD architecture

the SmartSSD CSD storage, can be efficiently transmitted via the P2P link. This innovative approach significantly mitigates the need for the conventional round-trip data traffic between the storage and the host CPU.

This architecture refinement translates into real benefits in terms of computational efficiency, particularly when dealing with resource-intensive operations such as data shuffling and encryption, both of which involve substantial I/O interactions. The reduction in I/O latency and data transfer overhead is instrumental in optimizing the overall performance.

4   Design and Implementation

In this section, we introduce the design of the relational database with accelerated shuffling. To the best of our knowledge, this is the first proposed approach that utilizes CSD to address the database security issue.

At its core, our implementation comprises three integral components: 1) modules pertaining to relational databases, excluding storage considerations, 2) the BAIT and storage module, and 3) a module dedicated to accelerating data shuffling processes. Within the domain of relational database modules, we have developed modules which encompasses an SQL parser, a system manager responsible for overseeing DDL (Data Definition Language) statements, a query manager responsible for handling DML (Data Manipulation Language) statements, and an indexing component. The BAIT and storage module have the primary responsibility of managing page I/O operations and shuffling processes. Consequently, this module exhibits a close association with SmartSSD CSD. It is noteworthy that while the source code of this module within our prototype may not exhibit distinctiveness from the relational database modules, it is conceptually delineated as a separate entity. Given that SmartSSD CSD incorporates embedded FPGA chips, we have also introduced an acceleration module to optimize the shuffling operation by harnessing the computational capabilities of the FPGA.

4.1   Matrix-Shuffling Algorithm

In our previous work, SCORD adopts multi-level shuffling primarily on account of security consideration. Multi-level shuffling is able to reduce the potential information leakage during the "segmented" shuffling process. By shuffling elements within the confines of their respective blocks, the degree of diffusion is restricted, which may unintentionally reveal the presence of the shuffling.

However, it is noteworthy that even with the multi-level shuffling, certain recognizable patterns such as clustered distribution may be found by displaying the indexes of tuples across multiple breached versions of one relation in the coordinate grid, especially in the column that most tuples are unique. Since the columns of one relation are shuffled using different permutation sequence at different times and the values stay unchanged (only positions changed), the tuple values in the first version of the breached relation can be used as the control group to assign the "indexes". Through the values, the "new indexes" can be deduced in the following versions of the breached relation. By plotting all the "indexes" on the coordinate grid with different color, an obvious clustered distribution can be observed, and the attackers are able to infer the existence of "group" operations.

As depicted in Fig. 3, a straightforward shuffling experiment involving a column with parameters $l = 250$ and $w = 5$ reveals obvious patterns in multi-level shuffling. In this visualization, gray dots, black dots, and blue dots indicate the "indexes" across different versions of one breached relation column. Consequently, an evaluation of the statistical distribution may readily unveil the presence of shuffling.

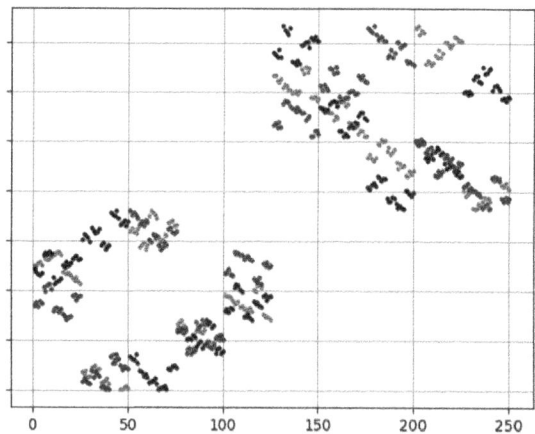

Fig. 3. Distribution of one column at different times. The horizontal axis represents the referential index of the data.

As a solution, we propose an enhanced *Matrix-Shuffling* algorithm based on SCORD's shuffling technique. Instead of employing multiple levels of shuffling, our algorithm adopts a matrix-based approach. The tuples within a given column, denoted as $x_0, x_1, ..., x_{l-1}$, are organized into an $l/w \times w$ matrix denoted as $\mathbb{X}$, with its transpose expressed as $\mathbb{X}^T$:

$$\mathbb{X} = \begin{pmatrix} x_0 & x_1 & \cdots & x_{w-1} \\ x_w & x_{w+1} & \cdots & x_{2w-1} \\ \vdots & \vdots & \ddots & \vdots \\ x_{(l/w-1)w} & x_{(l/w-1)w+1} & \cdots & x_{l-1} \end{pmatrix} \quad \mathbb{X}^T = \begin{pmatrix} x_0 & x_w & \cdots & x_{(l/w-1)w} \\ x_1 & x_{w+1} & \cdots & x_{(l/w-1)w+1} \\ \vdots & \vdots & \ddots & \vdots \\ x_{w-1} & x_{2w-1} & \cdots & x_{l-1} \end{pmatrix}$$

Within the matrix framework, each row undergoes shuffling according to one deranged permutation sequences, followed by independent shuffling of each matrix column. Considering data structure conventions, a matrix is commonly represented as a two-dimensional vector or array. Therefore, to execute column-wise shuffling on the matrix, a matrix transpose operation becomes necessary. The security analysis of the *Matrix-Shuffling* is conducted in Sect. 5.2. It is noteworthy that the column cannot be reorganized into a matrix all the time and the remainder $l\%w$ is suggested to do the shuffling in-situ.

Transposing a matrix and shuffling inside the matrix row can be computationally costly, particularly for a large matrix. In our design, SmartSSD CSD is indispensable for matrix operations, especially for a large matrix, due to its ability to perform computations directly within the storage device, which can minimize data movement between storage and the host system, thus reducing latency and bandwidth requirements.

4.2 Shuffling Offloading

In regular computer systems, the host processor engages in a sequence of communications with the storage devices, wherein data from the storage devices is read into the memory hierarchy, and subsequent computational operations are performed. Upon completion of these computations, the data is then written back to the storage devices from the memory.

When an SmartSSD CSD is installed in the system, the accelerated host computation may fall into one of the following two scenarios: 1) the host processor reads the data from the SmartSSD CSD storage and transfers them into the DRAM of the SmartSSD CSD, or 2) the processor inside the SmartSSD CSD communicates directly with the storage inside the SmartSSD CSD via the fast and private P2P link, enabling it to access and retrieve the data directly. In the former case, data must traverse through the host memory before reaching the FPGA DRAM. Data stored in SmartSSD CSD follows a path that includes traversal through the SmartSSD storage controller, the SmartSSD PCIe switch, the host memory, the SmartSSD PCIe switch once again, and finally, the FPGA DRAM. Conversely, the latter scenario involves direct communication between the processor and the storage inside the SmartSSD CSD, and there is no reliance on host resources for data transfer. Therefore, the P2P communication approach can significantly reduce latency and is less resource-intensive on the host system.

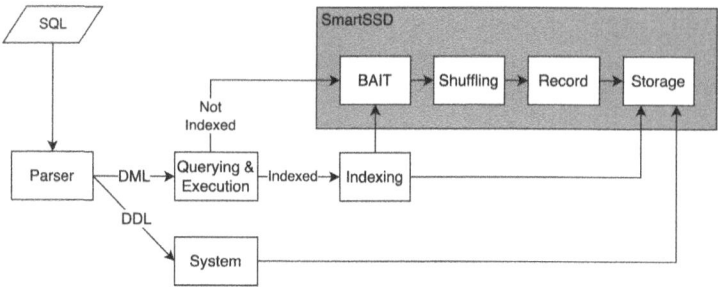

Fig. 4. System Architecture with SmartSSD CSD

Fig. 4 shows the interaction between the SmartSSD CSD and the host system in the context of SQL querying. The host system initiates this interaction by dispatching instructions and, where necessary, pertinent data essential for executing SQL statements, such as data to be inserted or updated. The SmartSSD CSD then starts to load data from the SmartSSD storage to the FPGA DRAM via the P2P communication link, with data being loaded in page-sized units. Once all the relevant pages have been successfully loaded into the FPGA DRAM, the de-shuffling and subsequent processing procedures are initiated. Upon the completion of these operations, the pages will undergo shuffling once again and then be stored back into the SmartSSD storage. At the same time, the processing results will be put to the output buffer and then returned to the host side. This approach empowers the SmartSSD CSD to effectively offload a significant portion of the I/O-intensive tasks, thereby conserving computational resources on the host for other concurrent operations.

4.3   Acceleration Module

The SmartSSD CSD contains one Xilinx Kintex Ultrascale+ KU15P FPGA with 1.143 million SLCs (system logic cells) and approximate 300,000 LUTs (Look Up Tables). These SLCs and LUTs are key components in FPGA that play a crucial role in accelerating applications as they can be configured to implement various digital logic functions. With the help of the FPGA, we designed the acceleration unit to accelerate the generation of deranged permutation.

As described in previous sections, a deranged permutation plays a vital role in the shuffling. Generating a permutation is faster than generating a deranged permutation, especially as the length of the sequence increases. Since highly efficient algorithms such as Fisher-Yates shuffle [12] can help to generate the permutation, the generation of the deranged permutation can be conducted as two steps:

- Uniformly generating a permutation sequence at random with Fisher-Yates shuffle at a time complexity of $O(n)$.
- Verifying whether the generated permutation meets the criteria of being deranged. If it does, the algorithm outputs this permutation as the result.

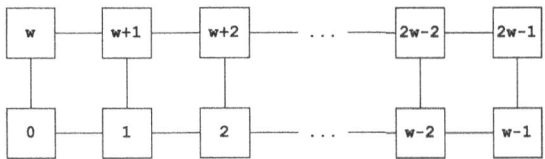

Fig. 5. Linear array model of parallel computations

If not, the process is repeated to generate more random permutations until a valid deranged permutation is obtained.

In general, the acceleration module covers the verification of whether the generated permutation is deranged based on a linear array model. We implemented a linear array based on the FPGA to accelerate the check of deranged permutation. The linear array model is consisted of $2w$ identical processors (w is the size of one block) and these processors are organized into two one-dimensional structure. Except for the first one, each $i - th$ processor, for $1 < i < w$, has local connection to processor $i - 1$. All first half w processors have the local connection to one designated processor in the other half. Every processor has a finite number of registers, and can execute basic instructions. This linear array model belongs to the SIMD class of computer architectures. During the execution of the derangement checking, processors work in synchronous manner and perform in parallel the same sequence of instructions. It is recommended that one selected processor, such as the first one, may have some additional features and can control the work of other processors. The structure of the model is shown in the Fig. 5.

During the verification, the sequence before shuffling would be stored in the processors 0 to $w - 1$, and a new permutation sequence before derangement verification would be stored in the processors w to $2w - 1$. As one pair of processors, the processor i and the processor $i + w$ would conduct the comparison in parallel to accelerate the verification of derangement.

5   Experiment Evaluation and Analysis

To evaluate the efficiency of shuffling with SmartSSD CSD, we implemented a prototype of the design on a workstation with SmartSSD installed.

Our prototype is implemented with C++ in AMD Vitis Unified Software Platform 2023.1. The experiments are conducted on a workstation machine with one Intel i7-10700F processor and 32GB of DDR4 memory. It is important to note that the SmartSSD CSD is designed for server environment, which is typically equipped with powerful cooling system and native U.2 interface. To facilitate our experiments, we used one U.2 to PCIe x4 adapter along with one additional 70mm × 70mm fan attached to the SmartSSD CSD. These measures were taken to prevent the FPGA and NAND components from reaching their thermal thresholds. The SmartSSD CSD driver and the AMD Vitis development environment were both installed on an Ubuntu 20.04.6 system, with an additional installation of an older version of the Linux Kernel to accommodate the development requirement.

In our performance evaluation, we employed an IMDB review dataset, as sourced from [1]. This dataset consists of numeric unique identifiers for reviews, accompanied by associated information such as reviewers, movies, ratings, and more. It contains over 5 million review records. To enhance its usability for our purposes, we converted the dataset from its original `json` format to `sql` format.

5.1 Shuffling Performance Evaluation

To simplify the performance evaluation process, we have made the following assumptions: 1) We have established that the shuffling process is initiated when the number of tuples l is a multiple of w (size of the block). This approach ensures that no remaining tuples are left during the conversion from a list of tuples to a matrix. 2) We assume that all columns within the relation have been indexed to maximize query performance.

To practically examine the performance of our design, we conducted a comparative analysis between two systems. In one system, SmartSSD CSD was enabled, and in the other, it was disabled. In the former, shuffling and storage operations were offloaded to the SmartSSD CSD, while in the latter, all processing was executed solely by the host CPU. We evaluate the system performance using a collection of insertion and update statements denoted as $\mathbb{Q}$. The total execution time without shuffling enabled is represented as $T_{OFF}^{\mathbb{Q}}$. Meanwhile, the execution time with shuffling enabled on a CPU-only platform is denoted as $T_{CPU}^{\mathbb{Q}}$, and the execution time with shuffling enabled with the involvement of SmartSSD CSD is denoted as $T_{CSD}^{\mathbb{Q}}$. We conducted evaluation tests encompassing both INSERT and UPDATE operations involving 50,000 and 100,000 statements. The average results of 10 rounds are shown in Table 3.

Table 3. Running Times Under Three Cases

Operation	# of Statements	$T_{OFF}^{\mathbb{Q}}$	$T_{CPU}^{\mathbb{Q}}$	$T_{CSD}^{\mathbb{Q}}$
INSERT	50,000	5.27 s	5.88 s	5.42 s
INSERT	100,000	9.31 s	10.42 s	9.69 s
UPDATE	50,000	242.18 ms	306.85 ms	254.19 ms
UPDATE	100,000	387.11 ms	498.23 ms	399.65 ms

Based on the results, the following observations can be made:

– **Minor Impact of Shuffling Process**: The time consumed by the shuffling process is found to be relatively small when compared to the overall execution time. Moreover, the integration of SmartSSD CSD demonstrates its capability to further minimize the time allocated to shuffling. This suggests that the

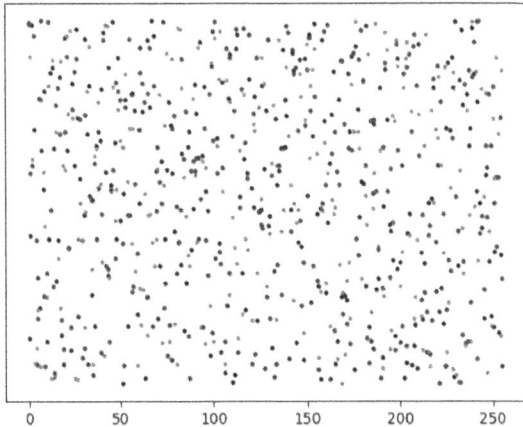

Fig. 6. Distribution of multiple versions of the breached tuple column after matrix shuffling.

shuffling operation, when accelerated by SmartSSD CSD, imposes minimal influence on the system's overall performance. Consequently, a greater portion of computational resources can be directed towards other critical operations, such as data processing and index building.

- **Substantial Performance Enhancement**: The execution of statements involving SmartSSD CSD participation exhibits a notable improvement in relative performance from 5 to 10 times faster comparing to scenarios where only the CPU is engaged in shuffling computations. This substantial enhancement underscores the significant reduction in the overhead time associated with shuffling when SmartSSD CSD is employed.

5.2 Distribution and Information Entropy Analysis

For the purpose of eliminating the distribution traces of shuffled tuples, we ameliorate the shuffling algorithm. We conduct the same experiment as Sect. 4.1, and the distribution is reflected in Fig. 6. Our shuffling based on matrix could clearly achieve a better randomness on the distribution of data points after shuffling. The matrix shuffling can conceal the distribution pattern splendidly even the attackers contrast the different shuffled versions on same tuple column. If the attackers acknowledge the fact that all the columns are shuffled, even worse, the attackers happen to know the all the correct columns in one tuple, it is unlikely that the attackers can restore the entire relation because each column has different deranged permutation sequence as the shuffle key.

Information entropy is one of the most important feature of randomness [22]. High entropy is crucial for security-critical applications, and high entropy means that the outcomes are more unpredictable and random. Let m be the procedure of shuffling; the formula for calculating the information entropy can be expressed as follows:

$$H(m) = -\sum_{i=1}^{n} p(x_i) \cdot \log_2(p(x_i))$$

where n represents the number of possible positions for the tuple to move during the shuffling, and $p(x_i)$ represents the probability of selecting each position x_i. Table 4 records the different information entropy with various column size under multi-level shuffling and matrix shuffling. We can clearly see that the matrix shuffling could provide a much larger information entropy for a better randomness.

Table 4. Information Entropy Under Different Shuffling Algorithm

Column Size	64	256	1024	1024
Multi-level Shuffling	24	64	160	4096
Matrix Shuffling	295.995	1683.996	8769.006	43250.047

6   Conclusion

In conclusion, we have proposed a solution for enhancing database security through the use of SmartSSD-accelerated cryptographic shuffling. Our approach accelerates the shuffling algorithm using the SmartSSD CSD, including software and hardware architectures, and further optimizes the algorithm by exploiting the built-in FPGA. We have conducted performance evaluations to show the performance improvements the SmartSSD CSD brings to the shuffling algorithm. Our proposed solution provides a promising direction for enhancing database security and protecting sensitive information. In the future, we will continue to explore how the features of large-scale storage systems can be integrated with database systems to enhance their performance and security.

References

1. IMDb Review Dataset - ebD. https://www.kaggle.com/datasets/ebiswas/imdb-review-dataset
2. Samsung SmartSSD. https://www.xilinx.com/applications/data-center/computational-storage/smartssd.html
3. SmartSSD computational storage drive installation and user guide (2021). https://www.xilinx.com/content/dam/xilinx/support/documents/boards_and_kits/accelerator-cards/1_3/ug1382-smartssd-csd.pdf
4. Cost of a Data Breach Report 2023. Technical report, IBM Security (2023)
5. Acharya, A., Uysal, M., Saltz, J.: Active disks: programming model, algorithms and evaluation. ACM SIGOPS Oper. Syst. Rev. **32**(5), 81–91 (1998)

6. Arasu, A., Eguro, K., Kaushik, R., Kossmann, D., Ramamurthy, R., Venkatesan, R.: A secure coprocessor for database applications. In: 2013 23rd International Conference on Field programmable Logic and Applications, pp. 1–8. IEEE (2013)
7. Boneh, D., Di Crescenzo, G., Ostrovsky, R., Persiano, G.: Public key encryption with keyword search. In: Cachin, C., Camenisch, J.L. (eds.) Public key encryption with keyword search. LNCS, vol. 3027, pp. 506–522. Springer, Heidelberg (2004). https://doi.org/10.1007/978-3-540-24676-3_30
8. Boneh, D., Waters, B.: Conjunctive, subset, and range queries on encrypted data. In: Vadhan, S.P. (ed.) TCC 2007. LNCS, vol. 4392, pp. 535–554. Springer, Heidelberg (2007). https://doi.org/10.1007/978-3-540-70936-7_29
9. Cash, D., et al.: Dynamic searchable encryption in very-large databases: data structures and implementation. Cryptology ePrint Archive (2014)
10. Coles, M., Landrum, R.: Transparent data encryption. In: Vadhan, S.P. (ed.) Expert SQL Server 2008 Encryption, pp. 127–150. Springer, Cham (2009). https://doi.org/10.1007/978-3-540-70936-7_29
11. Do, J., Kee, Y.S., Patel, J.M., Park, C., Park, K., DeWitt, D.J.: Query processing on smart SSDS: opportunities and challenges. In: Proceedings of the 2013 ACM SIGMOD International Conference on Management of Data, New York New York USA, pp. 1221–1230. ACM (2013). https://doi.org/10.1145/2463676.2465295, https://dl.acm.org/doi/10.1145/2463676.2465295
12. Fisher, R.A., Yates, F., et al.: Statistical Tables for Biological, Agricultural and Medical Research, Edited by RA Fisher and F. Yates. Oliver and Boyd, Edinburgh (1963)
13. Gaetjen, S., Knox, D., Maroulis, W.: Oracle Database 12c Security. McGraw-Hill Education Group (2015)
14. Geng, T., Alsuwat, H., Huang, C.T., Farkas, C.: Securing relational database storage with attribute association aware shuffling. In: 2019 IEEE Conference on Dependable and Secure Computing (DSC), pp. 1–8. IEEE (2019)
15. Geng, T., Huang, C.T., Farkas, C.: SCORD: shuffling column-oriented relational database to enhance security. In: Wang, G., et al. (eds.) UbiSec 2023. CCIS, vol. 2034, pp. 163–176. Springer, Cham (2023). https://doi.org/10.1007/978-981-97-1274-8_11
16. Grubbs, P., Lacharité, M.S., Minaud, B., Paterson, K.G.: Pump up the volume: Practical database reconstruction from volume leakage on range queries. In: Proceedings of the 2018 ACM SIGSAC Conference on Computer and Communications Security, pp. 315–331 (2018)
17. Hedam, N., Tychsen Clausen, M., Bonnet, P., Lee, S., Friis Larsen, K.: Delilah: EBPF-offload on computational storage. In: Proceedings of the 19th International Workshop on Data Management on New Hardware, pp. 70–76 (2023)
18. Kang, Y., Kee, Y.s., Miller, E.L., Park, C.: Enabling cost-effective data processing with smart ssd. In: 2013 IEEE 29th Symposium on Mass Storage Systems and Technologies (MSST), pp. 1–12. IEEE (2013). https://doi.org/10.1109/MSST.2013.6558444, http://ieeexplore.ieee.org/document/6558444/
19. Kapko, M.: Progress software's MOVEit meltdown: uncovering the fallout (2024). https://www.cybersecuritydive.com/news/progress-software-moveit-meltdown/703659/
20. Koo, G., et al.: Summarizer: trading communication with computing near storage. In: Proceedings of the 50th Annual IEEE/ACM International Symposium on Microarchitecture, pp. 219–231 (2017)

21. Lee, J.H., Zhang, H., Lagrange, V., Krishnamoorthy, P., Zhao, X., Ki, Y.S.: SMARTSSD: FPGA accelerated near-storage data analytics on SSD. IEEE Comput. Archit. Lett. **19**(2), 110–113 (2020)
22. Li, S.Y., Miguel Angel, B.H.: A novel image protection cryptosystem with only permutation stage: multi-shuffling process. Soft. Comput. **27**, 15319–15336 (2023)
23. Popa, R.A., Redfield, C.M., Zeldovich, N., Balakrishnan, H.: CRYPTDB: protecting confidentiality with encrypted query processing. In: Proceedings of the Twenty-Third ACM Symposium on Operating Systems Principles, pp. 85–100 (2011)
24. Priebe, C., Vaswani, K., Costa, M.: EnclaveDB: a secure database using SGX. In: 2018 IEEE Symposium on Security and Privacy (SP), pp. 264–278. IEEE (2018)
25. Riedel, E., Gibson, G., Faloutsos, C.: Active storage for large-scale data mining and multimedia applications. In: Proceedings of 24th Conference on Very Large Databases, pp. 62–73. Citeseer (1998)
26. Salamat, S., Haj Aboutalebi, A., Khaleghi, B., Lee, J.H., Ki, Y.S., Rosing, T.: NASCENT: near-storage acceleration of database sort on SmartSSD. In: The 2021 ACM/SIGDA International Symposium on Field-Programmable Gate Arrays, pp. 262–272 (2021)
27. Salamat, S., Zhang, H., Ki, Y.S., Rosing, T.: NASCENT2: generic near-storage sort accelerator for data analytics on SmartSSD. ACM Trans. Reconfigurable Technol. Syst. (TRETS) **15**(2), 1–29 (2022)
28. Soltaniyeh, M., Lagrange Moutinho Dos Reis, V., Bryson, M., Yao, X., Martin, R.P., Nagarakatte, S.: Near-storage processing for solid state drive based recommendation inference with SmartSSDs®. In: Proceedings of the 2022 ACM/SPEC on International Conference on Performance Engineering, pp. 177–186 (2022)
29. Song, D.X., Wagner, D., Perrig, A.: Practical techniques for searches on encrypted data. In: Proceeding 2000 IEEE Symposium on Security and Privacy. S&P 2000, pp. 44–55. IEEE (2000)
30. Tang, L., Li, T., Jiang, Y., Chen, Z.: Dynamic query forms for database queries. IEEE Trans. Knowl. Data Eng. **26**(9), 2166–2178 (2013)
31. Tavakoli, E.B., Beygi, A., Yao, X.: RPKNN: an OpenCL-based FPGA implementation of the dimensionality-reduced KNN algorithm using random projection. IEEE Trans. Very Large Scale Integr. VLSI Syst. **30**(4), 549–552 (2022)
32. Tomes, E., Altiparmak, N.: A comparative study of HDD and SSD raids' impact on server energy consumption. In: 2017 IEEE International Conference on Cluster Computing (CLUSTER), pp. 625–626. IEEE (2017)
33. Tu, S.L., Kaashoek, M.F., Madden, S.R., Zeldovich, N.: Processing analytical queries over encrypted data. In: Proceedings of the VLDB Endowment (2013)
34. Wang, J., Park, D., Kee, Y.S., Papakonstantinou, Y., Swanson, S.: SSD in-storage computing for list intersection. In: Proceedings of the 12th International Workshop on Data Management on New Hardware, pp. 1–7 (2016)
35. Wu, S., Li, Q., Li, G., Yuan, D., Yuan, X., Wang, C.: ServeDB: secure, verifiable, and efficient range queries on outsourced database. In: 2019 IEEE 35th International Conference on Data Engineering (ICDE), pp. 626–637. IEEE (2019)
36. Xue, K., Li, S., Hong, J., Xue, Y., Yu, N., Hong, P.: Two-cloud secure database for numeric-related SQL range queries with privacy preserving. IEEE Trans. Inf. Forensics Secur. **12**(7), 1596–1608 (2017)

Ensuring End-to-End IoT Data Security and Privacy Through Cloud-Enhanced Confidential Computing

Md Shihabul Islam$^{(\boxtimes)}$ ⓘ, Mahmoud Zamani ⓘ, Kevin W. Hamlen ⓘ, Latifur Khan ⓘ, and Murat Kantarcioglu ⓘ

The University of Texas at Dallas, Richardson, TX 75080, USA
{md.shihabul.islam,mxz173130,hamlen,
lkhan,muratk}@utdallas.edu

Abstract. IoT devices gather data from the most intimate and sensitive aspects of our lives, transmitting it to untrusted cloud services for further managing and automating tasks through interconnecting smart devices without human intervention. To safeguard sensitive and private IoT data, solutions based on Trusted Execution Environments (TEEs) could be utilized, providing end-to-end encrypted solution. Specifically, TEEs securely process sensitive data within a protected area of the processor, isolated from the main operating system and applications, ensuring data confidentiality and integrity. However, in this study, we demonstrate that the end-to-end encryption offered by TEE based solutions for IoT devices may not be entirely sufficient. We present the first attack against TEE-based IoT solutions that can deduce sensitive information, such as a motion sensor reading, merely by analyzing memory access patterns. Our findings show that we can identify the type of device with about 95% accuracy and determine the values sent by IoT devices, like temperature readings, with approximately 85% accuracy. To counter these vulnerabilities, we design a system that enhances data security for IoT solutions in the untrusted cloud, using techniques like data oblivious execution and padding. With these defenses, we observe significant reduction in accuracy of device type detection and value prediction to at most 27% and 19%, respectively.

Keywords: IoT · Data Security and Privacy · Confidential Computing

1 Introduction

The rapid expansion of Internet of Things (IoT) devices, expected to reach 1 trillion devices by 2035 [34], underscores their widespread use across applications, specially automation platform transforming homes, cities, healthcare, and numerous other domains. However, the generation and transmission of diverse data, including sensitive information, by these devices necessitate robust measures to preventing potential breaches and ensuring confidentiality of personal

© IFIP International Federation for Information Processing 2024
Published by Springer Nature Switzerland AG 2024
A. L. Ferrara and R. Krishnan (Eds.): DBSec 2024, LNCS 14901, pp. 71–91, 2024.
https://doi.org/10.1007/978-3-031-65172-4_5

and proprietary information. Despite cloud services implementing security features like authentication and access control, these have been insufficient and vulnerable to breaches previously [3,23]. For instance, data from over half a million CloudPets users was compromised, exposing emails, passwords, photos, and over two million voice recordings stored insecurely [17]. With 44% of users highly concerned about private data leaks from smart home IoT devices [14], it's clear that an additional layer of protection is crucial to fully ensure data confidentiality, integrity, and safeguard user privacy.

Confidential computing stands as a promising approach to protecting IoT data when it is processed in the cloud. At its core, confidential computing ensures that data remains encrypted not only during transit and at rest, but also during the processing phase, a state traditionally considered vulnerable. By making use of hardware based Trusted Execution Environments (TEE), confidential computing creates a protected space in memory, isolating the data and the operations on it from other processes, even those with higher privilege levels. For instance, Microsoft Azure cloud service supports confidential computing with TEEs, namely, Intel Software Guard Extensions (INTEL SGX). For IoT data, this means that sensitive information sent to the cloud for processing can remain encrypted, thereby bolstering the security and privacy of IoT systems.

Despite the security advantages offered by TEEs for IoT data, it is not clear whether they disclose sensitive information even if the underlying hardware is trusted. One significant area of concern involves side-channel information [15,41,44]. Even when the TEE is trusted, it is still susceptible to certain types of attacks, specifically, those predicated on analyzing memory access patterns [16,45]. Sophisticated attackers can monitor these patterns, even in the face of encryption, to infer sensitive information about the operations being performed or the data being processed. This poses a potential vulnerability for IoT systems, as an attacker could potentially discern, for instance, a user's behaviors, user's intimate devices, or a device's operational status from the patterns of data access in memory. Therefore, while TEEs and confidential computing represent crucial steps forward in securing IoT data, further investigation is required to understand the magnitude of such attacks on IoT systems.

In this work, to our knowledge, *we are the first ones to demonstrate attacks against* TEE-based IoT systems, specifically leveraging memory access patterns. By developing a machine learning (ML) based inference attack, we show that these access patterns, while seemingly innocuous, can inadvertently leak significant and sensitive information regarding the underlying IoT platform. This discovery underscores the need for additional layers of protection even within TEE-based systems to fully safeguard the privacy and integrity of sensitive IoT data, emphasizing that encryption alone may not be sufficient against certain kinds of attacks.

In this paper, leveraging the insights gathered by our ML-based inference attacks, we design a system, named SECIoTT (SECureing IOT data with Tee), that addresses the challenges associated with using TEE in the cloud to develop an end-to-end secure and privacy preserving IoT data management framework.

As a case study, we develop a rule-based programming platform for IoT automation as our application that runs in the TEE of an untrusted cloud and administers the automation for controlling and managing IoT devices and inter-device interactions. As a starting point, all data generated by IoT devices are encrypted and only decrypted inside the TEE, so that even the *cloud provider may not see the sensitive data*. In addition, we consider utilizing TEEs in the IoT/*edge* devices to ensure an end-to-end confidential computing platform, where IoT data is always protected both in transit and at rest.

To prevent access pattern leakage attacks on the TEEs and thus ensure the confidentiality of IoT data, we first explore using *data oblivious* solutions where we obscure the memory access patterns. Our results show that this may not be enough in certain cases due to different number of memory accesses processed for different devices (i.e., number of rules processed and number of events transmitted to control IoT behavior). To prevent this leakage, we explore *padding* different properties to obscure information, such as the number of rules used by each device and number of events dispatched to respective IoT devices to execute the automation. We provide empirical evidence that our ML-based inference attack can successfully utilize access pattern leakages to infer secret information of IoT data and automation framework from the TEE. For instance, we can detect device type (e.g., temperature sensor) and device measurement (e.g., temperature reading) with approximately 95% and 85% accuracy, respectively. Subsequently, through empirical evaluation, we demonstrate that a combination of oblivious execution and padding effectively thwarts our ML-based inference attack by obscuring access pattern information collected from the TEE, resulting in a reduction of device type and device measurement detection accuracy to approximately 27% and 19%, respectively. Furthermore, we conduct comprehensive experiments with real IoT devices to study the performance and scalability of SECIOTT regarding the IoT automation framework.

Our contributions can be summarized as follows:

- To our knowledge, we provide the first sensitive information inference attack against TEE-based IoT systems using memory access patterns.
- We introduce data oblivious execution to hide access patterns. We further show that it is not enough to reduce the attack success to the desired levels and develop padding techniques to hide sensitive information.
- Based on the insights provided by our novel sensitive inference attack, we design an efficient IoT data management and rule-based automation programming platform SECIOTT based on TEEs reliable for deploying on the untrusted cloud.
- We evaluate SECIOTT by simulating a smart home platform using 20 different and diverse real smart IoT devices in a realistic environment.

The rest of the paper is organized as follows. Section 2 explains the threat model. Section 3 introduces our proposed framework and its components. Section 4 details the adopted security measures to thwart side-channel attack. Section 5 describe the experiments and evaluation of the framework. Section 6 and Sect. 7 describe limitations with future work and related work, respectively. Finally, Sect. 8 concludes our work.

2   Threat Model

In this paper, we consider an adversary that seeks to gain insight into the private and sensitive information of the user, especially user's surrounding environment such as in a smart home system. The principal goal of the adversary is to access IoT device information, IoT data, user data, and automation rules stored in the untrusted cloud service provider and steal secrets for certain incentives. The adversaries may also eavesdrop on the communication channel between the client and the server to retrieve information.

We assume that the *only trusted* component that the user can depend on in the cloud is the TEE (e.g., INTEL SGX), whereas the other privileged software components and applications outside of TEE including the OS are *untrusted*. The adversary also has control over the memory and storage, hence, they may try to observe memory access patterns, cache usage, execution time, and any other resource usages by the program, and consequently infer application secrets [16]. However, we presume that adversaries are only curious about the private and sensitive user information and do not intend to endanger the communication between the IoT devices and the cloud or interfere with the automation process in the enclave. Moreover, we surmise that the user's IoT devices are equipped with TEE (e.g., ARM TRUSTZONE [42]) security features. We consider that adversaries cannot get root or physical access to the IoT devices to manipulate hardware. Denial-of-Service (DoS) attacks [26] and Rollback attacks [19] are out of the paper's scope.

3   System Design

We develop SECIOTT, a cloud-based *secure end-to-end encrypted* IoT *rule-based trigger-action* platform, especially designed to mitigate the leakage of sensitive IoT data and private user information. The principal idea is to utilize state-of-the-art TEEs, i.e. INTEL SGX in the untrusted cloud and ARM TRUSTZONE at the edge, and proper cryptographic techniques to alleviate data security and privacy issues.

3.1   Design Preliminaries

Here we briefly describe the overall architecture for SECIOTT, which is demonstrated in Fig. 1. We explore a smart home scenario[1] where various interconnected IoT devices like multipurpose sensors, smart bulbs, outlets, and security cameras facilitate home automation through shared resources. Users define

---

[1] We consider the smart home platform as our motivating application with a focus on the growing smart home device market, projected to reach 88 billion USD in consumer spending by 2025 [37].

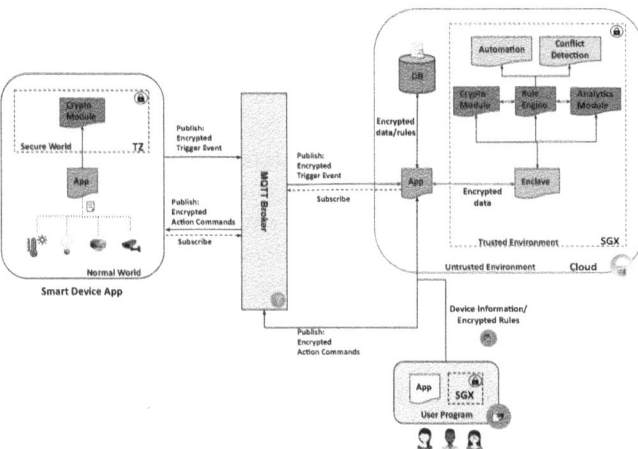

Fig. 1. SecIoTT architecture.

automation rules using a trigger-action programming paradigm, which are then sent to an untrusted cloud supported by INTEL SGX. We assume IoT devices are equipped with ARM TRUSTZONE and interact with cloud services using MQTT (Message Queuing Telemetry Transport) [21], which offers low bandwidth usage and a minimal code footprint. MQTT secures data during transmission via TLS/SSL and authenticates clients using protocols like OAuth. Importantly, our system is platform-agnostic and can be integrated into any smart environment, such as smart industry, that employs trigger-action-rule-based automation in the cloud.

3.2 Rule Deployment Strategy

To automate tasks using smart devices, users set up "rule" through a web or app interface in a "trigger-action" format, which links multiple devices for customized activities [2]. For example, SAMSUNG SMARTTHINGS allows users to connect SMARTTHINGS devices and manage automation through SmartApps using the *Rules API*, with the SMARTTHINGS cloud processing these rules [11]. Typically, a trigger-action rule includes conditions based on device states (e.g., active/inactive) or sensor readings (e.g., temperature) that trigger specific actions and commands (e.g., turn on/off) to control or activate devices as needed [36].

Typically, the cloud service provider initializes the required SGX enclave instances on the untrusted cloud platform for users. Following the successful attestation of the enclave as an authentic INTEL SGX platform, the server establishes a secure communication channel with the user for sharing sensitive details such as encryption/decryption keys and device information, as outlined in Appendix A.1. Users then encrypt and register their rules via this secure channel using the shared secret key. Our framework adopts the SMARTTHINGS JSON rule structure [27] to ensure compatibility with the established SMARTTHINGS system, with a sample rule displayed in Listing 4 in Sect. A.3. For encryption, we employ the Advanced Encryption Standard (AES) [9], specifically AES-GCM with a 128-bit key, ensuring data confidentiality and integrity in transit.

After receiving the rules, SECIOTT forwards them to the SGX *Enclave*, where they are decrypted using a shared secret key, parsed, re-encrypted with the enclave's own secret key, and then stored externally in a database. This ensures that user rules are never exposed in plaintext outside the enclave, maintaining their security at rest in the untrusted cloud environment. However, interactions between rules or with the physical environment may create unsafe conditions and increase the attack surface [10,40]. These potential conflicts are addressed in our *Rule-Conflict Detection* module, detailed in Sect. A.2.

3.3   Automation with IoT Data

In SECIOTT, TRUSTZONE securely stores sensitive device information, such as device IDs and encryption keys for payloads. It encrypts payload events (i.e., trigger event), which include device data and states (e.g., temperature, humidity, dust level, energy consumption), and transmits these via the MQTT publish method. Listing 5 in Sect. A.3 provides an example of such a trigger event.

Upon receiving trigger events, the untrusted cloud server sends them to the SGX enclave, where the payload is decrypted, and device and rule information is retrieved. The *RuleEngine* in the enclave then checks if the *trigger event* values match any registered *rule trigger conditions*. If a match occurs, the rule is *satisfied*, and the corresponding *action-command* is encrypted and sent to the designated actuator via MQTT to execute the automation. If there is no match, the event is discarded, and the system processes the next event. Actuators receive the encrypted *action-command event* via MQTT subscription, decrypt it in TRUSTZONE, verify the payload, and adjust their states accordingly.

In summary, SECIOTT processes the entire rule-based automation of the IoT platform within the INTEL SGX's secure memory container, the *enclave*. User-defined rules, device information, and data from smart devices remain encrypted outside the enclave and are only decrypted within it for processing. Encryption and decryption keys are securely stored inside the enclave to prevent data and rule manipulation by attackers. On the client side, sensitive device information and secret keys are safeguarded in the ARM TRUSTZONE to ensure protection at the edge. Thus, all data is secure at rest, supported by TEEs. Additionally, during transmission, client information is protected by the secure communication channels provided by SGX and MQTT.

4   Side-Channel Analysis: Attack and Defense

It is widely acknowledged that TEEs are vulnerable to side-channel attacks that adversaries exploit to steal sensitive information from an enclave process [22,45]. In this section, we analyze what plausible secret information of IoT and user data could be exposed through access pattern leakage of INTEL SGX. In addition, we discuss multiple defense strategies, including oblivious execution and padding, for SECIOTT to thwart the access pattern based side-channel.

4.1   Attack Analysis

Utilizing the enclave feature of INTEL SGX, we can prevent any unauthorized access to our rule-based framework. However, the adversaries with supervisor-level access and control of the untrusted system may observe side-channel information, such as instruction, memory, disk, cache usage, network access, and execution time patterns of different processes, or any other resource usages from the untrusted environment [22,45]. Utilizing the side-channel information, adversaries can exploit private contents of the enclave and infer enclave secrets [8]. In this work, we consider access pattern based side-channel attack based on program execution inside the enclave.

For this purpose, we utilize SGX-Step [38], an open-source framework that uses APIC timers to interrupt the target enclave after each instruction. This allows for the execution of attack-specific routines between handling the interruption and resuming enclave operations. By meticulously advancing through the enclave's execution step by step, SGX-Step can quantify side-channel data like the virtual address of instructions (i.e., instruction pointer) and execution duration of individual instructions. While the instruction pointer address is discernible in debug enclaves, we assume our attacker's access to these pointers to foster a more resolute and robust adversary model [24]. Utilizing sequences of these instruction pointer addresses, adversaries can observe the secret-dependent branching within the enclave's program, thus deducing concealed information. We use these virtual address sequence of all instructions in the enclave execution to evaluate our data oblivious approach.

As heterogeneous IoT devices may have diverse ruleset, processing the rules and events in the enclave may disclose various access patterns with distinguishing characteristics. Moreover, the number of rules or device events dispatched to the actuator devices may also leak certain information to the adversary. These patterns can be further analyzed to unravel secret properties of the rules and events, thus exposing underlying IoT data. Moreover, we retrieve encrypted rules from the database that corresponds to the *trigger-event* device. The number of retrieved rules may also leak certain information to the adversary, although they cannot comprehend the rule properties since it is only decrypted in the enclave. Furthermore, after rule *satisfaction*, the number of action-commands dispatched to the actuator devices are visible to the adversary. Although they cannot comprehend the command properties since it is encrypted before leaving the enclave. Yet, the quantity of retrieved rules or commands dispatched for each trigger event may leak information about the trigger or actuator devices.

In Fig. 2, we illustrate the severity of information leakage through virtual memory access patterns. Here, we collect access pattern information from SGX using SGX-Step while SECIoTT is securely performing automation with rules and events in the enclave. We consider the possibility that adversaries might deploy replay attacks to capture comprehensive access pattern data. Afterwards, we use ML-based classification models to deduce various properties, such as the type of triggering device, its measurements (values/states), and the count of rules linked to each device. The figure shows that when no defense is employed, the

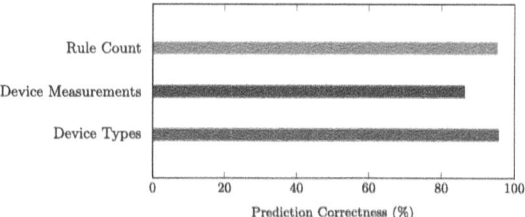

Fig. 2. Severity of information leakage via access pattern without any defense strategy.

access pattern leakage can successfully infer crucial information regarding user's IoT devices, data, and automation framework. Later in Sect. 5.2, we explain the attack in more detail and empirically show that these access pattern leakage can be alleviated using our proposed defense mechanism.

While adversaries cannot directly access data within the enclave, by observing side-channel information, they can obtain potentially valuable insights such as device types, device measurements/states, and rule counts. A substantial number of rules may suggest a more complex system and a user with sophisticated needs or preferences, thereby exposing areas of dependency on the system. Furthermore, users with an abundance of rules could be perceived as more valuable targets, possessing sensitive data or extensive automation control. Conversely, users with fewer rules may be perceived as less security-conscious, creating potential vulnerabilities for social engineering or other forms of attacks. Moreover, gaining insights into device types offers perceptions into user habits and allows attackers to customize attacks [30]. For example, knowing the types of devices in a smart home can allow attackers to craft convincing phishing emails or messages. In addition, correlating device types with IP addresses elucidates network topology, potentially revealing vulnerabilities in connections and communication patterns.

An adversary estimating device measurements/states or observing the pattern of rules can map user schedules, household environment, or even financial status, creating a comprehensive profile for further exploitation [13]. For example, rules involving motion sensors expose movement patterns and object presence; or sensor measurements (e.g., temperature, humidity) of trigger events provide insights into the physical layout and conditions of the home. Therefore, securing the properties and analysis of trigger-events and rules through mitigating side-channel leakage is crucial to prevent unintended privacy invasion, security breaches, and misuse of sensitive information.

4.2   Defense Strategy

Oblivious Execution. To obstruct side-channel from leaking sensitive and private information, one solution is to make the program *data oblivious* [22,31]. The intuition behind this approach is that any program logic that depends on the sensitive and private data should follow a data independent execution flow, so that the resource access patterns do not depend on the secret data. By enforcing this data-oblivious execution, the program effectively obfuscates any potential leakage of access pattern information, as the program's execution flow remains the same regardless of the specific data being processed. In addition, data independent execution is also effective against cache and page level attacks [4].

In our system, SECIOTT, we meticulously implement a comprehensive suite of data-oblivious strategies, such as control flow obfuscation, maintaining uniform access, code/memory padding, and masking code logic, to safeguard against access pattern leakage within the secure enclave environment. Control flow obfuscation and masking code logic ensure that the program's control and execution flow is independent of the sensitive data, making it challenging for adversaries to infer information from the observed execution patterns. Uniform instruction access guarantees that the memory access patterns of the program are consistent across different inputs and do not reveal any information about the underlying data. Additionally, code and memory padding are applied to further mask any discernible correlations between data processing activities and the observed patterns of resource utilization.

Usually, adversaries cannot observe the contents of the internal CPU registers outside of the trusted boundary [32]. This ability to conceal register contents is crucial, as operations involving these registers do not leave any discernible memory access patterns that could be exploited by attackers. Therefore, we utilize the registers to implement different oblivious data arithmetic operations using low-level assembly instructions to mask the underlying code logic without revealing the nature of the operations or the data being processed. Moreover, adversaries might detect conditions under which specific branches are executed, potentially revealing sensitive data through the patterns of branch accesses and causing branch prediction attacks. To hide the actual control flow of the program, we further utilize the registers to implement the oblivious flag-based conditional moves, where a register flag will determine which branch to take from all the possible values, which are also stored in registers [25]. An example of such an oblivious primitive, which checks an equality condition using registers, is demonstrated in Listing 1. This specific primitive is particularly useful for obliviously evaluating whether a trigger-event value meets the predefined threshold specified in a rule, thereby determining if the rule is satisfied without leaking information about the trigger or the rule's threshold value.

Throughout our implementation, we follow these data-oblivious strategies to hide the distinguishing patterns and make the program execution flow independent of different rules and events. As discussed in Sect. 4.1, both rules and events contain crucial and delicate information of users. Therefore, we make sure that managing rules, handling events, and processing rule automation in the *enclave* is performed in a data-oblivious manner. More specifically, we guarantee that the program execution flow is same for disparate rules and events during automation by following data-oblivious code structure and control flow ensuring identical instruction execution sequences.

Listing 2 and Listing 3 demonstrates the difference between a non-oblivious and oblivious function that compares the event value and rule's trigger value based on the "equality" condition during automation. In non-oblivious implementation, the function's execution flow directly depends on the comparison of the event value and the rule's trigger value. In contrast, the oblivious implementation leverages the oblivious primitive demonstrated in Listing 1, which utilizes CPU registers to perform the comparison in a data-independent manner. In addition, the oblivious function accesses all instructions and elements sequentially irrespective of the data being processed, thereby obfuscating control flow patterns. Therefore, even if an adversary were to observe the program's execution, they would be unable to deduce secret data, such as device trigger value or rule's threshold, based on the observed memory access patterns or control flow behavior.

Padding. As our ML-based inference attack shows that the number of rules and action-commands may disclose sensitive information, we try to obscure the leakage through *padding*. We obfuscate the number of encrypted rules retrieved from database by fetching a random number of dummy rules in addition to device-specific rules. Additionally, we try to obscure the leakage caused by the number of dispatched action-commands and the network pattern by sending a random number of dummy commands to dummy devices irrespective of any rules being *satisfied*. Therefore, the actual number of action-commands that are being sent to the actuator devices are hidden from the adversaries. To disambiguate the noise, we incorporate a *flag* in the payload to indicate whether a command or rule is a dummy or legitimate. While dummy rules are processed as usual within the enclave, they exclude the step of sending out action-commands. Conversely, when the actuator device receives action-commands, it decrypts the payload in the TRUSTZONE and acts only on the legitimate commands, ignoring any dummy ones. To prevent any information leakage from the randomization, we also provide a mode to retrieve a fixed number of rules and dispatch a fixed number of action-commands. By introducing a mix of real and dummy rules and events, it becomes difficult for attackers to distinguish genuine patterns from decoys.

Moreover, the discrepancy in the encrypted rule and event sizes may reveal important knowledge about the rules or IoT data. Adversaries could exploit this inconsistency by monitoring the rule and event sizes. To counter this, we standardize the size of the rules by padding them so that all appear uniform. We apply a similar strategy to the events as well. For simplicity of discussion, we denote the adoption of dummy data and padding of the data as *padding*. Padding can make it challenging for adversaries to infer sensitive information from observed patterns. It creates uncertainty and noise, making it difficult for adversaries to correlate rules or events confidently, hence preventing them from drawing accurate conclusions about user data or preferences. We empirically demonstrate the effectiveness of our data oblivious approach and padding later in Sect. 5.2.

```
int obliviousEqual(int val1,
↪    int val2, int out1, int
↪    out2){
     int out = 0;
     asm ("movl %1, %%eax;"
          "movl %2, %%ebx;"
          "movl %3, %%ecx;"
          "movl %4, %%edx;"
          "xorl %%eax, %%ebx;"
          "cmovz %%ecx,
          ↪    %%edx;"
          "movl %%edx, %0;"
          : "=r" ( out )  /* output
          ↪    */
          : "r" ( val1 ), "r" (
          ↪    val2 ), "r" (out1),
          ↪    "r" (out2)  /* input
          ↪    */
          );
     return out;
}
```

Listing 1. Obliviously checking equality condition

```
int
↪    equalityCondition(enum
↪    OpType op, int
↪    event_value, int
↪    rule_value){
     // checks if operator
     ↪    type is "equal"
     if(op == EQ){
          // checks if
          ↪    values are
          ↪    equal
          if(event_value ==
          ↪    rule_value)
               return 1;
     }
     return 0;
}
```

Listing 2. A non-oblivious function

```
int equalityCondition(enum OpType op,
↪    int event_value, int rule_value){
     int isEqualOp =
     ↪    obliviousEqual(op, EQ, true,
     ↪    false);
     int isEqualValue =
     ↪    obliviousEqual(event_value,
     ↪    rule_value, true, false);
     return
     ↪    obliviousAndOperation(isEqualOp,
     ↪    isEqualValue, true, false);
}
```

Listing 3. An oblivious function

5   Experiments and Evaluations

In this section, we evaluate SECIOTT within a smart home context by securely executing rules inside the TEE. Specifically, we assess how real-time processing of rules impacts the performance and scalability of the automation platform executed in cloud-based INTEL SGX. Additionally, we test the effectiveness of our ML-based inference attack through access patterns and demonstrate the efficacy of our proposed defense strategies in safeguarding IoT data confidentiality and integrity.

We develop the enclave application of the SGX cloud module by employing the INTEL SGX SDK APIs and libraries in *C*. We implement the on-device application with *C* language and OP-TEE [18] as the TEE of TRUSTZONE. In our prototype, we use a Python implementation to build the rules, where this approach could be easily extended with any other programming languages. In addition, we use MongoDB as the database to store rules.

5.1   Experimental Setting

Device Configuration. We use a system with 8-core i7-6700 (Skylake) processor containing INTEL SGX security features, operating at 3.4 GHz, running Ubuntu 18.04 with 64 GB RAM for simulating the cloud host machine. However, it is important to recognize that, our enclave solution is a prototype of a scheme that any cloud provider can integrate as a service with a TEE, including Microsoft Azure cloud. In addition, our attacks are applicable whether it is in cloud or on-prem as long as the attacker can observe the memory access pattern. For our evaluation, we use our rule-based IoT application and local cloud environment rather than any other existing IoT applications or cloud provider for the following two reasons: 1) IoT applications, not being open-source, could

Table 1. Smart devices used in the Smart Home Automation

Device Type	Brand	# Devices	Capability
Multi-purpose Sensor	SmartThings	1	Temperature Measurement, Contact Sensor
Multisensor	Aeotec	1	Temperature Measurement, Motion Sensor, Illuminance Measurement, Relative Humidity Measurement
Arrival Sensor	SmartThings	1	Presence Sensor
Water Leak Sensor	SmartThings	1	Water Sensor
Motion Sensor	SmartThings	1	Motion Sensor
Door/Window Sensor	Ecolink	1	Contact Sensor
Tilt Sensor	Ecolink	1	Contact Sensor
Smart Bulb	SmartThings, Phillips Hue	3	Light, Dimmer Control, Color Control
Smart Plug	SmartThings	2	Switch
Smart Outlet	SmartThings	2	Switch
Button	SmartThings	1	Button
Tracker	SmartThings	1	Presence Sensor
Security Camera	Arlo	1	Sound Sensor
Robot Vacuum	Samsung	1	Robot Cleaner Cleaning Mode, Robot Cleaner Movement
Smoke Detector	First Alert	1	Smoke
MultiSiren	Zooz	1	Alarm

not be adapted to work with SGX; 2) Any applications utilizing existing cloud-based confidential computing model (e.g., Microsoft Azure confidential computing cloud with INTEL SGX), could not be used as the side-channel information from these cloud vendors are inaccessible.

To assess SECIoTT in a realistic scenario and for security evaluation, we explore real embedded devices connected with real sensors and actuators as our IoT/edge device. We use two Raspberry Pi (i.e., RPi 3 Model B) and one STM32MP157C-DK2 from a popular STM32MP1 series of MPUs as our embedded devices. For sensors we use Raspberry Pi Sense HAT that can measure temperature, humidity, pressure, and orientation; as well as a few standalone motion and temperature sensors. For actuators we use LEDs. To assess the scalability and performance of SECIoTT, we simulate a smart home environment with 20 smart devices, shown in Table 1, from different categories with various capabilities. We set up all the devices in our lab and connect them with a SMARTTHINGS hub.

Ruleset and Dataset. For the experiments, we use rulesets of sizes $100, 1k, 10k,$ and $100k$ focusing on diverse combinations of sensors and actuators. For the automation, we accessed data from the SMARTTHINGS cloud using the SMARTTHINGS API and from the embedded devices over several days, experimenting with event sets of sizes $100, 1k, 5k$, and $10k$.

5.2 Security Evaluation

We evaluate our framework SecIoTT to verify its effectiveness against side-channel attacks by analyzing the program execution access patterns of the enclave program. The goal of this evaluation is to discover how our defense strategy (i.e., data-obliviousness and padding) obscures the distinguishing characteristics of the sensitive execution flow of SecIoTT. As explained in Sect. 4.2, we utilize SGX-Step [38] to collect virtual address sequence of all instructions in the enclave execution to evaluate our data oblivious approach. We define the address sequence for each incoming trigger event as a *trace*, which follows a *time-series* pattern. Utilizing machine learning techniques, we develop models based on these captured traces to identify unique features. Our hypothesis is that traces obtained using an effective oblivious approach should appear nearly indistinguishable, whereas traces without such methods should show clear differences. To validate this, we perform supervised classification on the collected traces, aiming to demonstrate that extracting information from the traces protected by the oblivious method resemble random conjecture.

We gather traces based on trigger events from different sensor devices and with 4 different settings: *oblivious with padding, oblivious without padding, non-oblivious with padding*, and *non-oblivious without padding*. We use in total 10 devices (i.e., motion, temperature, humidity, water, presence, contact sensor; smoke, carbon monoxide detector; illuminance measurement, button) and for each device we collect $1k$ traces. We perform supervised classification using Bi-directional LSTM [33] and XGBoost [6], where, given a trace, our goal is to predict the trigger device types, number of rules the trigger event satisfies, and the sensor value/state in the trigger event. We report here the highest accuracy from both these classifiers, although they exhibit very similar results. We select the first 80% of collected data as the training data and the remaining 20% as test data, while maintaining the same class ratio in both dataset.

Prediction of Device Type. The adversaries may want to identify the triggering device type, a knowledge that could provide insights into user's registered rules, surrounding environment, or even the value of the triggering device. Figure 3a displays the accuracy of the multiclass classification using traces collected from 10 different devices, resulting in 10 classes in our prediction model. Our model achieves an accuracy of about 95% in predicting the device type for the non-oblivious approach. However, when oblivious methods and padding are implemented, the variation in the address sequence is minimized, leading to a significantly reduced prediction accuracy of approximately 27%. Although this does not reach the level of random guess accuracy (i.e., 10%), it is significantly lower than that of the non-oblivious approach. Hence, SecIoTT prevents adversaries from observing data-dependent distinguishing characteristics from the side-channel traces that can assist them to discover the device type.

Prediction of Number of Rules. An adversary may attempt to determine the number of rules a trigger device event satisfies within the enclave, potentially revealing information about the user, as discussed in Sect. 4.1. For that purpose, we design a classifier that predicts the number of rules satisfied given a trace from a device event. We use the same 10 devices, settings, and classifiers from the previous experiment. Note that, our ruleset is balanced; that is for each trigger device the number of rules in the ruleset are same. However, depending on the value of the trigger event, different devices satisfy different number of rules. Figure 3b illustrates the classification results. Similar to the previous experiment, oblivious execution with padding achieves the lowest accuracy with approximately 34%, while no oblivious method and no padding obtains the highest accuracy with approximately 95%. Recall that, we incorporate dummy rules and action-commands in our *padding* scheme to make the actual number of processed rules and action-commands. Observations show that without the padding, the prediction accuracy is 49.5%, making padding a crucial component of the defense.

Prediction of Trigger Values. In this experiment, we aim to determine if adversaries can predict the state or value of a device/sensor's trigger event from the traces after *successfully identifying the trigger device type*. We developed classifiers for three different devices: a motion sensor, a temperature sensor, and a humidity sensor. The motion sensor's trigger values are classified as either *active* or *inactive*. For the temperature and humidity sensors, which produce continuous numerical values, we used a bucketing technique to categorize values into distinct bins. Each classifier predicts the specific bin containing the trigger value. We created five equally spaced bins for the class labels, with temperature ranging from 50 to 100 degrees Fahrenheit and humidity from 20% to 70%.

Figure 3c exhibits the binary classification result for the motion sensor device achieving lowest accuracy for traces collected with oblivious strategy, i.e. approximately 43.5%, less than random guess (50%). On the contrary, the non-oblivious approach gains highest accuracy, which means traces portray discernible features that isolates *active* from *inactive* traces. Similarly, Fig. 3d depicts the prediction accuracy of temperature and humidity values from traces, where accuracy score for the data oblivious approach is close to random guess (20%). Consequently, from these experiments, we can determine that our data oblivious approach, coupled with padding, successfully obscures the code execution pattern gathered from enclave execution, thereby hiding data-dependent alteration of the program and definitively mitigating the risks posed by access pattern based side-channel attacks.

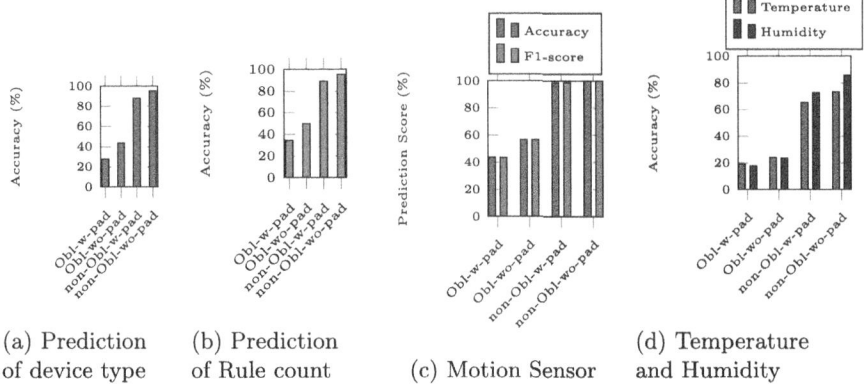

(a) Prediction
of device type

(b) Prediction
of Rule count

(c) Motion Sensor

(d) Temperature
and Humidity

Fig. 3. Multiclass classification (10 classes) result for predicting (a) device type and (b) number of rules satisfied in enclave. Prediction results of (c) Motion Sensor Values (*active, inactive*) and (d) Temperature and Humidity values (5 buckets).

5.3   Performance Analysis

Execution Time. Our goal of this evaluation is to measure the computational time overhead of SECIoTT after the integration of INTEL SGX and the side-channel defense techniques. We consider three cases for the experiment: 1) No SGX (*nosgx*), that provides no security guarantee of data; 2) with SGX but without any defense or encryption mechanism (*sgxnodef*), which may provide integrity of data but lacks confidentiality; and 3) with SGX including proposed oblivious strategy (*sgxdef*), that provides total security guarantee. We measure the average execution time of the rule-based automation with varying rule set size and a fixed device event set size of $10k$ by querying the devices listed in Table 1 for all the cases. Figure 4a illustrates the comparison of average execution time of the automation for the three cases. Overall, the introduction of SGX including proposed data-oblivious and padding techniques incur slight time overhead compared to the case with only SGX. Nevertheless, it ensures the data integrity and confidentiality, which the other cases fail to achieve, with a trivial trade-off in execution time.

Memory Consumption. We investigate how the enclave memory is affected by the proposed defense. We measure the heap memory of the enclave using the Enclave Memory Measurement Tool (EMMT) and perform the experiment with varying rule set sizes and a total of $10k$ devices events for two cases: 1) with SGX but without any defense (*sgxnodef*); and 2) with SGX including proposed defense strategy (*sgxdef*). Figure 4b illustrates the observed heap memory usage in the enclave in *kilobytes*. The implementation with the proposed defense uses more enclave heap memory than without defense. However, this increase is negligible as the framework does not store any rules or data permanently in the enclave, retrieving rules from the database only as needed.

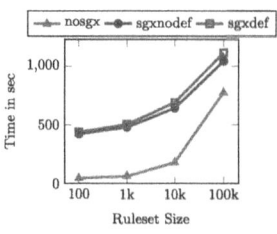

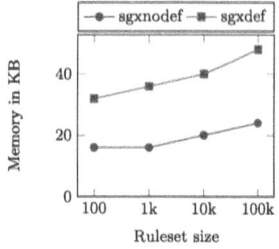

(a) Average execution time   (b) Enclave heap memory

Fig. 4. (a) Comparison of average execution time for three cases: no SGX ——, SGX without proposed defense ——, and SGX with proposed defense ——. A total of $10k$ device events are considered with varying Ruleset sizes of 100, $1k$, $10k$, and $100k$. (b) Comparison of Enclave heap memory (in KB) used for two cases: SGX but without defense —— and SGX with proposed defense ——. A total of 10000 device events are considered with varying Ruleset sizes of 100, $1k$, $10k$, and $100k$.

6   Limitations and Future Work

While our ML-based attack model performs effectively with heterogeneous IoT devices, in future, we plan to consider other ML-based models for our attack and multiple IoT environments. Furthermore, we plan to undertake a detailed investigation into other forms of side-channel information, such as execution timing patterns, to assess the extent of information leakage from secure IoT data execution in SGX. We also aim to examine alternative defense strategies, such as Oblivious RAM, to counter such attacks. This will include comparing these methods with our existing defenses and analyzing the trade-offs involved. Additionally, we intend to explore potential attacks targeting TRUSTZONE side of the system to further enhance our security measures.

7   Related Work

Recent studies address data privacy in IoT automation on rule-based Trigger-action platforms. Techniques like Filter&Fuzz [43] reduce trigger events sent to the cloud obscuring user behavior, while minTap [7] releases only necessary data to avoid over-privilege and unnecessary API access. However, vulnerabilities like exposure of rules and action events in the cloud persist. In contrast, SECIoTT conceals IoT automation execution and data, enhancing security and privacy in the untrusted cloud. Additionally, several defenses counter memory access pattern-based side-channel attacks on INTEL SGX. Obliviate [1] employs Oblivious RAM (ORAM) for file access within SGX, ZeroTrace [28] uses an oblivious memory controller with Path ORAM and Circuit ORAM, and Oblix [20] provides oblivious search indexes using ORAM. Despite ORAM's effectiveness, it suffers from high computational overhead, limiting its practicality [5,12,39]. Our experiments show that padding is necessary to fully conceal data access patterns, a functionality effectively incorporated by SECIoTT.

8   Conclusion

In this paper, we propose an end-to-end secure IoT *trigger-action rule*-based automation framework using TEE (e.g., INTEL SGX and ARM TRUSTZONE). We demonstrate that access pattern-based side-channel attacks against cloud-based TEE expose sensitive user information. To enhance TEE security, we propose countermeasures based on data oblivious execution and padding, significantly reducing IoT data prediction accuracy. We evaluate SECIoTT with real-time IoT data from smart devices in a practical smart home scenario.

Acknowledgments. The research reported in this work was supported in part by NSF awards DMS-2204795, OAC-2115094, CNS-2331424, ARL/Army Research Office award W911NF-17-1-0356, NIH award 5RM1HG009034-08, National Center for Transportation Cybersecurity and Resiliency (TraCR) and a gift from Cisco Inc.

A   Appendix

A.1   Secret Key Management

We outline the process of securely sharing and managing secret keys in SECIoTT to maintain data integrity and confidentiality among three entities: the user (with SGX), the IoT/edge device (with TRUSTZONE), and the untrusted cloud (with SGX). The IoT device securely stores an RSA private key in its TRUSTZONE, which cannot be altered. Initially, users set up their account and retrieve the public key from their devices. Following remote attestation between the user's and cloud's SGX, the public key of the IoT device is shared with the cloud. The IoT device sends an RSA signature to the cloud, where it is verified using the

Table 2. Rule Conflict Definitions

Conflict Type	Condition	Example
Shadow	$Trigger(R_B) \subseteq Trigger(R_A)$ or $Trigger(R_B) \cap Trigger(R_A) \neq \varnothing$, $Actuator(R_A) = Actuator(R_B)$, $Action(R_A) \equiv Action(R_B)$	R_A: if temperature $\leq 65°F$, air-conditioner = off R_B: if temperature $\leq 60°F$, air-conditioner = off
Execution	$Trigger(R_B) \subseteq Trigger(R_A)$ or $Trigger(R_B) \cap Trigger(R_A) \neq \varnothing$, $Actuator(R_A) = Actuator(R_B)$, $Action(R_A) \neq Action(R_B)$	R_A: if temperature $\leq 75°F$, air-conditioner = off R_B: if temperature $\geq 70°F$, air-conditioner = on
Environment Mutual (continuous toggle)	$Trigger(R_B) \neq Trigger(R_A)$, $Actuator(R_A) = Actuator(R_B)$, $Action(R_A) \neq Action(R_B)$	R_A: if it's raining, window = close R_B: if temperature $\leq 80°F$, window = open
Dependence Conflict (loop)	$Action(R_B) \mapsto Trigger(R_A)$ and $Action(R_A) \mapsto Trigger(R_B)$	R_A: if air-conditioner = on, window = close R_B: if window = close, air-conditioner = on
Incompleteness	$Action$ is incomplete	*Incomplete*: if temperature $\geq 75°F$, air-conditioner = on *Complete*: if temperature $\geq 75°F$, air-conditioner = on; else air-conditioner = off

* R_A = old rule, R_B = new rule   * *Trigger* $\Rightarrow$ Trigger condition
* *Actuator* $\Rightarrow$ Actuator device   * *Action* $\Rightarrow$ Action performed on Actuator

public key within the cloud's SGX enclave. The cloud enclave then generates and encrypts a secret session key using the IoT device's public key, which the device decrypts using its private key and stores securely in TRUSTZONE. All subsequent communications between the IoT device and the cloud are encrypted using this secret session key. Users can reset the key through the same procedure if needed.

A.2 Rule Conflict Detection

The interaction of rules designed to automate devices and tasks in environments like smart homes can create unsafe conditions and increase the attack surface, a risk that is often unnoticed in traditional platforms due to the complexity of potential device interactions and environmental variables [10,40]. For example, in a smart home, a conflict may arise between two rules: one *closing window shades when the user is absent* and another *opening them if the temperature drops below* 90° *F*. This conflict can lead to a race condition, potentially compromising user privacy without their knowledge. In this work, we develop a simple rule conflict detection scheme following [29,35] within SGX enclave. Rule conflicts are categorized into five types, detailed in Table 2. In SECIoTT, the detection process operates securely within the *RuleEngine*. It checks for conflicts when a user registers a rule, comparing it against existing rules that share the same environment or devices. If a conflict is detected, the user is alerted before the rule registration is finalized, allowing SECIoTT to preemptively address potential unsafe conditions and enhance the platform's robustness and practicality.

A.3 Sample Rule and Event JSON

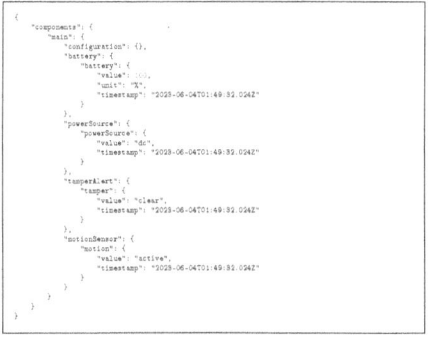

Listing 5. Sample Device Event in JSON.

Listing 4. Sample Rule in JSON.

References

1. Ahmad, A., Kim, K., Sarfaraz, M.I., Lee, B.: Obliviate: a data oblivious filesystem for intel SGX. In: NDSS (2018)
2. Bastys, I., Balliu, M., Sabelfeld, A.: If this then what? Controlling flows in IoT apps. In: Proceedings of the 2018 ACM SIGSAC Conference on Computer and Communications Security, pp. 1102–1119 (2018)
3. BlueVoyant: Third-party supply chain cyber risk CISO report, November 2020. https://www.bluevoyant.com/resources/ciso-report-download-form/
4. Chandra, S., Karande, V., Lin, Z., Khan, L., Kantarcioglu, M., Thuraisingham, B.: Securing data analytics on SGX with randomization. In: Foley, S.N., Gollmann, D., Snekkenes, E. (eds.) ESORICS 2017. LNCS, vol. 10492, pp. 352–369. Springer, Cham (2017). https://doi.org/10.1007/978-3-319-66402-6_21
5. Chang, Z., Xie, D., Li, F.: Oblivious RAM: a dissection and experimental evaluation. Proc. VLDB Endow. **9**(12), 1113–1124 (2016)
6. Chen, T., Guestrin, C.: Xgboost: A scalable tree boosting system. In: Proceedings of the 22nd acm sigkdd international conference on knowledge discovery and data mining. pp. 785–794 (2016)
7. Chen, Y., Alhanahnah, M., Sabelfeld, A., Chatterjee, R., Fernandes, E.: Practical data access minimization in {Trigger-Action} platforms. In: 31st USENIX Security Symposium (USENIX Security 2022), pp. 2929–2945 (2022)
8. Costan, V., Devadas, S.: Intel SGX explained. IACR Cryptol. ePrint Arch. **2016**(086), 1–118 (2016)
9. Daemen, J., Rijmen, V.: AES proposal: Rijndael (1999)
10. Ding, W., Hu, H.: On the safety of IoT device physical interaction control. In: Proceedings of the 2018 ACM SIGSAC Conference on Computer and Communications Security, pp. 832–846 (2018)
11. Fernandes, E., Jung, J., Prakash, A.: Security analysis of emerging smart home applications. In: 2016 IEEE Symposium on Security and Privacy (SP), pp. 636–654. IEEE (2016)
12. Grubbs, P., et al.: Pancake: frequency smoothing for encrypted data stores. In: 29th USENIX Security Symposium (USENIX Security 2020), pp. 2451–2468 (2020)
13. Hou, J., Qu, L., Shi, W.: A survey on internet of things security from data perspectives. Comput. Netw. **148**, 295–306 (2019)
14. Business Insider: The security and privacy issues that come with the Internet of Things, January 2020. https://www.businessinsider.com/iot-security-privacy
15. Intel: Intel® software guard extensions (intel® SGX) SDK for Linux* OS, June 2020. https://download.01.org/intel-sgx/sgx-linux/2.10/docs/Intel_SGX_Developer_Reference_Linux_2.10_Open_Source.pdf
16. Islam, M.S., Kuzu, M., Kantarcioglu, M.: Access pattern disclosure on searchable encryption: ramification, attack and mitigation. In: Ndss, vol. 20, p. 12. Citeseer (2012)
17. Larson, S.: Stuffed toys leak millions of voice recordings from kids and parents, February 2017. https://money.cnn.com/2017/02/27/technology/cloudpets-data-leak-voices-photos/index.html
18. Linaro: Open portable trusted execution environment (2022). www.op-tee.org
19. Matetic, S., et al.: {ROTE}: rollback protection for trusted execution. In: 26th USENIX Security Symposium (USENIX Security 2017), pp. 1289–1306 (2017)
20. Mishra, P., Poddar, R., Chen, J., Chiesa, A., Popa, R.A.: Oblix: an efficient oblivious search index. In: 2018 IEEE Symposium on Security and Privacy (SP), pp. 279–296. IEEE (2018)

21. MQTT: Message queuing telemetry transport (MQTT) (2022). http://mqtt.org/
22. Ohrimenko, O., et al.: Oblivious multi-party machine learning on trusted processors. In: 25th {USENIX} Security Symposium ({USENIX} Security 2016), pp. 619–636 (2016)
23. Paul, F.: Top 10 IoT vulnerabilities, April 2019. https://www.networkworld.com/article/3332032/top-10-iot-vulnerabilities.html
24. Puddu, I., Schneider, M., Haller, M., Čapkun, S.: Frontal attack: leaking {Control-Flow} in {SGX} via the {CPU} frontend. In: 30th USENIX Security Symposium (USENIX Security 2021), pp. 663–680 (2021)
25. Rane, A., Lin, C., Tiwari, M.: Raccoon: closing digital side-channels through obfuscated execution. In: 24th {USENIX} Security Symposium ({USENIX} Security 2015), pp. 431–446 (2015)
26. IBM X-Force Research: The weaponization of IoT devices, April 2017. https://www.ibm.com/downloads/cas/6MLEALKV
27. SAMSUNG: Samsung smartthings developers: working with rules (2021). https://smartthings.developer.samsung.com/docs/rules/overview.html
28. Sasy, S., Gorbunov, S., Fletcher, C.W.: ZeroTrace: oblivious memory primitives from Intel SGX. In: NDSS (2018)
29. Shah, T., Venkatesan, S., Ngo, T., Neelamegam, K., et al.: Conflict detection in rule based IoT systems. In: 2019 IEEE 10th Annual Information Technology, Electronics and Mobile Communication Conference (IEMCON), pp. 0276–0284. IEEE (2019)
30. Shahid, M.R., Blanc, G., Zhang, Z., Debar, H.: IoT devices recognition through network traffic analysis. In: 2018 IEEE International Conference on Big Data (Big Data), pp. 5187–5192. IEEE (2018)
31. Shaon, F., Kantarcioglu, M.: SGX-IR: secure information retrieval with trusted processors. In: Singhal, A., Vaidya, J. (eds.) DBSec 2020. LNCS, vol. 12122, pp. 367–387. Springer, Cham (2020). https://doi.org/10.1007/978-3-030-49669-2_21
32. Shaon, F., Kantarcioglu, M., Lin, Z., Khan, L.: SGX-bigmatrix: a practical encrypted data analytic framework with trusted processors. In: Proceedings of the 2017 ACM SIGSAC Conference on Computer and Communications Security, pp. 1211–1228 (2017)
33. Siami-Namini, S., Tavakoli, N., Namin, A.S.: The performance of LSTM and BiL-STM in forecasting time series. In: 2019 IEEE International Conference on Big Data (Big Data), pp. 3285–3292. IEEE (2019)
34. Sparks, P.: The route to a trillion devices. White Paper, ARM (2017)
35. Sun, Y., Wang, X., Luo, H., Li, X.: Conflict detection scheme based on formal rule model for smart building systems. IEEE Trans. Hum.-Mach. Syst. **45**(2), 215–227 (2014)
36. Ur, B., et al.: Trigger-action programming in the wild: an analysis of 200,000 IFTTT recipes. In: Proceedings of the 2016 CHI Conference on Human Factors in Computing Systems, pp. 3227–3231 (2016)
37. Vailshery, L.S.: Consumer spending on smart home related devices worldwide from 2019 to 2025, January 2021. https://www.statista.com/statistics/873607/worldwide-smart-home-annual-device-sales/
38. Van Bulck, J., Piessens, F., Strackx, R.: SGX-step: a practical attack framework for precise enclave execution control. In: 2nd Workshop on System Software for Trusted Execution (SysTEX), pp. 4:1–4:6. ACM, October 2017
39. Vuppalapati, M., Babel, K., Khandelwal, A., Agarwal, R.: {SHORTSTACK}: distributed, fault-tolerant, oblivious data access. In: 16th USENIX Symposium on Operating Systems Design and Implementation (OSDI 22), pp. 719–734 (2022)

40. Wang, Q., Datta, P., Yang, W., Liu, S., Bates, A., Gunter, C.A.: Charting the attack surface of trigger-action IoT platforms. In: Proceedings of the 2019 ACM SIGSAC Conference on Computer and Communications Security, pp. 1439–1453 (2019)

41. Wang, W., et al.: Leaky cauldron on the dark land: understanding memory side-channel hazards in SGX. In: Proceedings of the 2017 ACM SIGSAC Conference on Computer and Communications Security, pp. 2421–2434 (2017)

42. Winter, J.: Trusted computing building blocks for embedded Linux-based arm trustzone platforms. In: Proceedings of the 3rd ACM Workshop on Scalable Trusted Computing, pp. 21–30 (2008)

43. Xu, R., Zeng, Q., Zhu, L., Chi, H., Du, X., Guizani, M.: Privacy leakage in smart homes and its mitigation: IFTTT as a case study. IEEE Access **7**, 63457–63471 (2019)

44. Xu, Y., Cui, W., Peinado, M.: Controlled-channel attacks: deterministic side channels for untrusted operating systems. In: 2015 IEEE Symposium on Security and Privacy, pp. 640–656. IEEE (2015)

45. Yu, J., Hsiung, L., El'Hajj, M., Fletcher, C.W.: Data oblivious ISA extensions for side channel-resistant and high performance computing. In: The Network and Distributed System Security Symposium (NDSS) (2019)

Towards Atomicity and Composability
in Cross-Chain NFTs

Yong Zhi Lim[1]($\boxtimes$), Wen Qing Ong Perry[2]($\boxtimes$), and Jianying Zhou[1]($\boxtimes$)

[1] Singapore University of Technology and Design, Singapore, Singapore
`yongzhi_lim@mymail.sutd.edu.sg, jianying_zhou@sutd.edu.sg`
[2] Nanyang Technological University, Singapore, Singapore
`ongw0111@e.ntu.edu.sg`

Abstract. Non-Fungible Tokens (NFTs) have created new and interesting ways to own digital assets. Our work defines non-fungible tokens as a pair and addresses challenges in cross-chain NFT transfers, particularly on interoperability between siloed blockchains. We develop a methodology and propose a Greed Factor to prioritize various hashing methods for detecting duplicates. Our practical approach marks an advancement over existing methods by using an analysis of over 400,000 real-world NFTs across 50 collections, which is easily applicable to existing bridges. The experimental results show the limitations of current implementations, and we offer insights to strengthen their functionality and foster the future growth of NFTs.

Keywords: blockchain · cross-chain bridges · non-fungible token

1 Introduction

Non-Fungible Tokens (NFTs) have created new and interesting ways to own digital assets. Inspired by ERC-20 token standard, the introduction of the NFT or formally, ERC-721, allows ownership of distinguishable digital assets stored on the blockchain. This innovates and transforms how digital assets are actually owned while achieving the benefits of using the blockchain. Unfortunately, the current and existing hype with NFTs are mistakenly associated with a cost (e.g. cryptocurrencies) and are treated as a digitalized artwork or image, rather than an asset[1] which are highly trackable, and as its name implies 'non-fungible' - where it is treated as an unique entity on the blockchain. Although most NFTs currently reside solely on the *Ethereum* blockchain, there has been limited growth to enable interoperability between different blockchains.

To resolve transactions involving digital assets or NFTs between two different siloed blockchains, they simply rely on metadata or transactional data that is

---

[1] Digital assets or content not restricted to just images, but also includes audio, documents or videos.

© IFIP International Federation for Information Processing 2024
Published by Springer Nature Switzerland AG 2024
A. L. Ferrara and R. Krishnan (Eds.): DBSec 2024, LNCS 14901, pp. 92–100, 2024.
https://doi.org/10.1007/978-3-031-65172-4_6

stored on the source chain, retrieval of the digital asset (typically stored off-chain with the use of *InterPlanetary File System* (IPFS)), before transferring it to the destination chain. This poses a couple of concerns; 1) whether true *atomicity* is achieved by ensuring the existence of the NFT pair $\mathbb{N} < \mathbb{N}_o, \mathbb{N}_d >$, consisting of both on-chain transactional $\mathbb{N}_o$ and off-chain data storage $\mathbb{N}_d$ before the transaction is completed, and 2) whether true *composability* is achieved by ensuring that the NFT is able to interoperate once it is transferred to another blockchain. Although some *IPFS* attacks have been addressed in [4], it does not stop the originating minter from rug-pulling, removing or modifying the original resource, nor retain its ownership rights [6].

This paper is organized into 5 sections. Section 2 provides an analysis of how an NFT pair can be transferred from one blockchain to another, while providing a deep dive into its key interactions and possible threat vectors. Section 3 shows how we can protect the integrity of the NFT pair by redefining how checks could be performed during cross-chain transfers. Section 4 provides solutions on how to defend against proposed attack vectors and how its implementation is carried out. Finally, Sect. 5 concludes our work.

2   Analysis

2.1   Threat Model

The fundamental issue for cross-chain NFTs is to ensure that its transfer is correct and non-fraudulent. We assume that the owner's private keys ($K_{private}$), communications between the bridge ($\mathbb{B}$) and 2 blockchain channels ($\mathbb{S}_{BC}, \mathbb{D}_{BC}$) are timely and secure (see Fig. 1). The attacker should not have control of any oracle ($\mathbb{O}$) or obtained any private keys throughout the session while the NFT is in the process of a transfer. Any existing or already created assets should not apply.

Both [3,6] have identified issues with existing NFT implementations but have not addressed cross-chain transfers, completely missing, total loss or theft of NFTs. An attacker in our context, is likely motivated by greed, and will attempt to duplicate the most valuable NFTs in a specific collection $\mathbb{C}$ for financial gain across chains. Our work explores some of the underlying issues that occur in existing implementations during NFT transfers in cross-chain bridges. Here are our concerns: **(RQ1)** What are the existing weaknesses in NFTs during cross-chain transfer? **(RQ2)** What are the potential threat vectors in such an implementation involving the cross-chain transfer of a NFT? **(RQ3)** How can we further improve and defend against such cross-chain attacks?

To answer these questions, we first provide a simplified analysis for the exchange of assets between different blockchains. A typical flow of how a NFT asset, $\mathbb{N}$, is created and traversed between the two blockchains, the source blockchain $\mathbb{S}_{BC}$ and the destination blockchain $\mathbb{D}_{BC}$ before arriving to its proposed owner is shown in Fig. 1. We analyse the following components which provide an answer for **RQ1**:

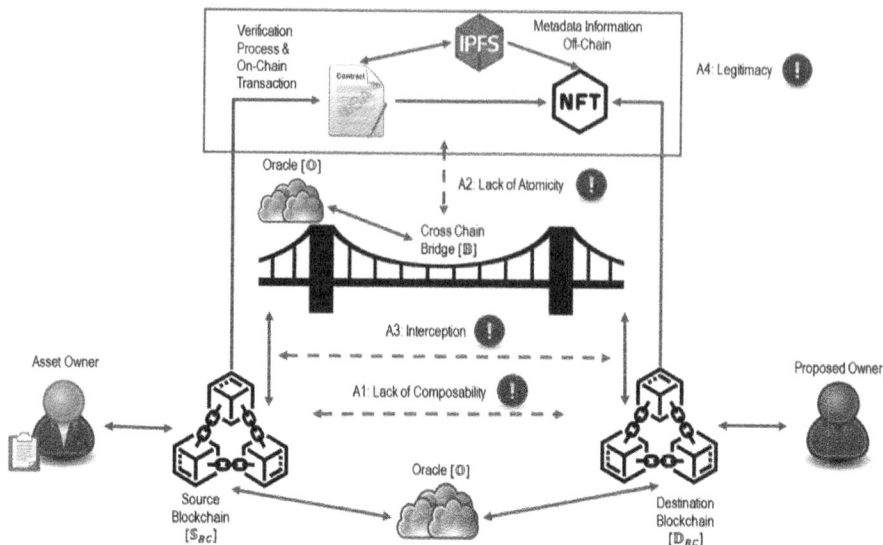

Fig. 1. Analysis of a simplified Cross-Chain NFT transaction, depicting how an asset is transferred from source to its destination blockchain involving its stakeholders

Stakeholders. The first and foremost part of our simplified ecosystem, we introduce two different stakeholders consisting of the asset owner as well as the prospective/future owner between two different blockchains. The asset owner first creates the NFT $\mathbb{N}$, then uploads the off-chain asset $\mathbb{N}_d$ (using a centralized or decentralized storage). Once the upload is complete, the smart contract is called on the source blockchain $\mathbb{S}_{BC}$ L2, before committing the creation of the on-chain asset $\mathbb{N}_o$ on L1. To complete the transfer of $\mathbb{N}$, the asset owner, who is the current owner of $\mathbb{N}$, first initiates the transfer of $\mathbb{N}$ from $\mathbb{S}_{BC}$ to $\mathbb{D}_{BC}$, before the proposed owner obtains $\mathbb{N}$.

Assets. The key component and next step of the NFT ecosystem, is accessing digital assets, which is usually stored off-chain by deploying *IPFS* or utilizing a centralized or trusted location. Creators can also choose to add attributes or traits in its smart contracts or metadata to uniquely identify the asset. After which, they add $\mathbb{N}$ to a collection $\mathbb{C}$ for a prospecting owner to mint them.

Blockchains and Bridge. Once the NFT $\mathbb{N}$ has been created, stakeholders can choose to publish or post their asset on the blockchain or NFT marketplace (NFTM) of their choice. Depending on how the bridges are configured, they are typically made transparent for the user, where an asset owner can transfer $\mathbb{N}$ by simply using a wallet, populating the recipient address in the target blockchain and paying the needed transaction fees (or *gas*) to complete the transfer. However, there are currently two issues in such bridges: the lack of true atomicity and legitimacy. Even though cross-chain bridges do exist, most of the listed bridges fails to consider such NFT assets as a pair, only involving the on-chain asset while conveniently forgetting the off-chain component - requiring only fields for $\mathbb{S}_{BC}$, $\mathbb{D}_{BC}$, $\mathbb{N}_o$, *signer* and destination address of the proposed owner's account (or wallet) in $\mathbb{D}_{BC}$. Bridges do not ensure that assets (be it stored off-chain or on-chain) exist, nor do they perform any form of checking to ensure that a duplicate exists on the destination blockchain as shown in Fig. 1. A malicious owner can simply perform rug pulls to pull the NFT out of the existence or create counterfeits which already has existed on another blockchain.

2.2 Threats

Although cross-chain bridge attacks have been previously studied in [7], they have not considered the off-chain component of digital assets, which should be part of the NFT pair $\mathbb{N} < \mathbb{N}_o, \mathbb{N}_d >$. To answer **RQ2**, we identify 4 possible threats (as shown in Fig. 1) that could occur with cross-chain NFTs:

A1: Lack of Composability. An originless $\mathbb{N}$ could exist. $\mathbb{D}_{BC}$ accepts $\mathbb{N}_d$ (without an on-chain asset $\mathbb{N}_o$) without verification from $\mathbb{S}_{BC}$. Checks to secure the asset should implement detection mechanisms at $\mathbb{D}_{BC}$ and $\mathbb{O}$.

A2: Lack of Atomicity. An assetless or missing off-chain asset $\mathbb{N}_d$ (either the metadata and/or asset) despite the existence of $\mathbb{N}_o$ in both $\mathbb{D}_{BC}$ and $\mathbb{S}_{BC}$. Checks should be implemented at $\mathbb{B}$ and $\mathbb{O}$.

A3: Interception. Stolen or missing $\mathbb{N}$ pair during transit or by impersonation. Checks should be implemented at $\mathbb{S}_{BC}$ and $\mathbb{D}_{BC}$ to ensure the correct identity of both the asset and proposed owners.

A4: Legitimacy. Non-unique or swapped $\mathbb{N}$ pair, where $\mathbb{N}_o$ no longer points to the original resource, with multiple $\mathbb{N}_o$ pointing to the same identical $\mathbb{N}_d$, or with its $\mathbb{N}_d$ switched out. Checks should be implemented at $\mathbb{D}_{BC}$ and $\mathbb{B}$.

Relooking at Fig. 1, we show that an attacker can potentially attack $\mathbb{N}_d$, which is separate from $\mathbb{B}$, $\mathbb{D}_{BC}$ and $\mathbb{S}_{BC}$. The issue here is that the attacker can potentially cause 1) loss [**A1, A2, A3**], 2) misdirected [**A3, A4**], 3) duplicated or swapped [**A4**] $\mathbb{N}_d$, causing the $\mathbb{N} < \mathbb{N}_o, \mathbb{N}_d >$ pair to be invalidated.

3 Methodology

It is without a doubt that such cross-chain NFTs transfers are uncommon, but exploratory work must be done to enable safer and future-proof cross-chain implementations [2]. As shown in Algorithm 1, an ideal NFT $\mathbb{N}$ should consist of a pair $\mathbb{N} < \mathbb{N}_o, \mathbb{N}_d >$, with $\mathbb{N}_o$ and $\mathbb{N}_d$ referring to the on-chain transactional data and the off-chain asset, respectively. Even though the newly minted NFT $\mathbb{N}$ creates a distinct transaction both on the source blockchain $\mathbb{S}_{BC}$ and its off-chain $\mathbb{N}_d$ counterpart, its creation is not made known publicly to the other blockchains.

This siloed ecosystem prevents this state of awareness to other blockchains. Similarly, whenever an identical asset is uploaded onto the off-chain, the exact same hash or *CID* (Content Identifier) is created - with no distinct way to

---

Algorithm 1. $\mathbb{S}_{BC}$ Source Blockchain

Require: $\mathbb{N} < \mathbb{N}_o, \mathbb{N}_d >$,
Ensure: existence of $\mathbb{N} < \mathbb{N}_o, \mathbb{N}_d >$
1: obtain $K_{private}$ from asset owner's account (or wallet)
2: verify existence of $\mathbb{N}_d$
3: smart contract calls to generate $sign(H|\mathbb{N}_o, \mathbb{N}_d|)$
4: check $\mathbb{B}$, $\mathbb{O}$ or $\mathbb{S}_{BC}$ for potential duplicates
5: **if** $\mathbb{N} < \mathbb{N}_o, \mathbb{N}_d >$ does not exist **then**
6:     $\mathbb{S}_{BC}$ mints $\mathbb{N}$
7: **end if**
8: prepare $\mathbb{N}$ for $\mathbb{B}$

Algorithm 2. $\mathbb{B}$ Bridge

Require: $\mathbb{O}$, $\mathbb{S}_{BC}$, $\mathbb{D}_{BC}$, address of proposed owner in $\mathbb{D}_{BC}$,
      $\mathbb{N} < \mathbb{N}_o, \mathbb{N}_d > \leftarrow K_{private}$
Ensure: existence of $\mathbb{N} < \mathbb{N}_o, \mathbb{N}_d >$
1: receives $\mathbb{N}$ from $\mathbb{S}_{BC}$
2: determine GF and prioritize queue
3: check $\mathbb{S}_{BC}$ for potential duplicates on $\mathbb{D}_{BC}$ or using $\mathbb{O}$ based on queue
4: determine if duplicate $\mathbb{N}_d$ belongs to specific $\mathbb{C}$
5: select and perform appropriate hashing H technique for $\mathbb{N}_d$
6: **if** $\mathbb{N}_o || \mathbb{N}_d$ does not exist **OR** $H|\mathbb{N}_d| >=$ threshold value **then**
7:     $\mathbb{B}$ informs $\mathbb{S}_{BC}$ of discrepancy
8: **end if**
9: prepare $\mathbb{N}$ for $\mathbb{D}_{BC}$
10: **if** lock & mint $\mathbb{N}$ **then**
11:     freeze $\mathbb{N}$ on $\mathbb{S}_{BC}$
12: **else if** burn & release $\mathbb{N}$ **then**
13:     burn $\mathbb{N}$ on $\mathbb{S}_{BC}$
14: **end if**

---

determine true ownership of the said digital asset [1]. As shown in Algorithm 1, checks are generated to ensure that the minted $\mathbb{N}$ pair not only already exists, but also to ensure that potential duplicates do not exist within the same $\mathbb{S}_{BC}$, $\mathbb{O}$ or $\mathbb{B}$ prior to transfer into $\mathbb{D}_{BC}$.

Once $\mathbb{B}$ receives $\mathbb{N}$ from $\mathbb{S}_{BC}$ in Algorithm 2, $\mathbb{B}$ should check for potential duplicates if $\mathbb{N}$ belongs in a specific collection $\mathbb{C}$ in $\mathbb{D}_{BC}$ and/or use of $\mathbb{O}$. This can be done by leveraging on image hashing instead of traditional file hashing techniques to pick out similar images. However, checking for such duplicates not within a specific collection each time causes a delay and we propose a GF (Greed Factor) to prioritize and rank this process. The GF is calculated and given by $GF(\mathbb{C}, \mathbb{D}, \mathbb{N}) = \log($Floor Price of $\mathbb{N}$ in $\mathbb{C} * ($Number of $\mathbb{N}$ Sales in $\mathbb{D}) * (\mathbb{C}$ Interval Volume in $\mathbb{D}/$Number of $\mathbb{N}$s in $\mathbb{C}) + 1)$.

The floor price of $\mathbb{N}$ is the lowest price for any $\mathbb{N}$ in a given $\mathbb{C}$ and gives a good understanding if $\mathbb{N}$ is trending. Since an attacker is likely to be motivated by the perceived pricing of $\mathbb{N}$, the chance that he/she will attack a particular $\mathbb{C}$ is

Algorithm 3. $\mathbb{D}_{BC}$ Destination Blockchain

Require: $\mathbb{N} < \mathbb{N}_o, \mathbb{N}_d >$
Ensure: existence of $\mathbb{N} < \mathbb{N}_o, \mathbb{N}_d >$,
    NFT $\mathbb{N} < \mathbb{N}_o, \mathbb{N}_d > \leftarrow K_{private}$
 1: receives $\mathbb{N}$ from $\mathbb{B}$
 2: **if** $\mathbb{N}$ has left $\mathbb{S}_{BC}$ **AND** existence of $\mathbb{N} < \mathbb{N}_o, \mathbb{N}_d >$ **then**
 3:     mint $\mathbb{N}$
 4:     inform $\mathbb{B}$ and $\mathbb{S}_{BC}$
 5:     delegate proposed owner for $\mathbb{N}$
 6: **end if**

---

greater. $\mathbb{C}$'s interval volume is given by $\mathbb{D}$, which is its 1, 7 or 30 day performance obtained from $\mathbb{O}$. GF is needed to rank NFT collections $\mathbb{C}$ for checking, the higher the GF, the greater the impact as it may be subject to misuse (e.g., duplication, rug pull, etc.). $\mathbb{B}$ can also perform checks to ensure the existence of the $\mathbb{N} < \mathbb{N}_o, \mathbb{N}_d >$ pair and prevent the transaction from happening. Once checks are passed, $\mathbb{B}$ prepares $\mathbb{N}$ for cross-chain transfer according to rules set by the owner programmed in the smart contract. Any discrepancies in $\mathbb{N}_d$ during transfer will affect the transaction and will be retrieved by $\mathbb{B}$ and $\mathbb{S}_{BC}$ prior to completing migration to $\mathbb{D}_{BC}$. To finish off the transfer, the existence of $\mathbb{N} < \mathbb{N}_o, \mathbb{N}_d >$ pair is checked again before minting and delegation of the proposed owner in Algorithm 3. With these implementations in place, this will solve **RQ3**.

4   Experimentation

To validate our use case, we deployed a bridge[2] using Polygon's *FxPortal* and performed a transfer of $\mathbb{N}$ between the *Ethereum Goreli* and *Polygon Mumbai* testnets according to Sect. 3, using *Hardhat* and *node.js*. Before $\mathbb{N}$ is transferred across different networks, bridge contract $\mathbb{B}$ handles the transfer and ensures that 1) $\mathbb{N}$'s ownership does indeed exist, 2) is updated correctly on $\mathbb{S}_{BC}$ (which detects **A2** and **A3**), 3) $\mathbb{N}$ does exist by inspecting and validating the `image` URI given in the metadata (which detects **A1** and **A2**) and 4) checking for duplicates of $\mathbb{N}$ in $\mathbb{D}_{BC}$ by calculating GF for each $\mathbb{C}$ (which detects **A4**).

Prior to $\mathbb{N}$'s transfer, $\mathbb{B}$'s NFT contracts are called to check if $\mathbb{N}$ is indeed minted in the $\mathbb{D}_{BC}$, burned in the $\mathbb{S}_{BC}$ and updates the owner (see Algorithm 2). The custody contract is to hold or release $\mathbb{N}$ after the $\mathbb{B}$'s NFT contract is deployed to verify $\mathbb{N}$'s owner and ID. Once $\mathbb{N}$ is transferred to $\mathbb{D}_{BC}$, the same checks are performed. After ensuring that $\mathbb{N}$ does not exist in $\mathbb{S}_{BC}$, $\mathbb{B}$ wraps $\mathbb{N}$ for acceptance into $\mathbb{D}_{BC}$. The original $\mathbb{N}_o$ is then burned on $\mathbb{S}_{BC}$ and minted onto $\mathbb{D}_{BC}$. The newly minted $\mathbb{N}$ is stored in the `mapToken` function and a reverse function `_processMessageFromChild` is maintained from the address of the wrapped $\mathbb{N}$ to the original $\mathbb{N}$ information. An event `TokenMappedERC721` is then emitted whenever $\mathbb{N}$ is newly minted. This event contains the original network ($\mathbb{S}_{BC}$), the

---

[2] https://github.com/limyz/nftBridge.

original token address, the new wrapped token address ($\mathbb{D}_{BC}$), and the name and symbol of the new token. Not only does this eliminate threat vectors **A2** and **A3**, but also helps to track and verify $\mathbb{N}$ transfers and its wrapper creation events. We have collected data on over 400k real-world NFT pairs $\mathbb{N} < \mathrm{N}_o, \mathrm{N}_d >$ on 50 different collections $\mathbb{C}$ with high market caps and trading volumes from *CoinMarketCap* and *OpenSea*. As an attacker, he/she is likely to target specific valuable $\mathbb{C}$ and attempt to recreate similar $\mathbb{N}_d$ from a single $\mathbb{C}$. Since checking for such duplicates can be resource consuming, we prioritize checks by using Greed Factor (GF) to determine and rank $\mathbb{C}$s that an attacker may likely copy or steal from.

Table 1. Duplicate detection of $\mathbb{C}$s measured with various hashing methods using `threshold=0`

	GF	hash-size	phash			ahash			dhash			whash-haar		
			8	16	32	8	16	32	8	16	32	8	16	32
Azuki	3.75	groups	27	0	0	354	19	4	31	1	0	468	81	22
		similarities	55	0	0	962	41	8	63	2	0	1749	232	47
		accuracy	99.45	100	100	90.38	99.59	99.92	99.37	99.98	100	82.51	97.68	99.53
Bored Ape Kennel Club	2.74	groups	869	27	0	1075	594	215	1155	222	16	806	872	429
		similarities	2884	58	0	6630	2014	543	4535	574	33	7915	3987	1393
		accuracy	69.34	99.38	100	29.51	78.59	94.23	51.78	93.9	99.65	15.84	57.61	85.19
Bored Ape Yacht Club	5.29	groups	536	9	0	861	263	8	477	6	0	599	35	5
		similarities	1400	18	0	5636	617	16	1336	12	0	4710	122	10
		accuracy	86	99.82	100	43.64	93.83	99.84	86.64	99.88	100	52.90	98.78	99.90
Cryptopunks	0	groups	1156	162	1	1317	1013	730	1453	819	470	1197	492	114
		similarities	3259	341	2	7036	3546	2191	4519	2056	1144	4850	1402	279
		accuracy	67.41	96.59	99.98	29.64	64.54	78.09	54.81	79.44	88.56	51.5	85.98	97.21
Cryptoads	2.17	groups	89	8	3	539	159	15	128	15	3	411	155	53
		similarities	198	16	6	3509	543	33	279	30	6	2078	543	144
		accuracy	97.16	99.77	99.91	49.65	92.21	99.53	96	99.57	99.91	70.18	92.21	97.93
Degods	4.04	groups	433	7	1	866	386	16	777	11	3	579	681	403
		similarities	1042	14	2	5600	1277	32	2446	22	6	7075	3965	1256
		accuracy	88.35	99.84	99.98	37.4	85.73	99.64	72.66	99.75	99.93	20.91	55.68	85.96
Doodles	3.12	groups	452	138	57	749	323	108	479	104	44	617	665	418
		similarities	1178	290	12	3168	762	229	1156	219	95	4448	2253	1048
		accuracy	88.22	97.1	99.88	68.31	92.38	97.71	88.44	97.81	99.05	55.51	77.47	89.52
Milady	3.11	groups	269	0	0	564	16	2	225	0	0	748	24	1
		similarities	627	0	0	1411	33	4	492	0	0	2052	68	2
		accuracy	93.57	100	100	85.53	99.66	99.96	94.95	100	100	78.95	99.3	99.98
Mutant Ape Yacht Club	4.36	groups	28	1	0	414	4	0	37	0	0	476	14	0
		similarities	61	2	0	1233	8	0	77	0	0	1721	28	0
		accuracy	99.06	99.97	100	81.01	99.88	100	98.81	100	100	73.49	99.57	100
Pudgy Penguins	3.12	groups	310	30	5	962	121	33	525	49	8	956	124	11
		similarities	666	60	10	2987	251	66	1274	99	17	3322	272	23
		accuracy	92.57	99.33	99.89	66.70	97.20	99.26	85.80	98.90	99.81	62.96	96.97	99.74

We have utilized 4 different image hashing methods, namely 1) perceptual (`phash`), 2) average (`ahash`), 3) differential (`dhash`) and 4) haar wavelet (`whash-haar`) hashing and attempt to detect duplicates $\mathbb{N}_d$ using the `undouble` library, with its accuracy calculated using the difference between the total number of $\mathbb{N}_d$ in $\mathbb{C}$ and similarities found. `undouble` detects (near-)identical images across an entire system or directory by using a multi-step process of pre-processing the images (grayscaling, normalizing, and scaling), computing the image hash, and the grouping of images based on a threshold value [5]. The threshold value, as depicted in Algorithm 2, is given by the difference of the image hash values to determine similarities, where $H|\mathbb{S}_{BC}[\mathbb{N}_d]| - H|\mathbb{D}_{BC}[\mathbb{N}_d]|$. The number of groups indicate how many groups of $\mathbb{N}_d$ similarities in $\mathbb{C}$ are found. If no hash can be computed for $\mathbb{N}_d$, it is deemed that $\mathbb{N}$ does not exist [**A1, A4**]. Since each $\mathbb{N}_d$ produces an unique input, it creates an unique output (hash) even though it may be closely similar. Table 1 provides the accuracy of each of the image hashing methods. As there is no formal way to measure security, *GF* ranks each $\mathbb{C}$ for potential duplication and processes them accordingly. Each $\mathbb{N}_d$ is calculated based on real-world data and collections which are more volatile (given in *GF*). A higher *GF* represents the likelihood of that $\mathbb{C}$ being likely to be duplicated. Since the curated function calls for checking possible duplicate $\mathbb{N}_d$ are written in *Python* and *Solidity*, this experimentation could be easily extended to existing implementations using the `web3` library and NFT bridges.

5   Conclusion and Future Work

Our work has revealed an insight into the state-of-the-art, its slow adoption, and the explosive growth of NFTs in recent years. The lack of study of how NFTs work interoperably in blockchains has siloed assets onto the *Ethereum* ecosystem, preventing what we believe is potentially limiting the future growth of NFTs. We have presented an analysis into the current state of NFTs and its cross-chain interactions, along with potential threat vectors that could occur with them. By redefining the NFT as a pair, we have developed a methodology, experimented with bridge contracts, and utilize different hashing techniques to protect against duplicated assets and alleviate unnecessary stress on both chains.

References

1. IPFS Documentation. https://docs.ipfs.tech/
2. What are Cross-Chain NFTs?. https://chain.link/education-hub/cross-chain-nft
3. Lee, P., Abubakar, M., Lo, O., Pitropakis, N., Buchanan, W.: Non-fungible token fraud: studying security issues and improvements for NFT marketplaces using hashing techniques (2023). https://doi.org/10.21203/rs.3.rs-2573810/v1
4. Prünster, B., Marsalek, A., Zefferer, T.: Total eclipse of the heart – disrupting the interplanetary file system. In: 31st USENIX Security Symposium (USENIX Security 22), pp. 3735–3752. USENIX Association, Boston, MA (2022). https://www.usenix.org/conference/usenixsecurity22/presentation/prunster
5. Taskesen, E.: erdogant/undouble: Python package undouble is to detect (near-)identical images. https://github.com/erdogant/undouble

6. Wang, Z., Gao, J., Wei, X.: Do NFTs' owners really possess their assets? A first look at the NFT-to-asset connection fragility. In: Proceedings of the ACM Web Conference 2023. WWW '23, pp. 2099–2109. Association for Computing Machinery, New York, NY, USA (2023). https://doi.org/10.1145/3543507.3583281
7. Zhang, J., Gao, J., Li, Y., Chen, Z., Guan, Z., Chen, Z.: Xscope: hunting for cross-chain bridge attacks. In: Proceedings of the 37th IEEE/ACM International Conference on Automated Software Engineering. ASE '22, Association for Computing Machinery, New York, NY, USA (2023). https://doi.org/10.1145/3551349.3559520

A Privacy-Preserving Graph Encryption Scheme Based on Oblivious RAM

Seyni Kane[2,3] and Anis Bkakria[1(✉)] (iD)

[1] IRT SystemX, Palaiseau, France
anis.bkakria@irt-systemx.fr
[2] Applied Crypto Group, Orange Innovation, 14000 Caen, France
[3] SAMOVAR, Télécom SudParis, Institut Polytechnique de Paris, Paris, France

Abstract. Graph encryption schemes play a crucial role in facilitating secure queries on encrypted graphs hosted on untrusted servers. With applications spanning navigation systems, network topology, and social networks, the need to safeguard sensitive data becomes paramount. Existing graph encryption methods, however, exhibit vulnerabilities by inadvertently revealing aspects of the graph structure and query patterns, posing threats to security and privacy. In response, we propose a novel graph encryption scheme designed to mitigate access pattern and query pattern leakage through the integration of oblivious RAM and trusted execution environment techniques, exemplified by a Trusted Execution Environment (TEE). Our solution establishes two key security objectives: (1) ensuring that adversaries, when presented with an encrypted graph, remain oblivious to any information regarding the underlying graph, and (2) achieving query indistinguishability by concealing access patterns. Additionally, we conducted experimentation to evaluate the efficiency of the proposed schemes when dealing with real-world location navigation services.

Keywords: Privacy Enhancing Technology · Graph Encryption Scheme · Oblivious RAM · Trusted Execution Environment

1 Introduction

Cloud computing provides on-demand storage and computing resources, motivating data storage outsourcing for cost, availability, and efficiency. Securely storing and processing data in the cloud is challenging. Structured Encryption (SE) [1] has been proposed to protect outsourced data by allowing encrypted structured data to be queried with a query-specific token, revealing no useful information about the query or data.

SE schemes include graph encryption schemes [6,10]. The latter have various applications, including private navigation systems [16], online social networks [9],

---

The research presented in this paper was conducted while the author was affiliated with IRT-SystemX.

A. L. Ferrara and R. Krishnan (Eds.): DBSec 2024, LNCS 14901, pp. 101–108, 2024.
https://doi.org/10.1007/978-3-031-65172-4_7

and modeling highly confidential infrastructure. The main challenge is designing schemes that are secure, expressive, and efficient. Cryptographic techniques like Fully Homomorphic Encryption (FHE) [5] and secure Multi-Party Computation (MPC) offer high security but low efficiency. Other approaches trade some security for better efficiency by allowing controlled leakage [6,10]. Leakage examples include the Access Pattern (AP), the set of nodes and edges accessed by a query, and the Query Pattern (QP), the set of queries issued. An adversary observing AP and QP can infer information about the graph structure, connectivity, content, and query characteristics.

Ghosh, Kamara, and Tamassia (GKT) proposed a graph encryption scheme supporting shortest path queries with a balance between efficiency and security [6]. While their approach partitions and encrypts subgraphs separately, achieving optimal preprocessing time and space, it still leaks Access Patterns (AP) and Query Patterns (QP). Falzon et al. presented an attack exploiting QP leakage on the GKT scheme [3].

Existing graph encryption schemes, due to AP and QP leakage, aren't suitable for highly sensitive graphs. One potential solution is using Oblivious RAM (ORAM) to hide AP and QP leakage from untrusted servers [7]. However, ORAM can lead to high communication complexity. Trusted Execution Environments (TEE) offer a more efficient solution by creating secure enclaves within an application's address space [4,12,13,17]. TEEs can enhance oblivious data structures by reducing communication complexity between clients and servers.

Our Contributions. In this paper, we introduce an innovative graph encryption scheme designed to facilitate shortest path queries without compromising any information pertaining to the underlying data or the query itself. Our approach builds upon the foundation of the GKT scheme [6], widely recognized as the state-of-the-art solution for graph encryption, and incorporates advanced privacy-preserving techniques such as Oblivious RAM (ORAM) [7] and hardware isolation mechanisms like Trusted Execution Environments (TEE) [2].

Our scheme integrates ORAM to eliminate AP and QP leakage and incorporates TEE to optimize communication complexity by creating a secure enclave within the untrusted server. The key contributions of our work are as follows: We introduce TOGES, the first graph encryption scheme that synergistically employs ORAM and TEE to conceal both AP and QP leakage from an untrusted server, proving the adaptively semantic security of our scheme along with AP and QP indistinguishability. Additionally, we implemented our scheme and conducted evaluations using a real-world location navigation dataset, demonstrating its practicality and efficiency. For detailed information on the implementation and evaluation, please refer to the extended version of the paper [8].

Organization of the Paper. The rest of this paper is organized as follows: Sect. 2 defines the system and security model that we consider for our construction. It also states the assumptions, threat model, and security goals that we aim to achieve. Section 3 describes our construction in two versions: a basic version utilizing ORAM-based graph encryption, and an enhanced version leveraging TEE-ORAM based graph encryption. Section 4 analyzes the security of our construction and provides a formal security proof. Section 5 concludes the paper and suggests some possible future directions.

2   System and Security Models

2.1   System Model

In the presented scenario, we examine a situation where a client C, constrained by limited resources, seeks to delegate both data storage and computation tasks to an untrusted cloud server SP equipped with a TEE. This two-party system comprises a client, the data owner, and a cloud server, responsible for hosting the data. Crucially, the client places trust in the TEE embedded within the server, treating it as an extension of its own infrastructure.

2.2   Assumptions and Attacker Model

Our scheme involves two parties: C, responsible for encrypting a graph-based dataset and uploading it to CS, and CS, which hosts a TEE and stores the encrypted tree ORAM. C submits queries and is assumed to be trustworthy, while CS loads data into the enclave for query processing, considering the potential for an attacker to control the operating system on CS. The TEE provides a secure execution environment by cryptographically safeguarding code and data on an untrusted server, although it is vulnerable to side-channel attacks. These attacks allow adversaries to extract sensitive information by observing processing effects without direct access to the information source, and the OS can generate a trace of the enclave's accesses. We do not address these side-channel attacks, assuming adversaries cannot extract information from the TEE. The adversary can monitor communications between the TEE and external entities, including C. The TEE manages the position map and handles queries from C securely, establishing a secure channel with C to protect their communication. We presume the TEE's trustworthiness, considering data and access within the TEE as secure, and assume CS to be an honest but curious entity that follows the protocol while attempting to acquire information about the outsourced data.

3   ORAM Based Graph Encryption (OBGE)

In this section, we present our construction in three stages: a basic non-recursive version as a model for the scheme, an enhanced version combining TEEs and recursive ORAM.

3.1   A First Construction

We describe our basic construction and provide detailed explanations of the algorithms. Our trivial construction employs binary tree storage on the server, similar to Path ORAM, to eliminate access and query patterns in GKT schemes.

Description of the Protocol. We propose a novel protocol OBGE = (Setup, Query, Reveal) for privacy-preserving graph encryption based on GKT [6], and Path ORAM [14]. Our protocol allows a client to outsource a graph to a server and query it efficiently without revealing any information about the graph structure or the query results. We use a symmetric-key encryption scheme SKE = (Gen, Encrypt, Decrypt) similar to GKT and a pseudorandom function P in our protocol.

Algorithm 1: Setup

    input : Graph $G = (V, E)$, security parameter λ, pseudo random
           function P, symmetric-key encryption SKE
    output: Encrypted tree ORAM T, Position map PM, keys
           $key := (K_1, K_2, K')$

    Initialize empty dictionary $SPDX, SPDX'$;
    Initialise empty position map PM, and empty stash ST;
    $sp \leftarrow$ SKE.Gen(λ);   ▷ Generate keys
    $K_1 \leftarrow$ SKE.Gen$(sp); K_2 \leftarrow SKE.Gen(sp); K' \leftarrow_R \{0,1\}^\lambda$;
    $key := (K_1, K_2, K')$;
    $SPDX :=$ ComputeSPDX(G);
    for $(lab, val) \in SPDX$ **do**
        $tk_{lab} \leftarrow P_{K'}(lab); tk_{val} \leftarrow P_{K'}(val)$; ▷ Encrypt labels and values
        $SPDX'[tk_{lab}] := (tk_{val}, ct)$, where $ct =$ SKE.Encrypt$_{K_1}(val)$;
    end
    for $tk \in SPDX'$ **do**
        $PM[tk] := x$, where $x \leftarrow_R \{0, 1, \ldots, 2^L - 1\}$;   ▷ Assign random
        positions
    end
    for $tk \in PM$ **do**
        $(tk', ct') \leftarrow SPDX'[tk]$;   ▷ Retrieve encrypted values
        $c \leftarrow SKE.Encrypt_{K_2}(block)$, where $block = (tk, (tk', ct'), x)$;
        Upload c on $\mathcal{P}(x)$;   ▷ Encrypt and upload encrypted block
    end
    return $T, PM, ST, key := (K_1, K_2, K')$;

Setup. The setup process is depicted in Algorithm 1. We consider a graph $G = (V, E)$. C creates an SP-matrix M_G, which stores the shortest paths between any pair of vertices in G. Then, the client performs the same steps as in GKT to obtain a dictionary $SPDX$, which maps each pair of vertices (u, v) to the next vertex w on the shortest path linking u to v. The client then transforms $SPDX$ into a new dictionary $SPDX'$ as follows. For each entry $(u, v) \mapsto w$ in $SPDX$, the client generates two tokens: one for the key, $tk \leftarrow P_{K'}(u, v)$, and one for the value, $tk' \leftarrow P_{K'}(w, v)$, using a pseudorandom function P. The client also encrypts the value using a symmetric key encryption scheme, $ct' \leftarrow$ SKE.Encrypt$(K_1, (w, v))$, where K_1 is a secret key. The client then sets $SPDX'[tk] := (tk', ct')$ to construct the new dictionary.

The client creates a binary tree ORAM T of size N, where $ZN > |V|^2$, and each node acts as a bucket capable of holding up to Z blocks. A block is represented by the tuple $(tk, (tk', ct'), x)$, where tk is the token identifying the block (linked to the query (u, v)), tk' is the token identifying the next hop on the shortest path between u and v, ct' is the encryption of (w, v) by SKE, and x denotes the leaf node identifier in the ORAM tree T. The client initializes an empty stash ST, an array capable of holding blocks, and a position map PM, consisting of two columns. The first column stores tokens identifying blocks, and the second column stores the leaf identifier currently associated with each block. For each token $tk \in SPDX'$, the client randomly assigns a leaf identifier $x \leftarrow_R 0, 1, \ldots 2^L - 1$ and records it in the position map PM.

C begins by initializing the tree T. For each $block = (tk, (tk', ct'), x)$, employ C to encrypt the block using K_2: $c \leftarrow$ SKE.Encrypt$(K_2, block)$. Next, retrieve the corresponding leaf identifier $x := PM[tk]$ from the position map. Subsequently, upload the encrypted block to the appropriate node along the path from the root to the leaf x in T, denoted as $\mathcal{P}(x)$. This process follows the same mechanism as described in PathORAM [14]. Consequently, by the end of the setup phase, buckets containing fewer than Z data blocks are populated with dummy blocks.

The client finishes the setup phase by uploading the ORAM tree T to the server. The client store the position map PM, the stash ST, and the keys. The Setup protocol is described in detail in Algorithm 1.

Algorithm 2: Query

   input : Query $q = (u, v)$, Encrypted tree ORAM T, K_2, K'
   output: Encrypted path $resp$

   C computes $tk \leftarrow P_{K'}(q)$;   ▷ Extract the leaf identifier
   $x := PM[tk]$, $status := \epsilon$, $resp := \epsilon$;   ▷ Initialization
   Set variable $curr := tk$;   ▷ Set current node to the leaf
   while $status \neq$ "$SearchEnd$" **do**
      C executes $block \leftarrow$ Access$(curr)$;   ▷ Access current block
      if $block = \bot$ **then**
         set $status :=$ "$SearchEnd$";
      else
         Parse $block$ as $(tk, (tk', ct'), x)$;
         set $resp := resp \cup ct'$ and $curr := tk'$;
      end
   end
   return $resp$;   ▷ Return encrypted path

Query. To query the shortest path between nodes u and v, the client computes the token tk of the query using $P_{K'}(u, v)$. And execute Access(tk) with tk, which return the decrypted $block = (tk, (tk', ct')) \leftarrow$ SKE.Decrypt$_{K_2}(c)$. If $block \neq \bot$, C parses $block$ and stores ct' in a variable $resp$, and recall Access(tk'), this time with token of the next hope tk'. Repeats this process from step 4 until the Access procedure return $\bot$. The Query protocol is described in detail in Algorithm 2.

Reveal. Once the client receive the cipher-text he decrypts the ciphertext with his key K_1 and obtains the plaintext, which is the requested path. The Reveal protocol is described in detail in Algorithm 3.

Algorithm 3: Reveal

input : $resp$, $key := K_1$
output: Decrypted path $p_{u,v}$

C Set variable $p_{u,v} := \epsilon$;   ▷ Initialization
Parse $resp$ as $ct_1 \cup ct_2 \cup \cdots \cup ct_k$;   ▷ Parse encrypted path
for $i = 1, \ldots k$ **do**
    $m_i \leftarrow$ SKE.Decrypt(K_1, ct_i) ;   ▷ Decrypt each node
    Set $m := m \cup m_i$;   ▷ Aggregate decrypted nodes
end
return M;

Theorem 1 (Correctness of OBGE). *If P is a secure PRF, SKE is correct and the ORAM is correct, then OBGE is correct.*

Proof. The correctness follows from the fact that the tokens generated by P have a negligible probability of colliding in a random function, and consequently, this negligible probability extends to the PRF P as well.

3.2   Enhanced Construction

To address the limitations of basic construction in handling large graphs, such as those used in location navigation, we introduce an enhanced approach employing a dual strategy. First, we implement a Trusted Execution Environment (TEE) on the server side, creating a secure enclave that provides a confidential, tamper-resistant execution environment. Within this enclave, a compact client module manages the client's stash and position map, acting as the Path ORAM controller to handle queries securely without exposing sensitive data. This eliminates the need for client-side storage, enhancing security and simplifying interactions. However, TEEs like Intel SGX have limited storage, typically 128 MB [15], which can be insufficient for larger graphs as the position map and stash sizes might exceed this capacity. To address this, we propose using a recursive ORAM data structure [11] to minimize data stored in the TEE, optimizing storage utilization while maintaining ORAM protocol security. This dual approach enhances scalability and efficient use of TEE resources, making it suitable for applications involving large graphs and datasets.

4   Security Analysis

In the following, we examine the security of OBGE. We demonstrate that our construction is adaptively-secure with respect to a well-defined leakage profile. We define our leakage profile below.

Setup Leakage. Our scheme has no setup leakage. Indeed, in the `Setup` phase of our scheme, we set the size of the ORAM tree $ZN > |v|^2$, so we do not reveal the size of the graph.

Query Leakage. In the query phase, the server only learns the leaf identifiers associated with blocks containing the path of the current query. And we replace these leaf identifiers with random ones on the tree T as we access the blocks. Moreover, we re-encrypt the accessed blocks using a semantically secure SKE and relocate them either in the stash or in a new branch on the tree associated with the new leaf identifier. The server also does not learn the access pattern, which is guaranteed by the underlying Path ORAM protocol. So the only thing the server learns during a query is the length of the requested path $|p_{u,v}|$, by counting the number of recursions.

We formalize the leakage function of our scheme as $\mathcal{L} = \mathcal{L}_Q(q = (u, v)) = |p_{u,v}|$. The security of our scheme is summarized in the following theorem, with the detailed proof available in the extended version of this paper [8].

Theorem 2 (Security of *OBGE*). *If P is a secure PRF, SKE is INDCPA-secure, ORAM is secure according to definition in [14] then OBGE, as described above, is adaptively $\mathcal{L}$-semantically secure, where $\mathcal{L}$ is the leakage function.*

5   Conclusion

In conclusion, the need for secure graph encryption schemes is paramount in various applications, including navigation systems, social networks, and infrastructure modeling. Existing methods, while efficient, suffer from vulnerabilities such as access pattern and query pattern leakage, which compromise security and privacy. To address these challenges, we propose a novel graph encryption scheme that leverages ORAM and TEE to conceal both access pattern and query pattern leakage. Our scheme builds upon the foundation of existing state-of-the-art solutions like the GKT scheme, enhancing security while maintaining efficiency. Through evaluations on real-world datasets, we demonstrate the practicality and effectiveness of our solution in protecting sensitive graph data while allowing secure shortest path queries.

Acknowledgments. This work were supported by the french national research agency funded project AUTOPSY (grant no. ANR-20-CYAL-0008). Additionally, part of this work was done as part of IRT SystemX project PFS (Security of Smart Ports).

References

1. Chase, M., Kamara, S.: Structured encryption and controlled disclosure. In: Abe, M. (ed.) ASIACRYPT 2010. LNCS, vol. 6477, pp. 577–594. Springer, Heidelberg (2010). https://doi.org/10.1007/978-3-642-17373-8_33

2. Costan, V., Devadas, S.: Intel SGX explained. Cryptology ePrint Archive (2016)
3. Falzon, F., Paterson, K.G.: An efficient query recovery attack against a graph encryption scheme. In: Atluri, V., Di Pietro, R., Jensen, C.D., Meng, W. (eds.) Computer Security – ESORICS 2022. ESORICS 2022. LNCS, vol. 13554. Springer, Cham (2022). https://doi.org/10.1007/978-3-031-17140-6_16
4. Fuhry, B., Bahmani, R., Brasser, F., Hahn, F., Kerschbaum, F., Sadeghi, A.-R.: HardIDX: practical and secure index with SGX. In: Livraga, G., Zhu, S. (eds.) DBSec 2017. LNCS, vol. 10359, pp. 386–408. Springer, Cham (2017). https://doi.org/10.1007/978-3-319-61176-1_22
5. Gentry, C.: Fully homomorphic encryption using ideal lattices. In: Proceedings of the Forty-First Annual ACM Symposium on Theory of Computing, pp. 169–178 (2009)
6. Ghosh, E., Kamara, S., Tamassia, R.: Efficient graph encryption scheme for shortest path queries. In: Proceedings of the 2021 ACM Asia Conference on Computer and Communications Security, pp. 516–525 (2021)
7. Goldreich, O., Ostrovsky, R.: Software protection and simulation on oblivious rams. J. ACM (JACM) **43**(3), 431–473 (1996)
8. Kane, S., Bkakria, A.: A privacy preserving graph encryption scheme based on oblivious ram. https://github.com/nserser/graph_encryption (2024). online; 28 May 2024
9. Lai, S., Yuan, X., Sun, S.F., Liu, J.K., Liu, Y., Liu, D.: Graphse^2: an encrypted graph database for privacy-preserving social search. In: Proceedings of the 2019 ACM Asia Conference on Computer and Communications Security, pp. 41–54 (2019)
10. Meng, X., Kamara, S., Nissim, K., Kollios, G.: GRECS: graph encryption for approximate shortest distance queries. In: Proceedings of the 22nd ACM SIGSAC Conference on Computer and Communications Security, pp. 504–517 (2015)
11. Patel, S., Persiano, G., Yeo, K.: Recursive ORAMs with practical constructions. Cryptology ePrint Archive, Paper 2017/964 (2017). https://eprint.iacr.org/2017/964, https://eprint.iacr.org/2017/964
12. Rachid, M.H., Riley, R., Malluhi, Q.: Enclave-based oblivious ram using intel's SGX. Comput. Secur. **91**, 101711 (2020)
13. Sasy, S., Gorbunov, S., Fletcher, C.W.: ZeroTrace: Oblivious memory primitives from intel SGX. Cryptology ePrint Archive (2017)
14. Stefanov, E., et al.: Path ORAM: an extremely simple oblivious ram protocol. J. ACM (JACM) **65**(4), 1–26 (2018)
15. Will, N.C., Maziero, C.A.: Intel software guard extensions applications: a survey. ACM Comput. Surv. **55**, 1–38 (2023)
16. Wu, D.J., Zimmerman, J., Planul, J., Mitchell, J.C.: Privacy-preserving shortest path computation (2016). arXiv preprint arXiv:1601.02281
17. Wu, Z., Li, R.: OBI: a multi-path oblivious ram for forward-and-backward-secure searchable encryption. In: NDSS (2023)

Privacy

DT-Anon: Decision Tree Target-Driven Anonymization

Sabrina De Capitani di Vimercati$^{(\boxtimes)}$ ⓘ, Sara Foresti ⓘ, Valerio Ghirimoldi, and Pierangela Samarati ⓘ

Computer Science Department, Università degli Studi di Milano, Milan, Italy
{sabrina.decapitani,sara.foresti,pierangela.samarati}@unimi.it,
valerio.ghirimoldi@studenti.unimi.it

Abstract. More and more scenarios rely today on data analysis of massive amount of data, possibly contributed from multiple parties (data controllers). Data may, however, contain information that is sensitive or that should be protected (e.g., since it exposes identities of the data subjects) and cannot simply be freely shared and used for analysis. Business rules, restrictions from individuals (data subjects to which data refer), as well as privacy regulations demand data to be sanitized before being released or shared with others. Unfortunately, such protection typically comes with a loss of utility of the released data, impacting the performance of the analytics tasks to be executed.

In this paper, we present DT-Anon, a target-driven anonymization approach that aims at protecting (anonymizing) data while preserving as much as possible the capability of a classification task operating downstream to learn from the anonymized data. The basic idea of our approach is to perform the anonymization process on partitions produced by a decision tree driven by the target of the classification task. Each partition is then independently anonymized, to limit the impact of anonymization on the attributes and values that work as predictors for the target of the classification task. Our experimental evaluation confirms the effectiveness of the approach.

Keywords: data anonymization · machine learning classifier · target-driven anonymization · decision tree

1 Introduction

Today's society is highly dependent on data, with huge (and ever increasing) amount of data generated, collected, and processed. Concepts such as big data, data analytics, and machine learning are today common terms for the layperson, witnessing their pervasiveness in every context of our daily life. As a matter of fact, the availability of massive amount of data, together with powerful and efficient computational infrastructures and services, and hence ability to extract knowledge from data, is at the heart of our smart society, bringing great benefits in different domains, from business to leisure.

© IFIP International Federation for Information Processing 2024
Published by Springer Nature Switzerland AG 2024
A. L. Ferrara and R. Krishnan (Eds.): DBSec 2024, LNCS 14901, pp. 111–130, 2024.
https://doi.org/10.1007/978-3-031-65172-4_8

More and more scenarios rely today on different parties contributing to the collection, sharing, and analysis of data. However, often data collections include information that cannot be freely shared (e.g., [2]). This is, for example, the case of data referred to individuals, whose privacy needs to be protected, as demanded by data regulations. The EU General Data Protection Regulation (GDPR), the California Consumer Protection Act (CCPA), and other similar regulations worldwide, demand protection for information referred to individuals (data subjects) whose identity and sensitive information should be properly protected by data anonymization solutions (with anonymized data being exempt from the obligations set out in the regulations).

Notwithstanding the great benefit of operating on data, it is therefore of utmost importance to ensure that the privacy of data subjects (to whom the data refer) be properly protected in such data sharing and analytics process. When data analysis is performed by external parties (other than the data controller responsible for the data), this implies that data should be properly anonymized before being shared with these parties. Unfortunately, anonymization, which by design causes information loss (for a privacy gain) in the data, can have a significant negative impact on the performance of the downstream data analytics task, with the known tension between privacy and utility.

In this paper, we consider a scenario characterized by multiple parties (data controllers) contributing data for a data analytics task. Our goal is to design a *target-driven anonymization* approach, that is, a data anonymization approach aware of the data analytics task operating downstream, and driven by it. We consider classification as data analytics task, and present a technique that both protects the privacy of data subjects and maintains the utility of the anonymized data with respect to the classification task to which data are fed. Figure 1 shows our reference scenario, with data controllers anonymizing their data before providing them for the global knowledge base on which the classification task operates. Our data anonymization approach, called DT-Anon, anonymizes data aware of the data classification task to which data are contributed, with the goal of minimizing the effect of protection on the data analysis to be executed.

The remainder of the paper is organized as follows. Section 2 illustrates the basic concepts of our approach. Section 3 describes the problem addressed and the rationale of our approach. Section 4 describes our target-driven anonymization. Section 5 illustrates our experimental evaluation. Section 6 discusses related work. Finally, Sect. 7 presents our conclusions. Appendixes A and B report a theorem on the approach and the DT-Anon pseudocode, respectively.

2 Basic Concepts

We consider anonymization of datasets to be contributed to a data analytics task. Wishing to operate with truthful information for data analytics [1], we assume anonymization to be carried out according to k-anonymity [4,12] enhanced with ℓ-diversity [11]. We assume datasets to be anonymized to be relational tables, where each relation R is characterized by a set $\{a_1, \ldots, a_n\}$ of attributes comprising:

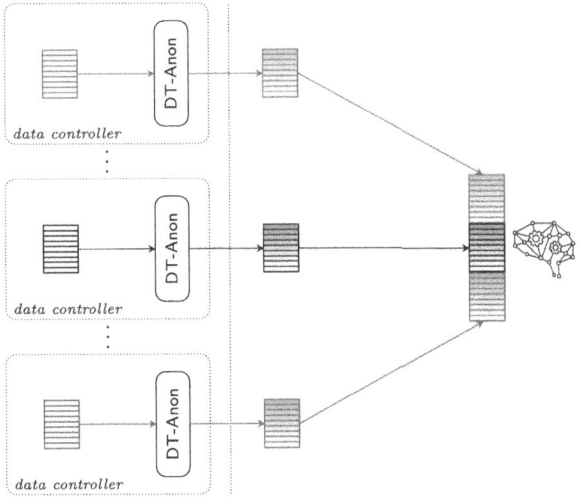

Fig. 1. Reference scenario

- *identifiers*: attributes identifying data subjects, that is, entities to which data refer (we consider these attributes to be removed before release and therefore discard them from our treatment);
- *quasi-identifier*: set of attributes that jointly can, through linking with other sources, possibly reduce the uncertainty about identities of data subjects;
- *sensitive*: attribute whose values, in association with (the identity of) data subjects, are considered sensitive and should therefore be protected.

Data anonymization through k-anonymity and ℓ-diversity implies generalizing values of the quasi-identifier attributes to ensure each combination of (generalized) values of quasi-identifying attributes appearing in the table to occur at least k times (k-anonymity), and each group of tuples with the same generalized quasi-identifying values to have at least ℓ well-represented sensitive values (ℓ-diversity). We consider generalization applied at the level of cell (in contrast to the whole attribute column), hence operating at finest possible grain to limit information loss [5]. We represent the generalized value for a set of values as the interval between the minimum and maximum values in the set for continuous (i.e., numerical) attributes, and as a set comprising all the values for categorical attributes. Also, for simplicity, in the following examples, we assume the well-represented criterion of ℓ-diversity to be enforced by requiring each group to include at least ℓ different values ($\ell = 1$ implies requiring only k-anonymity to hold with no restriction on the occurrences of the sensitive values). We refer to a transformed (generalized) version of a relation satisfying k-anonymity and ℓ-diversity as a (k, ℓ)-anonymous version of the relation. Figure 2 illustrates an example of original dataset and two possible $(2, 2)$-anonymous versions of it, considering Age and State to work as quasi-identifier and attribute Income to be sensitive.

	Age	State	Job	Income
t_1	51	MN	Gov	150
t_2	30	CA	Non-gov	150
t_3	35	CA	Non-gov	120
t_4	35	TX	Gov	300
t_5	62	MN	Gov	200
t_6	40	TX	Gov	300
t_7	62	CA	Gov	300
t_8	37	MN	Non-gov	150
t_9	36	TX	Gov	180
t_{10}	24	MN	Gov	100
t_{11}	64	CA	Gov	250

(a)

	Age	State	Job	Income
t_2	[30-35]	CA	Non-gov	150
t_3	[30-35]	CA	Non-gov	120
t_7	[62-64]	CA	Gov	300
t_{11}	[62-64]	CA	Gov	250
t_1	[51-62]	MN	Gov	150
t_5	[51-62]	MN	Gov	200
t_8	[24-37]	MN	Non-gov	150
t_{10}	[24-37]	MN	Gov	100
t_4	[35-40]	TX	Gov	300
t_6	[35-40]	TX	Gov	300
t_9	[35-40]	TX	Gov	180

(b)

	Age	State	Job	Income
t_1	[37-62]	MN	Gov	150
t_5	[37-62]	MN	Gov	200
t_8	[37-62]	MN	Non-gov	150
t_2	[24-35]	{CA,MN}	Non-gov	150
t_3	[24-35]	{CA,MN}	Non-gov	120
t_{10}	[24-35]	{CA,MN}	Gov	100
t_4	[35-36]	TX	Gov	300
t_9	[35-36]	TX	Gov	180
t_6	[40-64]	{CA,TX}	Gov	300
t_7	[40-64]	{CA,TX}	Gov	300
t_{11}	[40-64]	{CA,TX}	Gov	250

(c)

Fig. 2. An example of original relation (a) and of two $(2, 2)$-anonymous versions of the original relation (b)–(c)

3   Problem Definition and Sketch of the Approach

Our reference scenario is characterized by multiple *data controllers* contributing data for a data analytics task, operating on the collective information contributed by the different controllers. As data analytics task we consider *classification*. The goal of classification is to learn from classified data a model (classifier) able to predict the class (i.e., the value of a target attribute) associated with unseen data. Intuitively, classification learns dependencies of a given attribute (*target*) from other attributes (*predictors*) in the dataset. Datasets contributed by data controllers collectively represent the training data on which the classification task learns. We therefore consider a dataset $R(a_1, \ldots, a_n)$ to be released by a data controller contributing to the classification task to include, besides the quasi-identifier and sensitive attributes, also a

– *target* attribute, denoted τ, of interest for the classification task for which data are released.

Note that the target attribute cannot coincide with the sensitive attribute, by the definition of the problem at hand (as we aim at maintaining as much as possible the correct prediction of the target attribute while protecting instead inferences on the sensitive attribute).

Data may contain identifying, quasi-identifying, or sensitive information, and therefore should be anonymized before being released to external parties. As said, we assume anonymization with k-anonymity and ℓ-diversity, and hence the release to the classification analytics task of (k, ℓ)-anonymous datasets. Anonymization is enforced independently by each data controller, which could even operate with different values of k and ℓ, depending on the degree of protection wished.

The problem then becomes the possible negative impact of the anonymization on the ability of the classification task to learn dependencies relevant for

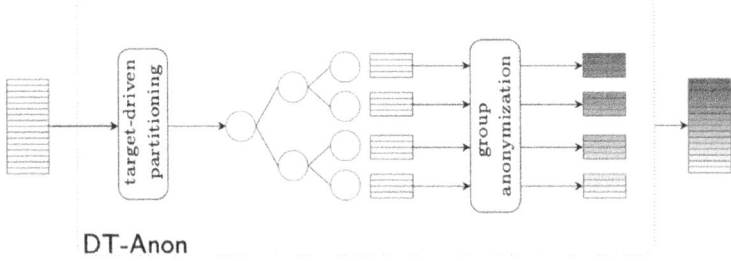

DT-Anon

Fig. 3. Overall working of our target-driven anonymization

the classification, that is, dependencies between attributes that represent good predictors of the target and the target. Intuitively, if predictor attributes are generalized in the anonymization, the classification will be operating with less information (suffering the information loss caused by generalization). Since different anonymous versions of a table can exist, corresponding to different groups of generalized tuples and/or generalization of different attributes/values in the quasi-identifier, our goal is the definition of an anonymization process aware of the data analytics task downstream and driven by it.

Basically, the problem we address is: *Given a relational table $R(a_1, \dots, a_n)$ where the set $\{a_1, \dots, a_n\}$ of attributes includes quasi-identifier attributes QI, a sensitive attribute s, and a target attribute τ, compute a (k, ℓ)-anonymous version of R that performs well for a classification task with target τ.*

In other words, we aim for an anonymization that preserves as much as possible the correlation among quasi-identifier and target values. With generalization as the technique for achieving anonymization, this implies to aim for a generalized version of the dataset that maintains as specific as possible the values of the predictor attributes in the quasi-identifier on which the target values depend more, generalizing instead values of other attributes from which the target attribute is less (or no) dependent. We do so by applying the anonymization process on subsets of the dataset, where each subset groups tuples that are equal or most similar with respect to predictor attribute values (so that generalization does not affect them or affect them with limited information loss).

Our approach, called DT-Anon, comprises two steps (Fig. 3):

- *target-driven partitioning* operates on the datasets by partitioning data producing groups driven by the target of the classification task. More precisely, groups are defined through a decision tree guided by the classification target. Intuitively, the decision tree partitions the tuples producing groups that will have equal or close/similar predictor values; hence avoiding their generalization or limiting the effect of a possible generalization on them.

– *group anonymization* applied on each group of tuples produced in the previous step (leaf nodes of the decision tree) with a classical anonymization approach.

Since the target-driven anonymization is enforced independently by each data controller before releasing data, in the following we illustrate our approach with reference to a single dataset. We will consider the presence of different of such anonymized datasets used for the same data analytics task in the experimental evaluation (see Sect. 5).

Example 1 (Running example). As running example, we consider the table in Fig. 2(a), where the pair ⟨Age,State⟩ is the quasi-identifier, Income is the sensitive attribute, and Job is the target for the classification task to which data are to be contributed (the example omits identifiers and other attributes since they are not relevant for the work). We will also refer to the two $(2, 2)$-anonymous versions of the table reported in Figs. 2(b)–(c).

4 Target-Driven Anonymization

We describe in more details the two phases of our target-driven anonymization.

4.1 Target-Driven Partitioning

The first step of our approach is the partitioning of tuples to produce groups of tuples that - when generalized - best preserve the correlation between (generalized) quasi-identifier and target values. Intuitively, this corresponds to produce groups that maintain in the same group tuples with the same (or as close as possible) values for predictor attributes in the quasi-identifier, so that the anonymization (generalization in particular) would not affect, or has limited impact, on them. Of course, predictors are not known, but should be learned from the data themselves.

Our approach to such target-driven partitioning is to use a machine learning algorithm based on a *decision tree* to make predictions. In other words, we create a model that predicts the value of the target attribute by learning *decision rules* from the other (quasi-identifier) attributes.

The construction of a decision tree starts from the root node that represents the whole dataset. The decision tree is then recursively built by splitting the dataset represented by a node into subsets that are represented as its child nodes. More precisely, for each node of the tree, a set of possible split values is identified for each attribute. The algorithm for the construction of the decision tree selects the attribute and the split value(s) that are most significant with respect to a specific criterion defined on the target [7] (e.g., information gain).

This process terminates when a stopping condition is satisfied (e.g., the values of the target attribute for the tuples in the leaf nodes are sufficiently uniform or all the attributes have been used for splitting). By construction, the attributes used in the splitting operations are those on which the target attribute depends more since they permit to partition the tuples in groups that are as similar as possible with respect to the target attribute. Furthermore, the tuples in these groups are also similar with respect to these attributes since they all satisfy the same decision rules (i.e., the if-then rules defined over the attributes used in the splitting operation).

We adapt this classical construction of a decision tree to our goal. First, we only use the quasi-identifier attributes for the splitting operations. Other attributes, including the sensitive attribute, are not considered. The rationale for using the quasi-identifiers only is that, as previously mentioned, the construction of the decision tree identifies the (quasi-identifier) attributes on which the target depends more and tuples with similar values for these attributes are grouped together, which is exactly what we need. The reason for not using the sensitive attribute (e.g., Income for our running example) is the need to ensure diversity of its values in each generalized group. The reason for not considering other attributes is that they are not affected by generalization and their values therefore are never impacted by the process. Using them not only would not help but could actually have negative effect, as it might prevent optimal consideration of quasi-identifier predictors in the partitioning (which would eventually result in more generalization on them). Second, we add the condition that a node can be split only if the resulting child nodes represent a partition of the parent relation with a sufficient number of tuples for satisfying the k-anonymity and ℓ-diversity requirements. (Intuitively, this requires the parent node to have at least $2k$ tuples to permit at least a binary split.) Each leaf node of a decision tree built considering these two changes is therefore a node that, by construction, represents a group of tuples of size at least k and with at least ℓ well-represented values for the sensitive attribute. This target-driven partitioning phase ensures to result in a (k, ℓ)-compliant decision tree formally defined as follows.

Definition 1 ((k, ℓ)-compliant decision tree). *Let $R(a_1, \ldots, a_n)$ be a relation with QI, s, and τ the quasi-identifier, sensitive, and target attributes in $\{a_1, \ldots, a_n\}$, respectively, and $DT(N, E)$ be a decision tree built over R for predicting target attribute τ, with N the set of nodes and E the set of edges. DT is a (k, ℓ)-compliant decision tree iff for each leaf node $n \in N$, the set of tuples R_n represented by n is such that $\mid R_n \mid \geq k$ and R_n includes tuples with at least ℓ well-represented values for s.*

Example 2 (Decision tree). Figure 4 illustrates an example of a decision tree built over the relation in Fig. 2(a). The root node coincides with the whole table that is split over attribute State. The resulting child nodes correspond to the set of

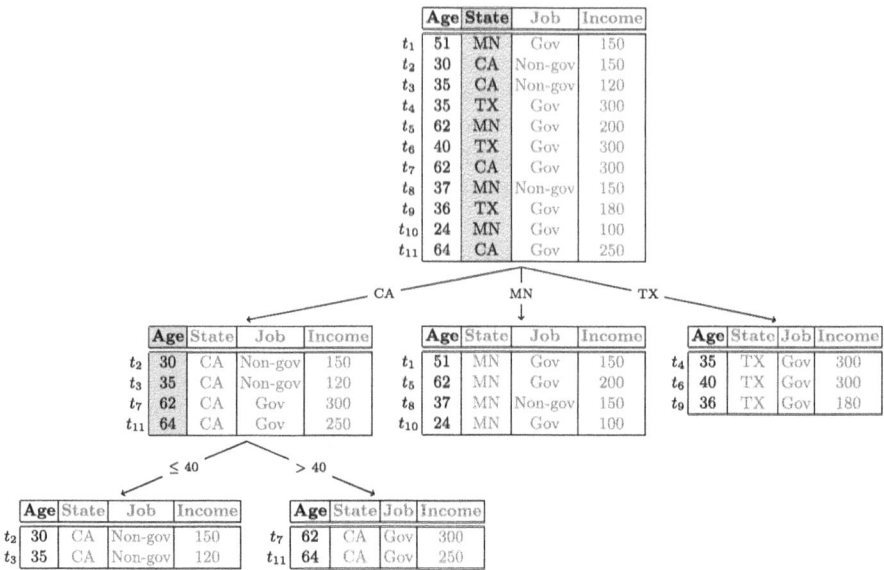

Fig. 4. An example of decision tree built over the relation in Fig. 2(a)

tuples related to employees working in California (CA), Minnesota (MN), and Texas (TX). For employees working in California, there is a further split that distinguishes between employees with age less than or equal to 40 and over 40. The leaf nodes are labeled with either 'Gov' or 'Non-gov' as job. For each node, attributes with a gray background are those used for splitting, and attributes with gray values are the attributes that cannot be used for splitting (i.e., sensitive attribute or attributes already used for split). In this tree, for example, the first (from left) leaf node corresponding to the set $\{t_2, t_3\}$ of tuples is associated with a decision rule of the form "IF State=CA AND Age$\leq$ 40 THEN Job is 'Non-gov'". Since each leaf node represents a group of at least two tuples with at least two different values for the sensitive attribute Income this is a $(2, 2)$-compliant decision tree.

4.2   Group Anonymization

The goal of the second phase is to independently anonymize each group of tuples represented by the leaf nodes of the (k, ℓ)-*compliant* decision tree built in the previous phase. The construction of the groups, clustering together tuples that are equal or close in values for predictors, ensures minimizing the impact of generalization on predictors. The problem becomes then to compute a generaliza-

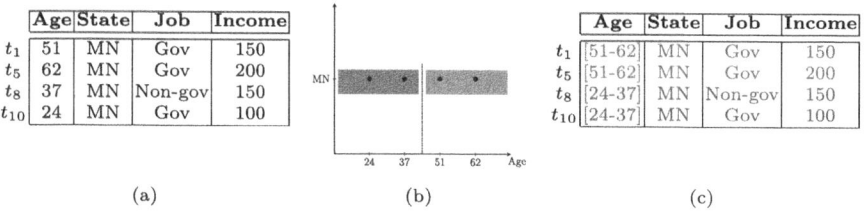

	Age	State	Job	Income
t_1	51	MN	Gov	150
t_5	62	MN	Gov	200
t_8	37	MN	Non-gov	150
t_{10}	24	MN	Gov	100

	Age	State	Job	Income
t_1	[51-62]	MN	Gov	150
t_5	[51-62]	MN	Gov	200
t_8	[24-37]	MN	Non-gov	150
t_{10}	[24-37]	MN	Gov	100

(a)   (b)   (c)

Fig. 5. An example of a relational table (a) with its spatial representation and partitioning (b), and the corresponding $(2,2)$-anonymous version (c).

tion that produces a (k, ℓ)-anonymous version of each leaf node while minimizing information loss. This can be achieved with classical approaches for k-anonymity and ℓ-diversity. In particular, we consider the application of Mondrian [10], a multi-dimensional algorithm that provides an efficient and effective approach for achieving k-anonymity (which we consider extended with ℓ-diversity). Mondrian leverages a spatial representation of the data, mapping each quasi-identifier attribute to a dimension, and each combination of values of the quasi-identifier attributes to a point in such a multi-dimensional space (multiple tuples with the same coordinates translate into a point with a multiplicity greater than 1). Mondrian then recursively partitions the multi-dimensional space in two sub-spaces by selecting a dimension (i.e., an attribute in the quasi-identifier) and a split point (i.e., a value in the domain of such an attribute), in such a way that each sub-space includes at least k points/tuples with at least ℓ different values for the sensitive attribute. The process terminates when any further partitioning would generate sub-spaces with less than k points (or the points in the sub-spaces would have less than ℓ different values for the sensitive attribute). Finally, all the tuples in each subspace are generalized to the same combination of (generalized) values for the quasi-identifier. The motivation for choosing Mondrian is that, while it being a well established reference in the field as efficient and effective approach, its approach to cutting multi-dimensional space to partition tuples is similar to how a decision tree works.

Example 3 (Anonymization). Consider the table in Fig. 2(a) and the decision tree in Fig. 4, where again the quasi-identifier is the pair ⟨Age,State⟩. Anonymization is applied independently on the four leaves. For the groups of two tuples no further split can be performed and hence only generalization is applied, reporting the age interval (instead of the specific values). For the group of four tuples, also reported in Fig. 5(a) and in the multi-dimensional space in Fig. 5(b), Mondrian will perform a split over attribute Age (the only one with different values) resulting in two groups to be generalized (Fig. 5(c)).

The anonymized version of the dataset is finally obtained through the union of the anonymized groups represented by the leaf nodes of the (k, ℓ)-*compliant* decision tree. Clearly, being each group of tuples (k, ℓ)-anonymous, also the union is, as formally captured by the theorem in Appendix A. For instance, the whole

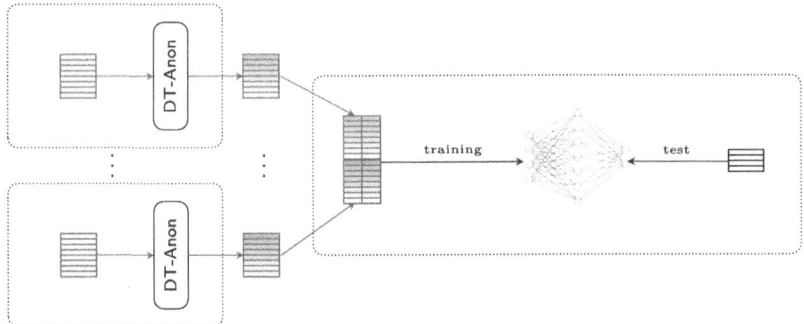

Fig. 6. Experimental scenario

dataset comprising the results of the anonymization of the different groups produced by the decision tree in Fig. 4 is the table in Fig. 2(b). Appendix B presents the algorithm used to compute a target-driven anonymization of a relation.

5 Experimental Results

We conducted a series of experiments to evaluate the effectiveness of DT-Anon in producing an anonymized dataset that can be used for training a classifier with good performance. In the following, we first describe the methodology applied and the datasets used in the experimental evaluation (Sect. 5.1), and then report and discuss the experimental results (Sect. 5.2).

5.1 Experimental Settings

Methodology. Our experimental evaluation has the goal of comparing the performance of a classifier when considering anonymization of the dataset of each single data controller running independently from the classification goal (i.e., following a classical approach) and when considering anonymization produced by DT-Anon. To evaluate the impact of anonymization on the classification task, we also evaluate the performance of the classifier trained over the original (raw) datasets.

Figure 6 illustrates the experimental scenario with multiple data controllers simulated by our experiments. We evaluated the classifier using different available datasets, partitioning the data in a training and a test set. To simulate the presence of multiple data controllers, we randomly partitioned the training set in different datasets (on which we then operated independently). In the paper, we report the results for the case of two data controllers, assuming two partitions of the original data on which the target-driven anonymization operated independently. For simplicity, we assumed k and ℓ (for which we considered different values) to be the same for all the data controllers.

As classifier, we considered a neural network, which implied transforming data into numeric variables to be fed in the neural network. Transformation, dependent on the type of attributes, worked as follows.

- *Categorical attributes*. Each categorical attribute is replaced with x binary attributes (one for each possible value of the attribute). Data values are then encoded through their representation via the binary attributes, setting to 1 the binary attribute(s) corresponding to their value(s). Scalar values will have only one binary attribute set to 1 (single-hot encoding) while sets of values (resulting from generalization) may have more than one attribute set to 1 (multi-hot encoding). For instance, with reference to the table in Fig. 2(c), attribute State will be represented using three binary attributes (State$_{CA}$, State$_{MN}$, and State$_{TX}$), and its generalized value in tuple t_2 encoded with the first two attributes set to 1 and the latter set to 0.
- *Numerical attributes*. Scalar values of numerical attribute a are standardized, meaning that each value v is substituted with $(v - \mu_a)/\sigma_a$, where μ_a is the mean of the values in the relation for attribute a and σ_a is their standard deviation. For interval values (resulting from generalization), transformation is preceded by replacing each interval value with its mid point. For instance, with reference to our running example, Age interval [62-64] in generalized tuples t_7 and t_{11} is replaced by 63.

The transformed anonymized datasets have then been used for training a classifier. We built a neural network with three hidden layers with 64, 32, and 16 neurons, respectively. We used the ReLu activation function and the Adam optimizer. These are the default parameters also used in the scikit-learn[1] implementation.

Evaluation Metrics. We evaluated the performance of the neural network (trained over anonymized datasets or raw datasets) by measuring the *accuracy* and the *$F1_{macro}$* score. Accuracy is the ratio between the number of correct predictions and the total number of predictions. It then measures the percentage of correct classifications that a trained machine learning model achieves. $F1_{macro}$ score is defined as the average of the class-wise F1 scores and is used for a multi-class classification problem. Formally, given a classification problem with a set C of classes, the $F1_{macro}$ score is defined as: $F1_{macro} = \frac{\sum_{c \in C} F1(c)}{n}$ where $F1(c)$ is the F1 score (i.e., the harmonic means of precision and recall) computed for class c.

Datasets. We performed experiments on different publicly available real-world datasets. We report here the results on datasets: *Bank* and *Nursery*, from the UCI machine learning repository and *Customer_segmentation*, from the Kaggle platform. The datasets, whose attributes are reported in Table 1, are as follows.

- *Bank* dataset[2] describes individuals using both numeric and categorical attributes (45,211 tuples). The dataset refers to direct marketing campaigns

---

[1] https://scikit-learn.org/stable.
[2] https://archive.ics.uci.edu/dataset/222/bank+marketing.

Table 1. Overview of the attributes in the considered datasets

Dataset	QI		Target (τ)	Sensitive (s)	Others
	Categorical	Numeric			
Bank	job marital education housing loan	age balance duration	y (2 values)	default (2 values)	pdays previous campaign poutcome contact day_of_week month
Nursery	parents, has_nurs form children housing finance health		class (5 values)	social (3 values)	
Cust_segm	gender ever_married graduated profession	age work_exp fam_size	segment (4 values)	spending_score (3 values)	var_1

based on phone calls related to a Portuguese banking institution. We consider as sensitive the binary attribute `default` that can assume two values stating whether the individual has credit in default. The target attribute is `y` that represents whether a client of the bank has subscribed a bank term deposit and has two possible values: 'yes' and 'no'.

- *Nursery* dataset[3] describes individuals using categorical attributes (12,960 tuples). It contains information derived from a hierarchical decision model that was realized to rank nursery school applicants. We consider as sensitive the categorical attribute `social` that represents the social conditions of the family of the applicant and can assume three values. The target attribute is `class` that represents the decisions and has five possible values: 'not_recom', 'recommend', 'very_recom', 'priority', and 'spec_prior'.
- *Customer_segmentation* dataset[4] describes individuals using both numeric and categorical attributes (13,330 tuples). The dataset includes information on the customers of an automotive company. We consider as sensitive the categorical attribute `spending` that represents the individual's spending score and can assume three values. The target attribute is `segment` that represents the market segment and has four possible values: 'A', 'B', 'C', and 'D'.

5.2 Results

Our experiments compare the performance of classifiers trained with anonymized datasets varying the privacy parameters k and ℓ. In the following, we use DT-Anon to refer to the neural network trained over a dataset anonymized with DT-Anon, and Anon to refer to the neural network trained over a dataset anonymized

---

[3] https://archive.ics.uci.edu/dataset/76/nursery.
[4] https://www.kaggle.com/datasets/kaushiksuresh147/customer-segmentation.

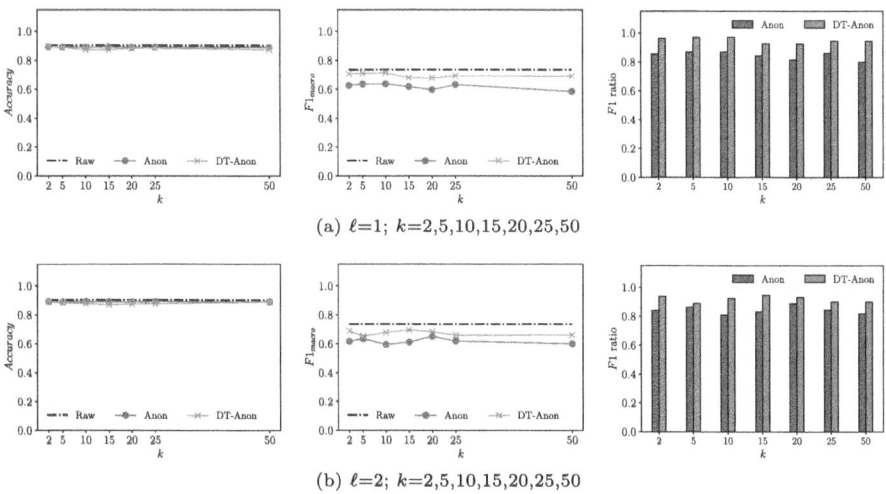

(a) $\ell=1$; $k=2,5,10,15,20,25,50$

(b) $\ell=2$; $k=2,5,10,15,20,25,50$

Fig. 7. Accuracy, $F1_{\mathrm{macro}}$, and $F1_{\mathrm{macro}}$ ratio varying k and ℓ for the *Bank* dataset

with a classical anonymization algorithm (which, in our implementation, is Mondrian revised to support the ℓ-diversity requirement). The performance of classifiers has been measured considering both the accuracy as well as the $F1_{\mathrm{macro}}$ score varying k (considering values 2, 5, 10, 15, 20, 25, 50) and ℓ (considering values 1, 2, 3, according to the number of distinct values for the sensitive attribute in the datasets). We consider as a baseline the neural network trained with the original (non anonymized) datasets. Figures 7, 8 and 9 illustrate the results of our experiments. In the figures, the orange (light gray in b/w) lines and bars refer to DT-Anon and the blue (dark gray in b/w) lines and bars refer to Anon. For each dataset and for each value of ℓ, we draw two line charts and one bar chart showing: accuracy, $F1_{\mathrm{macro}}$ score, and the ratio between the $F1_{\mathrm{macro}}$ score of DT-Anon (or Anon) and the $F1_{\mathrm{macro}}$ score of our baseline, respectively. This ratio shows how well a supervised learning model (our neural network) learns from the anonymized datasets with respect to how well it learns from the original raw dataset. In other words, it shows how much the anonymization of the dataset impacts the capability of the model to learn from (anonymized) data. As it is visible from the figures, DT-Anon performs better than Anon with almost all datasets and values of k and ℓ.

Accuracy. The experiments (line charts in the first column of Figs. 7, 8 and 9) show that the accuracy tends to decrease as the value of k and ℓ increases. This trend is due to the fact that higher values for k and ℓ require a higher amount of generalization, thus implying higher information loss that in turn produces a decrease in the effectiveness of the anonymized datasets in the learning process.

$F1_{\mathrm{macro}}$. The $F1_{\mathrm{macro}}$ score has a similar trend as the accuracy across all datasets (line charts in the second column of Figs. 7, 8 and 9). The experiments also

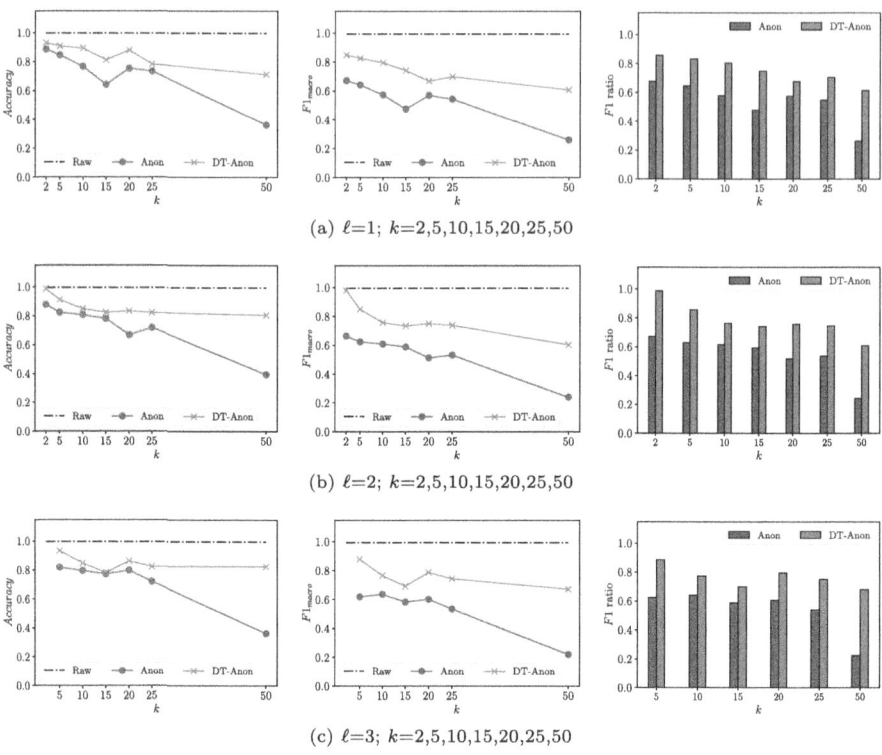

Fig. 8. Accuracy, F1$_{\mathrm{macro}}$, and F1$_{\mathrm{macro}}$ ratio varying k and ℓ for the *Nursery* dataset

show that, even though the enforcement of both the k-anonymity and ℓ-diversity requirements with $\ell > 1$ is stricter than the enforcement of the k-anonymity requirement only, the impact on the capability of the neural network to learn from the anonymized data remains similar to the cases where $\ell = 1$. Furthermore, the F1$_{\mathrm{macro}}$ score of DT-Anon is constantly higher than the F1$_{\mathrm{macro}}$ score of Anon.

F1$_{\mathrm{macro}}$*ratio.* The F1$_{\mathrm{macro}}$ ratio of DT-Anon for the *Bank* and *Customer_segmentation* datasets remains always higher than 0.82 (bar chars in Figs. 7, 8 and 9), again confirming that, with DT-Anon, the neural network preserves the capability of learning from the anonymized datasets. The F1$_{\mathrm{macro}}$ ratio of Anon is constantly lower. For the *Nursery* dataset the values are lower than those obtained for the other two datasets (from 0.60 with $k = 50$ and $\ell = 2$ to 0.98 with $k = 2$ and $\ell = 2$). Such values are, however, much higher than those obtained for Anon. This is probably due to the correlation between the quasi-identifier attributes and the target attribute in *Nursery*, which DT-Anon captures and preserves.

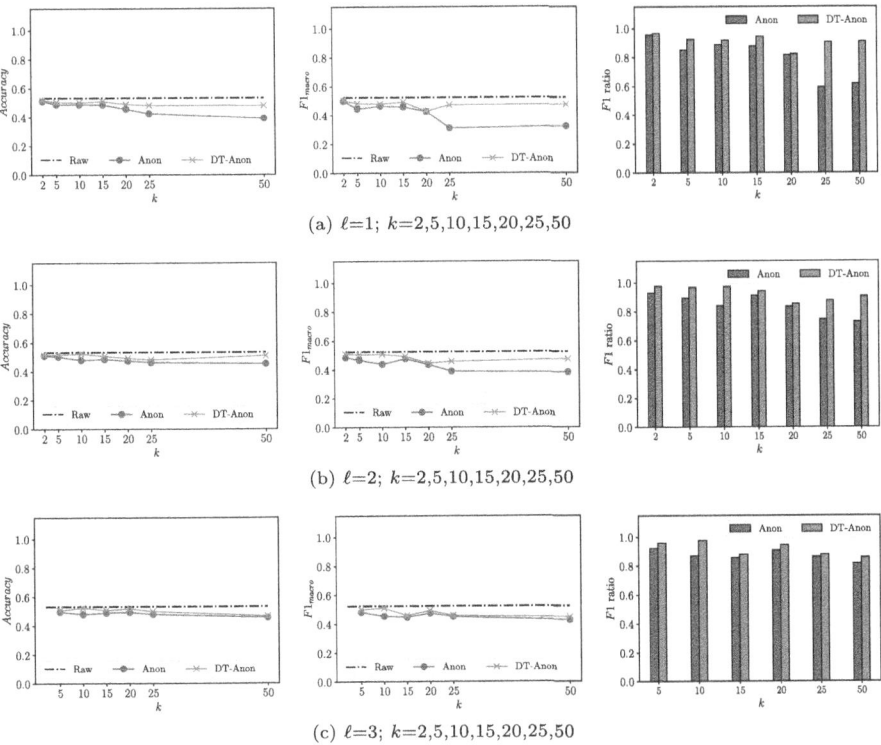

(a) $\ell=1$; $k=2,5,10,15,20,25,50$

(b) $\ell=2$; $k=2,5,10,15,20,25,50$

(c) $\ell=3$; $k=2,5,10,15,20,25,50$

Fig. 9. Accuracy, $F1_{macro}$, and $F1_{macro}$ ratio varying k and ℓ for the *Customer_ segmentation* dataset

6   Related Work

The problem of studying the effects of anonymization (e.g., k-anonymity, ℓ-diversity) on machine learning models has been the subject of several works (e.g., [3,6,9,13,14]).

Some of these proposals address the problem of evaluating the impact of different existing anonymization algorithms on the result of machine learning models (e.g., classifiers) and whether data anonymization can be enough to achieve privacy in machine learning (e.g., [13,14]).

Other proposals instead consider a problem similar to the one addressed in this paper and define an anonymization strategy that takes into account the subsequent use of the anonymized datasets (e.g., [3,6,8,9]). The work in [9] has introduced the problem of anonymizing data depending on a workload, which can be a classification or regression model or selection/projection predicates (i.e., selection or projection predicates that identify a subset of the data on which the anonymization is applied). The authors propose a variation of Mondrian where data are split in a way that minimizes the weighted entropy over the set of resulting partitions without violating the k-anonymity requirement. The main differences with our approach are that we consider a scenario with multiple data controllers and we can anonymize data using any anonymization algorithm. The work in [6] defines a method for learning how to generalize unseen data for classification analysis. It starts from an existing machine learning model and learns how unseen data should be generalized by training a *generalized model* (i.e., a decision tree) with data labeled with the existing model's predictions. The decision tree is then used to derive a set of generalization ranges obtained by combining the split values of each attribute from the tree's internal nodes. While sharing with us the idea of using a decision tree to build groups of "similar tuples", the problem addressed is completely different. Also the work in [8] builds a decision tree to determine the attributes that most influence the value of the target attribute. Leaf nodes of the decision tree with more than k data items are then anonymized by suppressing all attributes that are not used along the path from the root to the considered leaf node. Otherwise, a prune procedure is applied to obtain new leaf nodes of size at least k. The anonymization via suppression is then applied on these new leaf nodes. This proposal differs from our proposal in several aspects. The decision tree is build in a different way, the ℓ-diversity requirement is not considered, and the anonymization is enforced only through suppression, thus potentially reducing the information available for the classification task. In [3] the authors propose an approach for anonymizing data guided by relaxed functional dependencies. Such dependencies specify what subsets of attributes can be generalized and at which level, so to achieve a minimum level of anonymity (expressed through the k-anonymity requirement) while preserving data utility as much as possible. Data utility is measured in terms of classification accuracy and information gain. A set of generalization rules is extracted from the relaxed functional dependencies and then used for anonymizing datasets.

Other complementary solutions propose different anonymization strategies to improve the trade-off between privacy and utility also in machine learning scenarios (e.g., [15]).

7   Conclusions

We addressed the problem of anonymizing data in a scenario where multiple data controllers contribute to a classification task. We proposed DT-Anon, a data anonymization approach aware of and driven by the classification task downstream. DT-Anon enables data controllers contributing with their data to a classification task to anonymize their data while maintaining utility for the classification task. The experimental results confirm the ability of DT-Anon to limit, with respect to classical anonymization approaches, the information loss caused by data anonymization and hence its effect on the performance of the classification task. The paper leaves space for future work, including the consideration of other anonymization approaches (e.g., differential privacy) and the consideration of multiple data analytics tasks (e.g., tasks with different target attributes).

Acknowledgements. This work was supported in part by the EC under projects Chips JU EdgeAI (101097300) and GLACIATION (101070141), by the Italian MUR under PRIN project POLAR (2022LA8XBH), and by project SERICS (PE00000014) under the MUR NRRP funded by the EU - NGEU. Views and opinions expressed are however those of the authors only and do not necessarily reflect those of the European Union or the Italian MUR. Neither the European Union nor the Italian MUR can be held responsible for them.

A   Theorem and Proof

Theorem 1 ((k, ℓ)-anonymous dataset). *Let $R(a_1, \ldots, a_n)$ be a relation with QI, s, and τ the quasi-identifier, sensitive, and target attributes in $\{a_1, \ldots, a_n\}$, respectively, k and ℓ be the privacy parameters, and $DT(N, E)$ be a (k, ℓ)-compliant decision tree built over R for τ. The relation $\widehat{R} = \bigcup_{n \in N} \widehat{R}_n$, with $\widehat{R}_n$ the (k, ℓ)-anonymous version of relation R_n and $n \in N$ a leaf node of DT, is (k', ℓ)-anonymous, with $k' \geq k$.*

Proof. Let R_n be the partition of R represented by leaf node n in DT, and $\widehat{R}_n$ its (k, ℓ)-anonymous version. Table $\widehat{R}$ obtained merging the (k, ℓ)-anonymous partitions $\widehat{R}_{n_1}, \ldots, \widehat{R}_{n_m}$ of the leaf nodes $n_1, \ldots, n_m$ of DT includes m (k, ℓ)-anonymous partitions of R. Since each combination of (generalized) values for the quasi-identifier QI have either 0 or at least k occurrences in each $\widehat{R}_{n_i}$, such a combination will have either 0 or at least k occurrences also in $\widehat{R}$. Since a specific combination of (generalized) values for the quasi-identifier could appear in more than one anonymized partition, the number of occurrences of such a combination of values could be higher than k (i.e., $\geq j \cdot k$ if appearing in j partitions). Therefore, $\widehat{R}$ is k'-anonymous with $k' \geq k$. Furthermore, table $\widehat{R}$ still satisfies the ℓ-diversity property since the groups of tuples having the same combination of (generalized) values for QI can only grow (due to the presence of groups of tuples with the same value for QI in more than a partition of R). Hence, the number of well represented values can either grow or remain the same. $\square$

B   DT-Anon Algorithm

We now describe DT-Anon algorithm that enforces the target-driven anonymization. Figure 10 illustrates the pseudocode of the algorithm implementing the two phases of our approach.

Target-Driven Partitioning. Figure 10 illustrates the pseudocode of **BuildDT**, a recursive procedure used in the first phase of the DT-Anon algorithm for computing a (k, ℓ)-*compliant* decision tree. Procedure **BuildDT** receives as input a node n of the decision tree (which corresponds to the root node at its first invocation). The procedure first identifies the set $R_n \subseteq R$ of tuples that satisfy

Input

$R(a_1, \ldots, a_n)$: original relation
QI: quasi-identifier attributes in $\{a_1, \ldots, a_n\}$
s: sensitive attribute in $\{a_1, \ldots, a_n\}$
τ: target attribute in $\{a_1, \ldots, a_n\}$
k: anonymity requirement
ℓ: diversity requirement

Output

$\widehat{R}$: (k, ℓ)-anonymous version of R

DT-Anon
/* **Phase 1**: Compute a (k, ℓ)-*compliant* decision tree $DT(N, E)$ */
1: $N :=$ ROOT; $E := \emptyset$ /* set N of nodes and set E of edges of DT */
2: **BuildDT**(ROOT) /* ROOT node representing R */
   /* **Phase 2**: anonymize the leaves of DT */
3: $\widehat{R} := \emptyset$
4: **for each** leaf node $n \in N$ **do**
5:   $R_n = d_n(R)$
6:   $\widehat{R} := \widehat{R} \cup$ **Anonymize**(R_n)
7: **return**$(\widehat{R})$

BuildDT(n)
1: let $R_n = d_n(R)$ /* set of tuples in R satisfying decision rule d_n of n/*
2: **if** $|R_n| \geq 2k$ AND $\exists t_i, t_j \in R_n$: $t_i[\tau] \neq t_j[\tau]$ AND stop condition is not satisfied
3: **then** let *Split* be all possible splits of R_n
4:   **repeat**
5:     choose the most promising *split* in *Split*
6:     let N' be the set of nodes resulting applying *split* on R_n
7:     *found* := TRUE
8:     **while** *found*=TRUE AND $N' \neq \emptyset$ **do**
9:       let $n_i \in N'$ and $R_{n_i} = d_{n_i}(R_n)$
10:      **if** $|R_{n_i}| < k$ OR R_{n_i} has less than ℓ well-represented values for s
11:      **then** *found* := FALSE
12:      **else** $N' := N' \setminus \{n_i\}$
13:    *Split* := *Split* $\setminus$ $\{split\}$
14:   **until** *Split*=$\emptyset$ OR *found*=TRUE
15:   **if** *found*=TRUE
16:   **then** let N' be the set of nodes resulting applying *split* on R_n
17:     $N := N \cup N'$
18:     **for each** $n_i \in N'$ **do**
19:       $E := E \cup (n, n_i)$
20:       **BuildDT**(n_i)

Fig. 10. Pseudocode of the DT-Anon algorithm

the decision rule, denoted d_n, associated with node n (line 1). The procedure then verifies whether such a set of tuples can be further split (line 2). Existing decision tree algorithms split a node into child nodes until a stopping criterion is met or until the node represents a set of tuples with an homogeneous value for the target attribute. The **BuildDT** procedure adds a further check and verifies whether the set R_n of tuples represented by the considered node includes at least $2k$ tuples (i.e., at least a binary split can be enforced on the node). In this case, the procedure identifies all the possible candidate splits for n (line 3), considering the available attributes and partitions of their domains (like classical algorithms for building a decision tree). The procedure then evaluates, in decreasing order or effectiveness on the classification, the candidate splits checking whether the considered split *split* guarantees that the decision tree satisfies Definition 1 (lines 4–14). If *split* produces a (k, ℓ)-*compliant* decision tree, the split is enforced and the procedure recursively invokes itself on each (child) node resulting from the split (lines 15–20). As an example, suppose that the data owner wishes to compute a $(3, 1)$-anonymous version of the relation in Fig. 2(a). DT-Anon starts by invoking procedure **BuildDT** that, as shown in Fig. 4, splits the relation on attribute `State` in three child nodes. The second split of the first (from left) child node in Fig. 4 would be instead prevented because it generates two partitions with less than 3 tuples each.

Group Anonymization. The second phase of the DT-Anon algorithm in Fig. 10 consists in independently anonymizing the (sub)relations represented by the leaf nodes of the (k, ℓ)-*compliant* decision tree built in the first phase. For each leaf node $n \in N$ of the decision tree, the algorithm invokes an anonymization algorithm on relation R_n, and returns the (k, ℓ)-anonymous version of R_n, which is appended to the other (k, ℓ)-anonymous relations.

References

1. Abukmeil, M., Ferrari, S., Genovese, A., Piuri, V., Scotti, F.: A survey of unsupervised generative models for exploratory data analysis and representation learning. ACM CSUR **54**(5), 1–40 (2021)
2. Bhattacharjee, K., Islam, A., Vaidya, J., Dasgupta, A.: PRIVEE: a visual analytic workflow for proactive privacy risk inspection of open data. In: Proceedings of IEEE VizSec, Oklahoma City, OK, USA (2022)
3. Caruccio, L., Desiato, D., Polese, G., Tortora, G., Zannone, N.: A decision-support framework for data anonymization with application to machine learning processes. Inf. Sci. **613**, 1–32 (2022)
4. Ciriani, V., De Capitani di Vimercati, S., Foresti, S., Samarati, P.: k-anonymity. In: Yu, T., Jajodia, S. (eds.) Secure Data Management in Decentralized Systems, pp. 323–353. Springer, Boston (2007). https://doi.org/10.1007/978-0-387-27696-0_10
5. De Capitani di Vimercati, S., Foresti, S., Livraga, G., Samarati, P.: Data privacy: definitions and techniques. Int. J. Uncertain. Fuzziness Knowl.-Based Syst. **20**(6), 793–817 (2012)

6. Goldsteen, A., Ezov, G., Shmelkin, R., Moffie, M., Farkas, A.: Data minimization for GDPR compliance in machine learning models. AI Ethics **2**(3), 477–491 (2022)
7. Han, J., Kamber, M., Pei, J.: Data Mining: Concepts and Techniques. Morgan Kaufmann Publishers, Burlington (2006)
8. Kisilevich, S., Rokach, L., Elovici, Y., Shapira, B.: Efficient multidimensional suppression for k-anonymity. IEEE TKDE **22**(3), 334–347 (2010)
9. LeFevre, K., DeWitt, D., Ramakrishnan, R.: Workload-aware anonymization. In: Proceedings of KDD, Philadelphia, PA, USA (2006)
10. LeFevre, K., DeWitt, D.J., Ramakrishnan, R.: Mondrian multidimensional k-anonymity. In: Proceedings of ICDE, Atlanta, GE, USA (2006)
11. Machanavajjhala, A., Kifer, D., Gehrke, J., Venkitasubramaniam, M.: ℓ-diversity: privacy beyond k-anonymity. ACM TKDD **1**(1) (2007)
12. Samarati, P.: Protecting respondents identities in microdata release. IEEE TKDE **13**(6), 1010–1027 (2001)
13. Senavirathne, N., Torra, V.: On the role of data anonymization in machine learning privacy. In: Proceedings of IEEE TrustCom, Guangzhou, China (2020)
14. Slijepčević, D., Henz, M., Klausner, L.D., Dam, T., Kieseberg, P., Zeppelzauer, M.: k-anonymity in practice: how generalisation and suppression affect machine learning classifiers. COSE **111**, 102488 (2021)
15. Verdonck, J., De Boeck, K., Willocx, M., Lapon, J., Naessens, V.: A hybrid anonymization pipeline to improve the privacy-utility balance in sensitive datasets for ml purposes. In: Proceedings of ARES, Benevento, Italy (2023)

Visor: Privacy-Preserving Reputation for Decentralized Marketplaces

Tassos Dimitriou$^{(\boxtimes)}$

Computer Engineering Department, Kuwait University, Kuwait City, Kuwait
tassos.dimitriou@ieee.org

Abstract. Feedback mechanisms are crucial in e-commerce and collaborative systems, shaping trust, deterring dishonesty, and promoting good online behavior. Reputation, built through collective feedback, impacts vendor trustworthiness and support. However, centralized systems like eBay and Amazon pose privacy concerns as users must trust the central server to maintain accurate records and protect sensitive data. Decentralized solutions offer greater privacy and control over feedback, aiming to overcome the limitations of centralized ones. Yet, decentralized reputation systems face challenges such as susceptibility to Sybil attacks and more, which undermine user trust and system reliability.

In this work, we propose Visor, a decentralized reputation system addressing these challenges by leveraging randomizable signatures and zero-knowledge proofs to construct a reputation mechanism that safeguards privacy and prevents misuse. The security properties of Visor are formally demonstrated; the system guarantees integrity and ensures that users remain anonymous during feedback, while also maintaining unlinkability among pseudonyms and reviews associated with the same user. Finally, the system provides users with a comprehensive view of all transactions, thereby enabling public verifiability without reliance on any trusted third party.

Keywords: Reputation · Feedback · Decentralized Marketplaces · Privacy · Sybil resistance · Pointcheval-Sanders signatures

1 Introduction

Feedback mechanisms play a critical role in e-commerce and collaborative systems by helping establish trust, deter dishonest practices and incentivize good behavior. Reputation reflects users' collective opinions about provided services, products and resources, accumulated over time through appropriate feedback. As a result, vendors with higher reputation are trusted and supported, while those with lower reputation risk rejection and isolation. Ultimately, the effective management of reputation within e-commerce marketplaces is crucial for fostering trust, encouraging proper behavior and maintaining the marketplace's functionality.

© IFIP International Federation for Information Processing 2024
Published by Springer Nature Switzerland AG 2024
A. L. Ferrara and R. Krishnan (Eds.): DBSec 2024, LNCS 14901, pp. 131–150, 2024.
https://doi.org/10.1007/978-3-031-65172-4_9

Unfortunately, several e-commerce systems including eBay, Amazon and others, have a centralized character, thus offering only little privacy assurance: the user must trust the central server for upholding the accuracy of reputation records and protecting sensitive data from unauthorized disclosure or commercial exploitation. If these centralized systems get compromised, the leak of user profiles, interaction histories, and individual ratings can undermine the system's trustworthiness. Furthermore, existing systems often expose user identities during feedback, deterring honest users from submitting ratings for fear of reprisal or retaliation [18]. Thus, these systems must be trusted for correctness, availability and privacy assurance.

These limitations underscore the need for a solution that ensures robust anonymity without necessitating trust in a centralized party. Our work aims for a *decentralized* system, where users can leave reviews without disclosing their identities or relying on a central authority for supervision. By decentralizing the reputation system, users gain greater control over their privacy and security, overcoming the shortcomings associated with centralized systems and promoting trustworthy behavior among participants.

However, in the absence of adequate security measures, reputation systems are susceptible to various attacks [14, 16]. For instance, ballot-stuffing attacks are used to artificially inflate the reputation of underperforming vendors. In whitewashing, an entity may correct its bad reputation by exiting the current system and reintroducing itself in order to escape the consequences of bad reputation. These attacks and more, are instances of Sybil attacks in which a single malicious entity may create multiple identities in order to manipulate reputation [9]. For example, an attacker maintaining a collection of fake users can use them to provide negative feedback targeting a single vendor. Alternatively, a malicious vendor can attract users by using some of its pseudonyms to increase its own reputation and elevate its status. As a result, the goal of creating decentralized and privacy-preserving reputation systems presents significant challenges; these systems must not only show correct reputation profiles but also ensure unlinkability and safeguard user identity during feedback. Fundamentally, there appears to be an inherent tension between detecting or preventing Sybil attacks and maintaining reviewer anonymity: *How can the authenticity of feedback be ensured without knowledge of its origin?*

Contributions: In this work we propose Visor, a decentralized reputation system that targets e-commerce marketplaces. Such marketplaces consist of Service Providers (SPs) aiming at building their reputation by selling goods or services to clients. Our goal is to ensure that a vendor's reputation reflects the collective ratings of clients who rate SPs by providing appropriate feedback. We pay particular attention to providing privacy and making sure that unless explicitly disclosed by the client, no adversary can link a review to the client.

Visor is based on the use of randomizable signatures as detailed in [22] along with non-interactive zero-knowledge proofs to construct *review tokens* that can be used by clients to rate SPs. The use of randomizable signatures allows a token constructed by a service provider for a specific, yet anonymous, user to be used at a later time to rate the SP. Our approach may appear straightforward, however there are several technical difficulties that complicate the development of a system that is both provably secure and efficient at the same time. For example, proving that a token can only be used by a user who may be using one pseudonym to obtain the token and a different one when leaving a review is non-trivial. Similarly, the system should prevent an anonymous user to use a token more than once when rating an SP. Other technical challenges arise from developing appropriate zero-knowledge proofs to ensure the validity of statements and the privacy of operations.

To summarize, we make the following contributions

– We design Visor, a decentralized system that allows clients to rate SPs in a privacy-preserving manner. To validate user identities, Visor incorporates an initial registration phase which is necessary to safeguard against Sybil attacks, as demonstrated in [9]. Nevertheless, even this phase can be completely decentralized as explained in Sect. 3. Once this phase is over, no other party is involved apart from clients and SPs interacting with each other.
– We employ zero-knowledge proofs not only to guarantee anonymity but also to ensure the correctness and efficiency of various operations.
– We demonstrate that Visor effectively preserves privacy by providing formal proofs regarding the security and privacy offered by our system.
– We utilize a simple bulletin board as an append only ledger to ensure the validity and accountability of reviews by any entity, aiming for simplicity. This provides a comprehensive view of all transactions and facilitates public verification without depending on any trusted third party.

Organization: The rest of the paper is organized as follows. In the next section we review background work on reputation systems. In Sect. 3, we introduce our trust model and formalize the properties we expect from our system. Section 4 details the operations of Visor; its security properties are then analyzed in Sect. 5. Finally, Sect. 6 concludes this work.

2 Related Work

Reputation systems gather and share participants' performance scores, which are then made available to all users within the system. Such reputation profiles are very important in the context of marketplaces and trading systems as they help promote trustworthy behavior among users. However, to ensure the reliability of the system, it is imperative that reviews cannot be linked to participants.

Several privacy-respecting reputation systems have already appeared in the literature. While these systems offer privacy, they typically rely on a trusted third party (TTP). Androulaki et al. [1] devised a system in which users assign ratings by spending reputation coins managed by a TTP (acting like a "bank"), which is responsible for maintaining these reputation coins. Bethencourt et al. [5] introduced a pseudonym-based system where reputation is proportional to "votes" created and received from others. Unlike our approach, however, this system relies on trusting the registration authority for privacy. In [15], two TTPs are used to compute reputations. The first receives encrypted ratings of users along with tokens of validity while the second aggregates results and publishes the ratings. This system enforces privacy as long as the TTPs do not collude with each other. One of the TTPs is eliminated in [21], however the TTP still acts as an intermediary between raters and ratees, therefore transactions are not privacy preserving. A different approach is used in the work of [13] and [7], where each user solicits feedback from other participants in order to interact with another user. These protocols ensure the confidentiality of the submitted ratings, however the IDs of the participating entities are not hidden.

To eliminate dependency on a TTP, many works have a decentralized character in which there is no central location for aggregating or disseminating reputation. Bag et al. [2] developed PrivRep, a decentralized reputation system in which feedback providers use homomorphic encryption to publish their encrypted ratings on a bulletin board. Then, a reputation engine, operated by the marketplace owner, calculates scores using secure multiparty computation. Although privacy is enforced, users are unaware if their ratings were included in the calculation of reputation scores as participating entities are determined by the reputation engine, an approach that may lead to collusion.

Bazin et al. [3] use tokens created by clients to rate service providers after an interaction, combining blind signatures and zero-knowledge proofs to build the tokens. Service providers are responsible for managing the ratings destined for them and reporting correct reputation scores, however this requires the existence of "trackers", a set of entities that are needed to oversee the behavior of SPs and guarantee the reliability of the system. Beaver [24] does not use any TTPs and achieves a consistent view of reputation through consensus in a public blockchain. It requires the presence of an anonymous payment system like Zerocash [23] to facilitate anonymous transactions and payments between parties. Sybil resistance in Beaver depends on the use of appropriate transaction fees to mitigate malicious behavior.

In CLARC [4], anonymous credentials are combined with reputation in a system which allows issuers to issue credentials that can be used by clients to interact with SPs. After the interaction, clients can rate the service provided by the SPs. This is a decentralized system where the TTP has the role of a group manager needed to track misbehaving entities. This is not required in Visor. In a recent work, Dimitriou [6] developed a decentralized system which is based on a distributed ledger like Bitcoin's blockchain to allow participants to rate each other and make verifiable claims regarding their reputation.

Clients in Visor can generate as many pseudonyms they like when interacting with service providers. This concept is central in ensuring the privacy of users and protecting them against retaliation by SPs as shown in [18]. This concept has been central in many of the aforementioned works like [1,5]. RuP [19] is an early system where users can transfer their reputation into a new pseudonym, however this requires the involvement of a TTP. In AnonRep [25], users can change their pseudonyms only after a group of servers (TTPs) perform a verifiable shuffle of user pseudonyms. Similarly, in [11] the creation of new pseudonyms requires coordination with other users to prevent linkability. In our work, we show how pseudonym generation can be performed securely, eliminating the need for any intermediary and mitigating potential Sybil and other attacks against the reputation system.

3   Trust Model

Our model for an anonymous marketplace consists of a *registrar* R and *users* who wish to purchase digital goods and services from various *service providers* (*SP*s). Although, technically, users can be both "clients" and "sellers", in our model a user will refer to a client who wants to purchase an item or service from an SP and leave an anonymous review. A seller (service provider) on the other hand provides services that can be rated by users. At this mode, the long term ID of the seller is visible hence its reputation can be retrieved from all past reviews. Users can interact with as many pseudonyms as they like in order to hide their real ID.

The primary function of the registrar R is to ensure uniqueness of long term identities and prevent Sybil attacks or attacks that try to manipulate reputation. Consequently, R is only trusted to maintain reputation integrity rather than guaranteeing privacy and its role is strictly confined to issuing certificates. While the existence of R seems to contradict the decentralized nature of Visor, we should stress that this service can be decentralized as demonstrated recently in [17], where a system (CanDID) was developed to realize the concept of decentralized identity. Within Visor, the CanDID committee, consisting solely of decentralized nodes, can now undertake the role of the registrar. This committee conducts deduplication (ensuring identity uniqueness) while preserving privacy. Consequently, nodes within the committee are unable to access a user's real-world identity, and are also unable to determine user activities (see also [17] for more details). In the sequel, R will denote this decentralized set of nodes that issue user certificates and ensure uniqueness of identity. After user registration, R can also go *offline* as its presence is not required anymore in subsequent interactions.

Figure 1 depicts the primary use case anticipated in our system. After registration, a user can interact with an SP using a one-time pseudonym. The SP verifies that the user is a legitimate one and hands out a review token that allows the user to leave an anonymous review. Hence, it is important to take precautions to ensure that a user cannot be de-anonymized while submitting a review. The review is submitted to a public bulletin board B at some later time to prevent association with the previous interaction. As a result, anybody can calculate the rating of the SP from the reviews shown in B. If a user attempts to double-spend a review token, then its identity will be revealed.

We are assuming that the bulletin board B has an append-only structure. Thus, for example, service providers or malicious users cannot change or remove

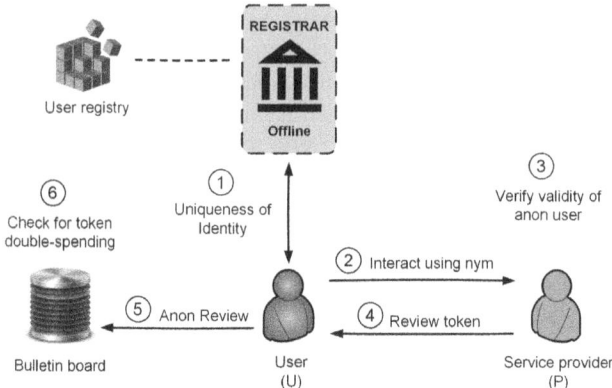

Fig. 1. System architecture

reviews. Furthermore, our system algorithms will ensure that anything posted on the board can be checked for validity, thus we can assume that only valid messages/reviews can be found in B.

As previously stated, users possess multiple pseudonyms; however, all of these must be linked to a *unique* long-term secret ID. Consequently, users should *not* be able to generate different identities that could be exploited to launch attacks against the reputation system. The properties we expect from Visor are:

- *Validation of reviews.* Anybody can publish a review in the bulletin board, however all reviews can be tested for validity.
- *Registration unforgeability.* Users cannot possess registration tokens that they have not been issued to them. Thus, users cannot participate in the system without a valid registration token issued by R.
- *Feedback authenticity.* SPs should be protected from false accusations (fake review tokens). Thus, a user cannot leave a review unless they have been issued a review token for that purpose.
- *User linkability in case of misbehavior.* A user who tries to use the same review token twice can be identified.
- *Non-frameability.* A malicious user cannot generate reviews for a given token that would trace to an honest user. Similarly, a malicious user cannot successfully show possession of an honest user pseudonyms.
- *Privacy.* Users should be assured of privacy, ensuring that their individual interactions cannot be exploited for identification. Thus, reviews for the same SP by two different users should be indistinguishable. Similarly, several SP reviews by the same user should also be indistinguishable.

In addition to the aforementioned security and privacy objectives, Visor is designed to prevent the following general attacks against a reputation system:

- *Sybil and related attacks*: Malicious users should not be able to generate different identities that can be used to influence rating scores. As the registrar's responsibility is to guarantee that each user (or service provider) can register only once, this effectively prevents Sybil attacks. This also prevents *whitewashing* (or re-entry) attacks, in which malicious SPs may attempt to erase their bad reputation and obtain a fresh start in the system. Finally, *self-promotion* refers to an SP's effort to enhance their own reputation. This attack, also known as *ballot stuffing*, is prevented in our system since an SP cannot issue a review token on itself. Additionally, a user will be identified in case of submitting more than one review using the same token.[1]

– *Denial of reputation updates.* The decision to accept or reject a reputation update should not depend on the SP's discretion. Visor's decentralized design ensures that reviews are always available and visible to all.

Operational Assumptions – Side Channels

Although Visor makes sure that user operations stay anonymous, there can be other metadata that can be used to compromise user privacy. For example, if the IP address of a user is exposed during communications with an SP or during feedback, it represents one such side channel and the user can be de-anonymized. Thus, we assume that users interact through *anonymous connections* such as those provided by a network like TOR [8].

Another side channel is the time a review is posted by the user. If this happens directly after an SP issues a review token to the client, then the client can easily be de-anonymized since only a small subset of users might have received a token during that interval. This set of possible clients matching a specific feedback is called the *anonymity set*. Determining the right tradeoff between the size of the anonymity set and delay of submitting a feedback is an interesting research direction, however it is out of scope of this work. A simple method would be for the user to see how many reviews have been submitted on average during a given interval.T (this is possible since reviews and time of submission are available in the bulletin board) and if this is acceptable the user can submit the review after this time T. So, here we will be assuming that feedback is submitted long after the review token has been given to the user.

3.1   Building Blocks

Pedersen Commitments. A commitment scheme enables a user to commit to a message m without revealing it to a receiver. Such a scheme is considered secure when it has both the "binding" and "hiding" properties. The binding property ensures that once a commitment to m is made, a malicious committer cannot change their decision. On the other hand, the hiding property ensures that the receiver cannot recover m. We will employ the Pedersen commitment scheme [20] which can be extended to commit to a number of messages

---

[1] Of course nothing prevents an SP for asking a collaborating client to vote positively for a given transaction. This is possible in any reputation system and cannot be prevented solely using cryptographic means.

$m, m_1, \ldots, m_n$. If $g, h, h_1, \ldots, h_n$ are generators of a group G of prime order q, to commit to $m, m_1, \ldots, m_n$, the user picks random $r \in Z_q$ and computes $\mathsf{Comm}(m, m_1, \ldots, m_n, r) = g^r h^m \Pi_{i=1}^n h_i^{m_i} \mod q$.

Pointcheval and Sanders Signature Scheme. An important building block of our protocol is the Pointcheval and Sanders (PS) Signature scheme [22]. A signature created by this scheme consists of only two group elements, independent of the number of messages to be signed. Another important characteristic is the ability of the signature to be randomized: when presented with a valid PS-signature $\sigma = (\sigma_1, \sigma_2)$ on a message m, anyone can generate another valid signature on the same message by choosing a random number t and computing $\sigma' = (\sigma_1^t, \sigma_2^t)$. This will enable service providers to check the legitimacy of users without being able to identify them. Additionally, a user can re-randomize a review token and leave a review without being linked to the original token. All this is possible since the scheme can be used to sign committed messages. Below, we describe the basic algorithms of the PS scheme [22].

- PS.Setup(1^k): Given a security parameter k, this algorithm outputs $pp \leftarrow (p, \mathbb{G}_1, \mathbb{G}_2, \mathbb{G}_T, e)$, where all bilinear groups must be of type 3. In the following, $\mathbb{G}_1^* = \mathbb{G} \setminus \{1_{G_1}\}$ and $\mathbb{G}_2^* = \mathbb{G} \setminus \{1_{G_2}\}$, which are the sets of the generators.
- PS.Keygen(pp): This algorithm selects $g \xleftarrow{\$} \mathbb{G}_1^*$, $\tilde{g} \xleftarrow{\$} \mathbb{G}_2^*$ and $(x, y, y_1, \ldots, y_r) \xleftarrow{\$} \mathbb{Z}_p^{r+2}$, computes $(X, Y, Y_1, \ldots, Y_r) \leftarrow (g^x, g^y, g^{y_1}, \ldots, g^{y_r})$ and $(\tilde{X}, \tilde{Y}, \tilde{Y}_1, \ldots, \tilde{Y}_r) \leftarrow (\tilde{g}^x, \tilde{g}^y, \tilde{g}^{y_1}, \ldots, \tilde{g}^{y_r})$, and sets $\mathsf{sk} \leftarrow X$ and $\mathsf{pk} = (g, Y, Y_1, \ldots, Y_r, \tilde{g}, \tilde{X}, \tilde{Y}, \tilde{Y}_1, \ldots, \tilde{Y}_r)$.
- $\sigma' \leftarrow$ PS.Sig(sk, C): A user seeking a signature on $(s, m_1, \ldots, m_r)$ presents a commitment $C \leftarrow g^t Y^s \prod_{i=1}^r Y_i^{m_i}$ to the signer. Both parties engage in a proof of knowledge of the commitment's opening. If the signer is convinced, he picks a random $u \xleftarrow{\$} \mathbb{Z}_p$ and returns $\sigma' \leftarrow (g^u, (XC)^u)$. Subsequently, the user can unblind the signature by computing $\sigma \leftarrow (\sigma_1', \sigma_2'/\sigma_1'^t)$.
- PS.SigVerify(pk, $(s, m_1, \ldots, m_r), \sigma$): This algorithm parses σ as (σ_1, σ_2) and checks whether $\sigma_1 \neq 1_{G_1}$ and $e(\sigma_1, \tilde{X} \cdot \tilde{Y}^s \cdot \Pi_{j=1}^r \tilde{Y}_j^{m_j}) = e(\sigma_2, g)$. If both conditions are satisfied it returns 1, otherwise 0.

Zero-Knowledge Proofs of Knowledge. A zero-knowledge (ZK) proof is a protocol by which a prover P can prove to a verifier V that a given statement is true, while avoiding revealing anything beyond the statement's validity. A ZK protocol must possess the properties of completeness, soundness, and zero-knowledge [12]. Completeness ensures that any truthful prover can successfully convince the verifier. Soundness suggests the existence of an algorithm that can extract a witness to the statement's validity from any potentially deceptive and successful prover P^*. Zero-knowledge ensures that a simulator can generate a transcript indistinguishable from a real interaction between a potentially untrustworthy verifier V^* and the prover P.

We will denote by $\pi = PoK\{(r, u, v) : C = g^r h^u \wedge I = g^v\}$ a proof of knowledge where the prover aims to convince the verifier that it knows $r, u,$

and v such that $C = g^r h^u$ and $I = g^v$. The variables r, u, v within parentheses represent private values, while C and I denote public information known to the verifier.

The proof can be made non-interactive by applying the Fiat-Shamir (FS) heuristic [10] in which the challenge of the verifier is generated by hashing the common input and the values committed by the prover in the beginning of the proof. A proof of knowledge can be transformed into a signature of knowledge for some message m by including the message in the hash of the FS transformation. Such a signature of knowledge is denoted by $\pi_{sig} = PoK\{(r, u, v) : C = g^r h^u \land I = g^v\}(m)$.

3.2   Visor Algorithms

Below, we outline the main operations expected from Visor. For simplicity, we do not list every single input that may be required by these algorithms. More details can be found in Sect. 4. The system consists of a registrar R, users U, service providers P, a bulletin board B and the following operations:

- Setup(1^λ): Mainly creates the parameters for Pointcheval-Sanders (PS) signature scheme by calling $params \leftarrow$ PS.Setup(1^λ).
- RegistrarKeyGen($params$): Generates a key pair for the registrar by calling $(sk_R, pk_R) \leftarrow$ PS.Keygen(params).
- UserKeyGen($params$): Executed by user U, it generates a secret key $sk_U \in Z_p$ and a corresponding public key $pk_U = g^{sk_U}$. During registration, pk_U will serve as the user's unique identity I.
- SPKeyGen($params$): Executed by a service provider P to generate a key pair $(sk_P, pk_P) \leftarrow$ PS.Keygen(params), which will be used to sign review tokens. A user with such a token can then rate the services of P by leaving an anonymous review. The signature on the token can be verified by anybody who has a certificate $cert_P$ assuring the authenticity of pk_P. The certificate is created by the registrar during registration of the service provider.
- IssueRegToken($params, U, R$): This is a protocol executed between a user U and the registrar R. First, a call to UserKeyGen($params$) generates the user's key pair (sk_U, pk_U), where pk_U will function as the user's public identity I, which will be stored in the registrar's database. I will be revealed and used as proof of cheating in the case a user tries to use the same review token twice. Afterwards, the user creates a commitment C on sk_U and proves knowledge of sk_U in both C and I. If the registrar is satisfied, it proceeds to sign C. The registration token σ_U is given by PS.Sig(sk_R, C). The user now can show randomized instances of σ_U when interacting with service providers.
- NymGen($params, U, sk_U$): Executed by U, this process generates a new pseudonym nym_U, linked to sk_U, which can be employed by U when interacting with a service provider. NymGen can be invoked as many times as necessary to create additional pseudonyms. It returns a Pedersen commitment on sk_U.
- IssueRevToken($params, U, nym_U, C_{ser}, R$): This is a protocol executed between a user U and the service provider P. The user uses a pseudonym

nym_U while $C_{ser} = g^S Y^{sk_U} Y_1^s Y_2^a$ is a commitment on a serial number s of the token. The serial number will be released during feedback to prevent the token being used twice. Additionally, the value a will be used in the construction of a double spending tag ds_{tag} to allow for user identification in case of double spending. The user sends σ_u, nym_U and C_{ser} along with a proof of knowledge that all are well formed. If the provider is satisfied, it returns a review token $\tau \leftarrow$ PS.Sig(sk_P, C_{ser}). After verifying the signature, U can use τ to leave an anonymous feedback for P.

– Feedback($params, U$): A user, in possession of a review token τ^*, can use this method to leave a review m for service provider P. The user releases the serial number s embedded in τ^* along with a double-spending tag $ds_{tag} = h \cdot sk_U + a$. It also creates a signature of knowledge $\pi_{sig}(m)$ on the values embedded in τ^*, ds_{tag} and signs the review m. The feedback, $\langle \tau^*, s, m, P, T, h, ds_{tag}, \pi_{sig} \rangle$ is then posted in the bulletin board through an anonymous channel. Anybody can then verify the validity of τ^* and $\pi_{sig}(m)$.

– IdentDS(ds_{tag_1}, ds_{tag_2}): Given two double-spending tags ds_{tag_1} and ds_{tag_2}, this algorithm returns the public key pk_U of the user who tried to double-spend a review token, otherwise it returns $\perp$.

3.3   Security Definitions

To define security, we first consider an adversarial user $\mathcal{A}$ who has the potential to act dishonestly and deviate from the prescribed protocols. In particular, we consider an adversary that may attempt to break unforgeability, feedback authenticity and non-frameability. This is captured by the property of *Legitimacy* (Le) along with *Double-spending Detection* (DsD) which are defined in Appendix A. We then look at adversaries whose objective is to identify a user while getting registration/review tokens or submitting a review. The *Privacy* (Priv) definition and appropriate oracles can also be found in Appendix A.

4   Protocol Details

4.1   Setup

The system parameters $(p, \mathbb{G}_1, \mathbb{G}_2, \mathbb{G}_T, e)$ are generated, where all bilinear groups must be of type 3 by executing $params \leftarrow$ PS.Setup(1^k). The registrar's key pair (sk_R, pk_R) is also generated by calling $(sk_R, pk_R) \leftarrow$ PS.Keygen(params).

4.2   Service Provider Registration

A service provider P creates a key pair (sk_P, pk_P) by calling $(sk_P, pk_P) \leftarrow$ PS.Keygen(params). The secret key sk_P will be used to sign review tokens that will enable clients to rate the SP. To verify the validity of the signature, clients must know this is coming from a legitimate SP. For this reason the registrar must sign the SP's public key. This signature on the public key and any additional data constitutes the provider's certificate $cert_P$ which is published in the bulletin board. There is no need for secrecy in the creation of the certificate as service providers are known to all parties.

4.3   User Registration

To prevent Sybil attacks, Visor requires that users register with R prior to participating in the system. This is a one-step process that can also be decentralized as explained in Sect. 3. This procedure will also create the necessary evidence required so that users cannot abuse the system.

At the heart of user registration lies the IssueRegToken$(params, U, R)$ procedure executed between U and R. Here, U creates a long-term secret $sk_U \in Z_q$ and sends $I = g^{sk_U}$ to the registrar, where I serves as the user's identification while sk_u remains confidential. Ensuring the uniqueness of I is important in detecting double-spending attempts; if a user attempts to use a review token more than once, I will be exposed, thereby revealing the user's identity. Additionally, the user selects a random $r \in Z_q$ and computes a commitment $C = g^r Y^{sk_U}$. Following this, the user requests a signature on C from R using the PS signature scheme. Thus, the user must convince R that both the identity I and the commitment C are created by a user who knows the secret sk_U. For this purpose, the user sends I, C, and a proof π_0, where

$$\pi_0 = PoK\{(r, sk_u) : I = g^{sk_U} \wedge C = g^r Y^{sk_U}\}.$$

If the proof passes verification, the registrar returns a signature $\sigma' = $ PS.Sig(sk_R, C) on the commitment C. The user checks the validity of the signature using PS.SigVerify(pk_R, sk_U, σ') and unblinds σ' by computing $\sigma_U \leftarrow (\sigma'_1, \sigma'_2/\sigma'^r_1)$. At this point, the user is considered registered, σ_U acts as a registration token, and U can show randomized instances of σ_U to other participants when interacting with them.

The public identity I is kept in the registrar's database along with other supporting documentation about U. Thus, the role of the registrar is to ensure that every user has a unique secret and has registered only once. This effectively prevents Sybil attacks.

4.4   Constructing the Review Token

Clients in our system can interact with service providers in an anonymous way (recall Fig. 1, Step 2). To achieve this, they must be capable of generating pseudonyms linked to their long-term secret key.

Operation NymGen$(params, U, sk_U)$ allows U to generate pseudonyms when interacting with SPs. Each pseudonym takes the form of a Pedersen commitment, expressed as $nym_U = g^t h^{sk_U}$, where t is a randomly selected number and sk_U denotes the user's long-term secret key. The process returns (nym_U, t) to the user, providing both forward anonymity and controlled linkability. This allows the user the option to either generate a new pseudonym or retain the existing one for each transaction conducted. The user additionally selects a random number s to serve as a "serial" number for the token, then proceeds to create the commitment $C_{ser} = g^d Y^{sk_U} Y_1^s Y_2^a$, for random d and a, where a will be used in double-spending identification.

Prior to issuing the review token, the service provider P must be convinced that U is a legitimate user. In particular, the user must prove (i) knowledge of d, s, a in C_{ser}, and (ii) that nym_U and the registration token $\sigma_U = (\sigma_1, \sigma_2)$ are bound to sk_U. For this, the user first picks random $r, w \in Z_p$ and computes $\sigma' = (\sigma_1^r, (\sigma_2 \cdot \sigma_1^w)^r)$. Then it sends nym_U, C_{ser} and the randomized signature $\sigma' = (\sigma_1', \sigma_2')$ to the SP along with a zero-knowledge proof of knowledge π_1 of t, d, s, a, b, w and sk_U such that

$$\pi_1 = PoK\{(t, d, s, a, w, sk_U) : nym_U = g^t h^{sk_U} \wedge$$
$$C_{ser} = g^d Y^{sk_U} Y_1^s Y_2^a \wedge$$
$$\mathsf{PS.SigVerify}(\mathsf{pk}_R, sk_U, \sigma') = 1\}$$

If the proof verifies, the provider P uses its secret key sk_P under the PS signature scheme to create the token $\tau' \leftarrow \mathsf{PS.Sig}(\mathsf{sk}_P, C_{ser})$, which is a signature on the tuple (d, sk_U, s, a). The user can now unblind τ' by computing $\tau \leftarrow (\tau_1', \tau_2'/\tau_1'^d)$ and verify it is a valid signature by checking whether $e(\tau_1, \tilde{X} \cdot \tilde{Y}^{sk_U} \tilde{Y}_1^s \tilde{Y}_2^a) = e(\tau_2, g)$. The user now has a valid review token that can use to provide feedback.

4.5 Feedback

The user submits the review at some random time after the actual interaction took place. This increases the anonymity set and prevents the provider P from associating a review with a particular interaction. The feedback has the form

$$\langle \tau^*, s, m, P, T, h, ds_{tag}, \pi_{sig} \rangle,$$

where

1. τ^* is a randomized variant of τ,
2. s is the serial number embedded in the token,
3. m is the actual review or rating about the service provided by P,
4. P is the global ID of the provider,
5. T is the time/timestamp of the feedback,
6. $h = H(\tau^*, s, m, T)$, where H is a collision-resistant hash function[2],
7. $ds_{tag} = h \cdot sk_U + a$ denotes a double-spending tag, and
8. The signature $\pi_{sig}(m)$ on the message m is a signature of knowledge about the token attributes sk_U, s, a.

The signature π_{sig} is used to ensure that only the intended user who knows sk_U can leave a review m, while s is released so that the token (or its randomized variants) can only be used once. In case U tries to utilize the token twice, ds_{tag} can be used to recover the secret sk_U of the user, in effect identifying and penalizing U, as will be explained in the next section. To create π_{sig}, the user

---

[2] Given τ^*, T, the user should not be able to find two different m_1, m_2 that produce the same h.

picks random $r, w \in Z_p$ and computes $\tau^* = (\tau_1^r, (\tau_2 \cdot \tau_1^w)^r)$. The signature of knowledge π_{sig} proves knowledge of w, sk_U, a (recall that s is released) such that

$$\pi_{sig} = PoK\{(w, sk_u, a) : ds_{tag} = h \cdot sk_U + a \quad \wedge$$
$$\mathsf{PS.SigVerify}(\mathsf{pk}_P, sk_u, s, a, \tau^*) = 1\}(m).$$

The feedback $\langle \tau^*, s, m, P, T, h, ds_{tag}, \pi_{sig} \rangle$ for provider P is posted in the bulletin board and anybody can check (i) the signature of P on τ^*, and (ii) the validity of π_{sig} and hence of the review.

4.6   Identifying Double-Spenders

Visor effectively thwarts double-spending attempts through the use of token serial numbers. As the token serial s is associated with the signature created by the provider and posted in the bulletin board, double-spending should not be attempted.

If, however, a user tries to provide a second review using the same (even if randomized) token, its identity can be revealed. In that case, there will be two feedback tuples $\langle \tau_1^*, s, m_1, T_1, h_1, ds_{tag_1}, \pi_{sig_1} \rangle$ and $\langle \tau_2^*, s, m_2, T_2, h_2, ds_{tag_2}, \pi_{sig_2} \rangle$ corresponding to the same token such that $ds_{tag_1} = h_1 \cdot sk_U + a$ and $ds_{tag_2} = h_2 \cdot sk_U + a$.

From these two equations, it is a simple matter to recover sk_U and hence the identity $I = g^{sk_U}$ of the user. This is proof of cheating, given the hardness of the discrete log problem.

5   Security Analysis

Below we give a series of theorems that prove the security of the reputation system developed.

Theorem 1. *Given that the Discrete Logarithm problem is hard and the Pointcheval-Sanders [22] signature scheme is secure, the reputation system achieves registration unforgeability, feedback authenticity and non-frameability.*

Proof: We are assuming the correctness of NIZK proofs π_0 and π_1, which necessitate the prover to possess the committed knowledge. We now consider the various cases under which adversary $\mathcal{A}$ can succeed in winning the legitimacy game (Definition 1 – Appendix A).

1. In the first case, $\mathcal{A}$ possesses a valid registration token without executing the RegGetToken protocol. This suggests that $\mathcal{A}$ can either demonstrate to R ownership of I_U – thus revealing knowledge of the secret identity sk_U in π_0 – or fabricate a token without R's involvement. The former scenario contradicts the hardness of the discrete logarithm problem and the correctness of π_0 as a proof of knowledge. The latter scenario contradicts the premise that the PS signature scheme [22] is secure against forgeries.

2. In the second case, $\mathcal{A}$ possesses a valid review token including sk_U and $pk_U = I_U = g^{sk_U}$ without executing the RegGetToken protocol. The RevGet-Token oracle models the IssueRevToken protocol in which the service provider P issues a review token. Similarly to the previous case, if $\mathcal{A}$ holds a valid review token without calling the RegGetToken or the RevGetToken queries, this means that either $\mathcal{A}$ can demonstrate ownership of a valid registration token which presupposes knowledge of sk_U in nym_U and in π_1, or $\mathcal{A}$ can fabricate a review token without P's involvement. As in the first case, the former scenario contradicts the hardness of the discrete logarithm problem and the soundness of the proof system, while the second contradicts the security of the PS signature scheme [22].

3. In the third case, $\mathcal{A}$ successfully executes Review for a user that has not been involved in RegGetToken or RevGetToken queries. As highlighted in the previous two cases, $\mathcal{A}$ should not acquire a valid token linked to a public identity that is not under $\mathcal{A}$'s control. Since the Review oracle models the Feedback algorithm, $\mathcal{A}$ cannot make such a call without knowing the value sk_U as this would contradict the fact that π_1, π_{sig} are sound ZK proofs.

Theorem 2. *The reputation scheme defies double-spending attempts.*

Proof: We are assuming the correctness of NIZK proofs π_1 and π_{sig}, which necessitate the prover to possess the committed knowledge. In particular, correctness suggests that each token contains a unique serial number that cannot be modified without detection. We now consider the three cases under which $\mathcal{A}$ can win the double-spending detection game (Definition 2 – Appendix A). Clearly, the last case is ruled out by the correct operation of IdentDS.

1. In the first case, there are two double spending tags with the same serial but different public keys. This suggests that either the adversary was able to use the same commitment C_{ser} to two different IDs which would contradict the binding property of the commitment or use a fresh commitment inside the PS signature created by the provider which would contradict the unforgeability of the PS signature scheme.

2. In the second case, the adversary uses the same review token, under the same user ID, with two different serial numbers. This again suggests that either the adversary was able to modify the commitment or use a fresh commitment inside the PS signature. Thus again, either the binding property is violated or the unforgeability of the signature scheme.

In summary, the mechanism for detecting double-spending ensures that review tokens can only be used once, as each token is bound to a unique token serial. The proof π_1 asserts that this is indeed the case. Additionally, the commitment C_{ser} and the proof also enforce the same secret key sk_U, the randomness a as well as the freshly chosen challenge value h to be used in the calculation of the $ds_{tag} = h \cdot sk_U + a$. Hence, double-spending a token always reveals the same token serial s and involves different challenges h_1, h_2 with overwhelming probability, thus extracting sk_U. □

Theorem 3. *Given that the Discrete Logarithm problem is hard, and both the commitment and the NIZK schemes are hiding and zero-knowledge respectively, the reputation scheme meets the privacy property.*

Proof: We consider the following cases under which an adversary $\mathcal{A}$ might be able to win the privacy game by distinguishing between two users U_0 and U_1 (Definition 3 – Appendix A).

1. In the first case, the adversary extracts the secret key sk_{U_b} during registration. Thus, by computing $I = g^{sk_{U_b}}$ it is easy to distinguish between U_0 and U_1. However, this is infeasible due to the hardness of the Discrete Logarithm problem and the hiding nature of the commitment scheme.
2. In the second case, the adversary obtains the user key sk_{U_b} from the attributes committed in nym_U or C_{ser} during the interaction with P. Thus, again $\mathcal{A}$ can win challenge in the privacy game. However, this is infeasible due to the hiding nature of the commitment scheme.
3. In the third case, the adversary extracts the secret key sk_{U_b} from the proofs π_0, π_1 and π_{sig} posted during registration or during the construction and submission of the review token. However, this is infeasible due to the ZK property of the NIZK proof system. $\qquad\square$

6   Conclusions and Future Research

In this work, we introduced Visor, a reputation framework designed to operate in a decentralized manner, preserve anonymity and withstand various known attacks against reputation systems. In our system, service providers issue tokens to users, enabling them to submit reviews anonymously. These tokens are based on randomizable signatures and Non-Interactive Zero Knowledge proofs, effectively thwarting Sybil attacks and preventing user de-anonymization. The system operates through a simple bulletin board where clients can leave reviews and access the reputation of service providers. As a result, there is no reliance on third-party entities to maintain reputation records. Furthermore, the validity of tokens and reviews can be verified by any interested party, while malicious users can be identified in case of misbehavior.

Looking ahead, a potential direction for enhancement involves the replacement of the bulletin board with a public ledger, such as Bitcoin's blockchain, to make certain operations more efficient. For instance, in such a system, service providers may find it easier to demonstrate their reputation, and clients may have improved access to service provider reputation profiles. However, this extension is left as future work.

A   Security Definitions

System Security. We consider adversarial objectives concerning two properties: legitimacy and detection of double-spending. The adversary $\mathcal{A}$ may interact with an honest registrar R and honest providers P as many times as necessary. $\mathcal{A}$ has access to the following oracles:

- Upon invoking RegGetToken(U), $\mathcal{A}$ initiates the IssueRegToken protocol with an honest R, under the condition that there are no ongoing or completed RegGetToken requests for U at the time. Upon successful completion, $\mathcal{A}$ will possess a secret sk_U and the registration token σ_U.
- Upon invoking RevGetToken(U), $\mathcal{A}$ initiates the IssueRevToken protocol with an honest P. Upon successful completion, $\mathcal{A}$ will obtain a review token τ_U.
- Review Upon invoking Review (U, τ), $\mathcal{A}$ initiates the Feedback protocol for a given review token τ.

The property of *Legitimacy* (Le) attempts to capture the goals of registration unforgeability, feedback authenticity and non-frameability. This property, as described in Definition 1, models the actions of an adversary who could potentially: (i) possess a registration token not generated by R or not associated with any U involved in a successful RegGetToken call, or (ii) possess a review token not generated by P or not associated with any U involved in a successful RevGetToken call.

Definition 1. (Legitimacy) *A reputation system upholds the Legitimacy property if, for any probabilistic polynomial-time adversary $\mathcal{A}$ in the experiment of Fig. 2, the advantage $\mathsf{Adv}_{\mathcal{A}}^{\mathsf{Le}}$ of $\mathcal{A}$ is negligible in λ, where*

$$\mathsf{Adv}_{\mathcal{A}}^{\mathsf{Le}}(\lambda) := Pr[\mathsf{Exp}_{\mathcal{A}}^{\mathsf{Le}}(\lambda) = 1].$$

Experiment $\mathsf{Exp}_{\mathcal{A}}^{\mathsf{Le}}(\lambda)$:
$(pk_R, sk_R) \leftarrow \mathsf{RegistrarKeyGen}(1^k)$
$(pk_U, sk_U) \leftarrow \mathsf{UserKeyGen}(params)$
$b \leftarrow \mathcal{A}^{\mathsf{RegGetToken, RevGetToken, Review}}(pk_R, pk_U)$
The experiment returns 1 iff

1. $\mathcal{A}$ possesses a valid registration token including with sk_U and $pk_U = I_U = g^{sk_u}$ that were not obtained from any RegGetToken query; or
2. $\mathcal{A}$ holds a valid review token including sk_U and $pk_U = I_U = g^{sk_u}$ that were not obtained from any RevGetToken query; or
3. $\mathcal{A}$ successfully executes a Review query in such a way that the honest $\mathcal{P}$ believes that the query involves a genuine public key pk_U, with no previous successful execution of RegGetToken or RevGetToken associated with it.

Fig. 2. Legitimacy experiment.

The Double-spending Detection (DsD) property, as defined in Definition 2, ensures that two reviews resulting in the same token serial have originated from the same user.

Definition 2. (Double-spending Detection) *A reputation system upholds the Double-spending detection property if, for any probabilistic polynomial-time adversary $\mathcal{A}$ in the experiment of Fig. 3, the advantage $\mathsf{Adv}_{\mathcal{A}}^{\mathsf{DsD}}$ of $\mathcal{A}$ is negligible in λ where*

$$\mathsf{Adv}_{\mathcal{A}}^{\mathsf{DsD}}(\lambda) := Pr[\mathsf{Exp}_{\mathcal{A}}^{\mathsf{DsD}}(\lambda) = 1].$$

Experiment $\mathsf{Exp}_{\mathcal{A}}^{\mathsf{Dsd}}(\lambda)$:

$(pk_R, sk_R) \leftarrow \mathsf{RegistrarKeyGen}(1^k)$

$b \leftarrow \mathcal{A}^{\mathsf{RegGetToken, RevGetToken, Review}}(pk_R, pk_U)$

The experiment returns 1 iff $\mathcal{A}$ successfully executes two Review queries on the *same* token, indicating double-spending. However, when employing the IdentDS algorithm, at least one of the following is true:

- The user public-keys pk_U^1 and pk_U^2 revealed are different, or
- The token serial numbers s_1 and s_2 released in these queries are different, or
- When applied to these two transactions, IdentDS outputs $\perp$.

Fig. 3. Double-spending detection experiment.

User Privacy. To address privacy concerns, we introduce an adversarial party $\mathcal{A}$, whose objective is to identify a user while getting registration/review tokens or submitting a review. The feedback mechanism is designed to prevent the disclosure of any user-identifiable information, and reviews should remain unlinkable. To define the actions of the adversary, $\mathcal{A}$ is allowed access to the following oracles:

- SysS(R): A call to this procedure enables $\mathcal{A}$ to commence the system setup process and produce the system parameter params along with a public key for R, who is under the control of $\mathcal{A}$.
- RegU($\mathcal{U}$): A call to this procedure enables $\mathcal{A}$ to generate a new user U through an IssueRegToken call executed between $\mathcal{A}$ and U. Upon successful termination, $\mathcal{A}$ will acquire U's public key pk_U along with any resulting transaction data.
- CorU(U): A call to this procedure enables $\mathcal{A}$ to compromise an honest user U and acquire U's secret key sk_U and registration token σ_U.
- CorP(P): A call to this procedure allows $\mathcal{A}$ corrupt an honest service provider P and obtain P's secret key sk_P.
- RevT($U(\sigma)$): A call to this procedure allows $\mathcal{A}$ run the IssueRevToken protocol between U and $\mathcal{A}$ controlling a provider P. Upon successful termination, $\mathcal{A}$ will acquire any resulting transaction data.
- ReviewP($\mathcal{U}(\tau, \sigma)$): A call to this procedure allows $\mathcal{A}$ run the Feedback protocol with an honest U, possessing τ and σ. Upon successful termination, $\mathcal{A}$ will acquire any resulting transaction data.

– Challenge($\mathcal{U}_0, \mathcal{U}_1$): A call to this procedure allows $\mathcal{A}$ initiate a IssueRevTo-ken or Feedback protocol by suggesting two honest users U_0 and U_1. The IssueRevToken protocol is run between U_b for $b = \{0, 1\}$ and $\mathcal{A}$, where $\mathcal{A}$ acts as the provider $\mathcal{P}$. The RegU queries for these two users should be asked, under the requirement that both users generate new one-time pseudonyms to be employed in various operations.

User Privacy (Definition 3) is ensured through the game shown in Fig. 4. The goal is to prevent the adversary from linking users to their tokens or reviews under the assumption that all actions are controlled by $\mathcal{A}$. Initially, the adversary registers multiple users, obtains review tokens, and submits feedback. Following this learning phase, $\mathcal{A}$ enters a Challenge phase with two users U_0 and U_1, chosen by $\mathcal{A}$. In this phase, a user U_b from the chosen pair, determined by a random bit b unknown to $\mathcal{A}$, submits a review using its token. Ultimately, $\mathcal{A}$ outputs a value b'. The scheme ensures privacy if the adversary cannot determine b with probability significantly better than $1/2$.

Definition 3. (User Privacy) *A reputation system upholds the User privacy property if, for any probabilistic polynomial-time adversary $\mathcal{A}$ in the experiment of Fig. 4, the advantage of $\mathcal{A}$ is defined by*

$$\mathsf{Adv}_{\mathcal{A}}^{\mathsf{Priv}}(\lambda) := Pr[\mathsf{Exp}_{\mathcal{A}}^{\mathsf{Priv}}(\lambda) = 1] = 1/2 + \epsilon,$$

where ϵ is negligible in λ.

Experiment $\mathsf{Exp}_{\mathcal{A}}^{\mathsf{Priv}}(\lambda)$:

$b \leftarrow \mathcal{A}^{\mathsf{SysS, RegU, CorU, RevT, Review, Challenge}}(1^\lambda)$

The experiment returns 1 iff $\mathcal{A}$ passes the following phases:

- *Setup phase:* $(pk_\mathcal{R}) \leftarrow \mathcal{A}^{\mathsf{SysS}}(1^\lambda)$
- *Learning phase:*
  transRecord $\leftarrow \mathcal{A}^{\mathsf{RegU, CorU, CorP, RevT, Review}}(pk_\mathcal{R})$
- *Challenge phase:*
  transRecord(U_b) $\leftarrow \mathcal{A}^{\mathsf{Challenge}}(U_0, U_1)$

Finally, $\mathcal{A}$ outputs $U_{b'}$ that is equal to U_b.

Fig. 4. User Privacy experiment.

It is implicitly assumed that $\mathcal{A}$ lacks any access to the internal memory of users. Otherwise, it could easily win the game described above. As stated in Sect. 3, all transfers are made through anonymous communication channels so that users are not identified by their IP addresses or other metadata.

References

1. Androulaki, E., Choi, S.G., Bellovin, S.M., Malkin, T.: Reputation systems for anonymous networks. In: Borisov, N., Goldberg, I. (eds.) PETS 2008. LNCS, vol. 5134, pp. 202–218. Springer, Heidelberg (2008). https://doi.org/10.1007/978-3-540-70630-4_13
2. Bag, S., Azad, M.A., Hao, F.: A privacy-aware decentralized and personalized reputation system. Comput. Secur. **77**, 514–530 (2018)
3. Bazin, R., Schaub, A., Hasan, O., Brunie, L.: Self-reported verifiable reputation with rater privacy. In: Steghöfer, J.-P., Esfandiari, B. (eds.) IFIPTM 2017. IAICT, vol. 505, pp. 180–195. Springer, Cham (2017). https://doi.org/10.1007/978-3-319-59171-1_14
4. Bemmann, K., et al.: Fully-featured anonymous credentials with reputation system. In: Proceedings of the 13th International Conference on Availability, Reliability and Security, pp. 1–10 (2018)
5. Bethencourt, J., Shi, E., Song, D.: Signatures of reputation. In: Sion, R. (ed.) FC 2010. LNCS, vol. 6052, pp. 400–407. Springer, Heidelberg (2010). https://doi.org/10.1007/978-3-642-14577-3_35
6. Dimitriou, T.: Decentralized reputation. In: Proceedings of the Eleventh ACM Conference on Data and Application Security and Privacy, pp. 119–130 (2021)
7. Dimitriou, T., Michalas, A.: Multi-party trust computation in decentralized environments in the presence of malicious adversaries. Ad Hoc Netw. **15**, 53–66 (2014)
8. Dingledine, R., Mathewson, N., Syverson, P.F., et al.: Tor: the second-generation onion router. In: USENIX Security Symposium, vol. 4, pp. 303–320 (2004)
9. Douceur, J.R.: The sybil attack. In: Druschel, P., Kaashoek, F., Rowstron, A. (eds.) IPTPS 2002. LNCS, vol. 2429, pp. 251–260. Springer, Heidelberg (2002). https://doi.org/10.1007/3-540-45748-8_24
10. Fiat, A., Shamir, A.: How to prove yourself: practical solutions to identification and signature problems. In: Odlyzko, A.M. (ed.) CRYPTO 1986. LNCS, vol. 263, pp. 186–194. Springer, Heidelberg (1987). https://doi.org/10.1007/3-540-47721-7_12
11. Florian, M., Walter, J., Baumgart, I.: Sybil-resistant pseudonymization and pseudonym change without trusted third parties. In: Proceedings of the 14th ACM Workshop on Privacy in the Electronic Society, pp. 65–74 (2015)
12. Goldreich, O.: Foundations of Cryptography: Volume 1, Basic Tools. Cambridge University Press, Cambridge (2006)
13. Hasan, O., Brunie, L., Bertino, E., Shang, N.: A decentralized privacy preserving reputation protocol for the malicious adversarial model. IEEE Trans. Inf. Forensics Secur. **8**(6), 949–962 (2013)
14. Hoffman, K., Zage, D., Nita-Rotaru, C.: A survey of attack and defense techniques for reputation systems. ACM Comput. Surv. (CSUR) **42**(1), 1–31 (2009)
15. Kerschbaum, F.: A verifiable, centralized, coercion-free reputation system. In: Proceedings of the 8th ACM Workshop on Privacy in the Electronic Society, pp. 61–70 (2009)
16. Koutrouli, E., Tsalgatidou, A.: Taxonomy of attacks and defense mechanisms in P2P reputation systems-lessons for reputation system designers. Comput. Sci. Rev. **6**(2–3), 47–70 (2012)
17. Maram, D., et al.: CanDID: can-do decentralized identity with legacy compatibility, sybil-resistance, and accountability. In: 2021 IEEE Symposium on Security and Privacy (SP), pp. 1348–1366. IEEE (2021)

18. Minkus, T., Ross, K.W.: I know what you're buying: privacy breaches on eBay. In: De Cristofaro, E., Murdoch, S.J. (eds.) PETS 2014. LNCS, vol. 8555, pp. 164–183. Springer, Cham (2014). https://doi.org/10.1007/978-3-319-08506-7_9
19. Miranda, H., Rodrigues, L.: A framework to provide anonymity in reputation systems. In: 2006 Third Annual International Conference on Mobile and Ubiquitous Systems: Networking & Services, pp. 1–4. IEEE (2006)
20. Pedersen, T.P.: Non-interactive and information-theoretic secure verifiable secret sharing. In: Feigenbaum, J. (ed.) CRYPTO 1991. LNCS, vol. 576, pp. 129–140. Springer, Heidelberg (1992). https://doi.org/10.1007/3-540-46766-1_9
21. Petrlic, R., Lutters, S., Sorge, C.: Privacy-preserving reputation management. In: Proceedings of the 29th Annual ACM Symposium on Applied Computing, pp. 1712–1718 (2014)
22. Pointcheval, D., Sanders, O.: Short randomizable signatures. In: Sako, K. (ed.) CT-RSA 2016. LNCS, vol. 9610, pp. 111–126. Springer, Cham (2016). https://doi.org/10.1007/978-3-319-29485-8_7
23. Sasson, E.B., et al.: Zerocash: decentralized anonymous payments from bitcoin. In: 2014 IEEE Symposium on Security and Privacy, pp. 459–474. IEEE (2014)
24. Soska, K., Kwon, A., Christin, N., Devadas, S.: Beaver: a decentralized anonymous marketplace with secure reputation. Cryptology ePrint Archive (2016)
25. Zhai, E., Wolinsky, D.I., Chen, R., Syta, E., Teng, C., Ford, B.: AnonRep: towards tracking-resistant anonymous reputation. In: 13th USENIX Symposium on Networked Systems Design and Implementation (NSDI 2016), pp. 583–596 (2016)

Attack

Resiliency Analysis of Mission-Critical System of Systems Using Formal Methods

Mahmoud Abdelgawad[✉] and Indrakshi Ray

Department of Computer Science, Colorado State University, Fort Collins, CO 80523, USA
{m.abdelgawad,indrakshi.ray}@colostate.edu

Abstract. A System of Systems (SoS) is a formation of heterogeneous systems that work together effectively to accomplish critical missions. These mission-critical SoS ensure the safety and security of any nation. Attacks on mission-critical SoS can have devastating outcomes. We need to design missions that are resilient to attacks. SoS engineering must specify, analyze, and understand where adverse events are possible and how to mitigate them while a mission-critical SoS is deployed. This work uses an end-to-end methodology for analyzing the resiliency of mission-critical SoS. The methodology starts with representing SoS mission as a workflow. It then converts the workflow into formal representation using Coloured Petri Nets (CPN). Threat models are extracted from the mission specification, which are used to construct the CPN representations of the attacks. The mission specifications and the attacks are composed to analyze the mission resiliency. The analysis identifies if the mission succeeds, fails, and is incomplete. We apply the methodology to an SoS consisting of a military vehicle and route-reconnaissance drones working together to monitor a national border and respond immediately to any physical threats. The result demonstrates how to restrict the mission to improve the resiliency of SoS formation. Such methodology is important for the early design stage of resilient mission-critical SoS.

Keywords: System of Systems (SoS) · Mission-critical Systems · Resiliency Analysis · Workflow · Formal Methods · Coloured Petri Nets (CPN)

1 Introduction

A System of Systems (SoS) is a formation of heterogeneous systems that interact for a global goal [4]. These systems operate independently, but none can accomplish the goal on its own; thus, they work together towards broader objectives. SoS characteristics include being independently managed, geographically distributed, evolving to meet mission needs, emerging from a combination of behaviors and properties of the system elements, and being influenced by a stimulus environment [2,4]. SoS has become common in technology domains, including transportation, water and energy resources, and defense. The SoS standardization [7] also considers Smart Grids, Smart Cities, and Internet-of-things (IoT) as part of one or more SoS. This paper is scoped to the SoS domains in defense where an operational environment, the objectives are established based on critical missions [5,11,19]. Most military SoS formations, for instance, ships,

© IFIP International Federation for Information Processing 2024
Published by Springer Nature Switzerland AG 2024
A. L. Ferrara and R. Krishnan (Eds.): DBSec 2024, LNCS 14901, pp. 153–170, 2024.
https://doi.org/10.1007/978-3-031-65172-4_10

aircraft, satellites, and ground vehicles, are equipped with independent system elements (i.e., sensors, weapons, communications) that must fulfill mission-critical objectives. These mission-critical systems are defined as systems whose one system element failure may significantly impact the mission-related functions [6,15]. Hence, resilience analysis of the mission-critical SoS is essential. The resiliency of mission-critical SoS is identical to cyber survivability defined for warfighter systems [17]. Mission-critical SoS must fulfill survivability requirements to continue a mission in the face of attacks. Therefore, SoS system engineering must specify and analyze a mission before deployment to evaluate its resilience and gauge what failures can be tolerated.

A mission can be described as a workflow consisting of various tasks executed by subjects and connected via different types of control-flow operators [20]. Mission also has a set of objectives. If all the objectives are satisfied, then the mission has *succeeded*. If some, but not all, objectives are satisfied, the mission is *incomplete*. When no objectives have been satisfied, the mission has *failed*. Many types of workflow resilience are defined: static, decremental, and dynamic resilience [18]. Static resiliency refers to a situation in which subjects become unavailable before the workflow executes, and no subjects may become available during the execution. Decremental resiliency expresses a situation where subjects become unavailable before or during the execution of the workflow, and no previously unavailable subjects may become available during execution. This work addresses the decremental resilience. Analyzing the workflow resiliency manually is error-prone and labor-intensive. Towards this end, there is a need for automated analysis of a mission and evaluating potential attacks and their effects. We use Coloured Petri Nets (CPN) [10] for the purpose of analysis. CPN is suitable for modeling process interaction and provides primitives for defining data types (i.e., colors). It is also associated with a high-level programming language (i.e., CPN-ML [8]) and Integrated Development Environment (i.e., CPN Tools [16]) that are practical to simulate processes interaction, manipulate data values, and verify state transitions of systems.

We designed a methodology that analyzes various aspects of mission resiliency for mission-critical SoS. The methodology leverages the workflow definition to describe a mission. It also utilizes a set of transformation rules [1] to convert the workflow to CPN. The methodology starts with specifying a mission and formally representing it as a workflow. The mission workflow is converted into the formal CPN specification which we refer to as CPN mission. The methodology systematically identifies threat models from the mission specification. The threat models are converted into formal CPN specifications, which we refer to as CPN attacks. These CPN attacks are used to demonstrate various attack scenarios. The CPN attacks are composed with the CPN mission, and the resulting system analyzed. It analyzes the mission's resiliency as it identifies when the mission succeeds, fails, and is incomplete.

The methodology is applied to an SoS formation consisting of a military vehicle required to monitor a national border and respond immediately to any threats. It also includes two unmanned aerial vehicles (i.e., drones) that perform surveillance and reconnaissance for the mission. A military base exchanges information between the military vehicle and drones to ensure a mission's success. The SoS mission decremental resiliency defined in [18] is analyzed. The application shows that the methodology is practical for mission analysis and helps improve mission-critical SoS resilience.

2 SoS Mission Description

This section describes a mission-critical SoS as a formation of a military vehicle and two rotary-winged drones. The military vehicle and drones exchange data through a Line Of Sight (LOS) communication. The military vehicle also exchanges data with an operation center site (i.e., a military base) through a Beyond Line Of Sight (BLOS) via satellite communication.

The mission requires the military vehicle to monitor the national border between the official ports of entry and respond to any threat. We assume the mission will occur in a region of interest with gravel roads and rough terrain. Human elements, the military vehicle drivers (e.g., soldiers), usually plan several routes to reach a Point of Interest (POI). In many cases, however, it is challenging to anticipate route hazards they may encounter (e.g., driving down a steep hill). Thus, drones play the role of surveillance and reconnaissance for the military vehicle to keep a safe distance from dangerous situations. The drones fly 5 mi ahead of the military vehicle at low altitudes using Global Positioning System (GPS) to inform locations. The drones accelerate and decelerate to keep a 5-mile distance from the military vehicle. They are also equipped with cameras and infrared sensors for transmitting day and night visions. If the drones fly during the daytime, they will use the cameras; if it is nighttime, they will use the infrared sensors. The drones keep flying as long as the battery is enough to return to the military vehicle for recharging. We assume the battery level contains 100 units, enough to fly for 4 h. If the drone is 5 mi from the military vehicle, it needs 2 battery units to return to the vehicle for recharging. The military vehicle is estimated to take 30–45 minutes to reach the POI if the human elements drive at average speed (60–70 miles per hour). The military vehicle usually reaches the POI, evaluates the situation, and returns to the starting point (i.e., station) without engagement. If the situation at the POI is evaluated for engagement, the human elements must receive instruction from the military base to engage (e.g., open fire against a threat). The mission requires the military vehicle to be fueled with a full petrol tank (i.e., gasoline) and connected to the military base. Actuating the cameras and infrared sensors, each takes one battery unit. Before flying, the drone's status is checked (LOS communication is up, cameras and infrared sensors are on, and the battery is fully charged). The military vehicle's status (full gasoline tank and connected to the military base) is also checked during deployment to accomplish a mission. The mission is successful if the military vehicle and drones reach the POI in time, report to the military base about the POI situation, and return to the station safely. The mission is considered incomplete if drones cannot collect data but can fly back to the vehicle, and the vehicle reaches the POI but cannot report to the military base. The mission fails when the drones cannot return to the vehicle (e.g., lost), the vehicle cannot reach the POI, or the vehicle or the drones do not return to the station.

Now, we need to analyze whether it is possible for the mission to succeed and in which attack scenarios the mission fails. An attack scenario changes the military vehicle and drones' attributes (i.e., mutable attributes that can be manipulated) and renders this SoS formation unable to perform a task. For instance, an attack scenario defuses the drones' cameras. It degrades its scanning capability, leaving the military vehicle unable to figure out the situation ahead, which may cause a mission to fail. Another attack scenario disrupts the BLOS communication, rendering the military vehicle disconnected from the military base and unable to report the threat situation, which may cause an incomplete mission. The methodology systematically covers possible attack scenarios that cause SoS degradation, and each attack scenario must be analyzed to determine if mission can be successful, failed, or incomplete with respect to the attack. A mission is resilient to an attack scenario if, after the attack occurs, it still meets its objectives.

3   Resiliency Analysis Methodology for Critical-Mission SoS

As illustrated in Fig. 1, the methodology consists of 7 steps, described below.

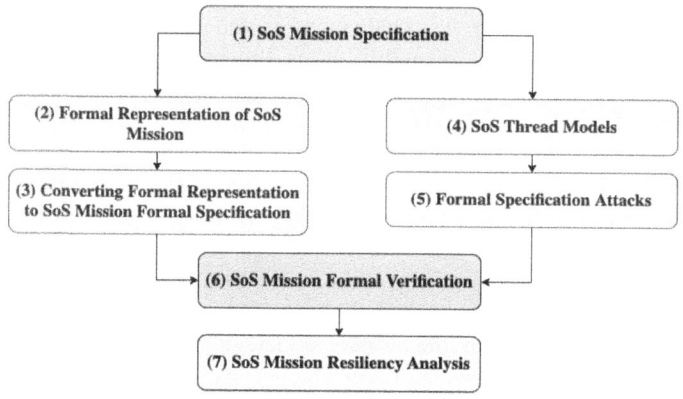

Fig. 1. Methodology for Mission Resiliency Analysis

Step 1 - SoS Mission Specification: It defines the mission *subjects* as active entities that execute tasks. Each *subject* has a type and a set of variables corresponding to its *attributes*. Some attributes of a subject (such as id) have values that cannot be changed – these are referred to as *immutable attributes*. All other attributes whose values can be changed are referred to as *mutable attributes*. The values of the attributes constitute a subject state. The state of a subject determines whether it can execute a given task. Step 1 also defines a set of *tasks*, which are atomic units of the mission. The mission starts with an initial task and ends with a final task. A mission includes a set of intermediate tasks between the initial and final tasks that may be grouped into sub-sets of tasks, referred to as sub-workflow in the formal representation. A mission also has a set of *objectives* to be accomplished.

Step 2 - Formal Representation of SoS Mission: This step represents the mission as a workflow. It decomposes the mission into many sub-workflows and connects them into a main workflow to represent the entire mission. It then instantiates the workflow by initializing each subject and its attributes, assigning each subject to tasks, and assigning initial conditions.

Step 3 - Converting Formal Representation to SoS Mission Formal Specification: The mission workflow is transformed into a formal specification for verifying the correctness and analyzing its resiliency.

Step 4 - SoS Threat Models: The threat models are derived from the mission specification. In this work, we focus on threats that may change the values of a subject's mutable attributes. When a subject's attribute value changes, it impacts the mission.

Step 5 - Formal Specification of Attacks: Each attack changes the mutable values of a subject. An attack requires preconditions to be satisfied so that it can happen. The effect of the attack is described as a post condition. Thus, each attack is specified with a precondition and post condition.

Step 6 - SoS Mission Formal Analysis: This step composes the Mission CPN with the Attack CPN. The Attack CPN can be attached only to the places in the Mission CPN where the Attack CPN's preconditions are satisfied. The state space and state transitions are verified and analyzed for each attack scenario attached to the mission formal specification.

Step 7 - SoS Mission Resiliency Analysis: SoS mission resiliency can be analyzed, including decremental and dynamic resiliency [18]. The resiliency analysis highlights constraints that must be enforced to make the mission more resilient.

4   SoS Mission Specification to Formal Representation

This section applies the first two steps of the methodology to the SoS example described in Sect. 2 to translate it into formal CPN specification.

4.1   SoS Mission Specification

We manually analyze the mission description to define the subjects, their attributes, tasks, and mission objectives. Only mutable attributes are considered. The SoS mission specification is summarized as follows:

- set of subjects = {*vehicle*, [*drones*]}
- set of *vehicle* attributes = {*location, GPS, speed, fuel, LOS, BLOS*}
- set of *drone* attributes = {*location, GPS, fly, acceleration, battery, camera, IR sensor, LOS*}

The tasks are:

1. check the *vehicle* status: *fuel* is full, LOS_v communication is up, and *BLOS* communication is up,
2. check the [*drones*] status: *battery* is set to 100 units, *camera* is on, *IR sensor* is on, and *LOS* communication is up,
3. drive the *vehicle* toward the POI with an average speed of 60–70 miles per hour,
4. fly [*drones*] toward the POI and keep 5 mi distance from the *vehicle*,
5. if it is daytime, each *drone* uses *camera* to scan the route and the POI area and provide day vision to the *vehicle*,
6. if it is nighttime, each *drone* uses *IR sensor* to scan the route and the POI area and provide night vision to the *vehicle*,
7. at the POI area, evaluate the situation, and report the military base in real-time,
8. if the threat requires engagement, the human element must receive instruction from the military base to engage, and

9. the *vehicle* and *[drones]* return to the station with data collected and evaluation reported.

The objectives = {*move toward POI*, *collect data*, *evaluate the situation*, *respond to threads*, *return to the station*}. Notice that we italicized the mission elements, which are used to formalize the SoS mission representation.

4.2   Formal Representation of the SoS Mission

We use the workflow defined in [1] to represent a mission. A workflow consists of tasks connected through operators. The syntax of the workflow is defined as follows.

Definition 1 (Workflow). *A workflow is defined recursively as:*

$$W = t_i \otimes (t \,|\, W_1 \otimes W_2 \,|\, W_1 \# W_2 \,|\, W_1 \& W_2 \,|\, if\{C\}\, W_1\, else\, W_2 \,|\, while\{C\}\{W_1\}) \otimes t_f$$

where sequence, exclusive choice, and, conditioning, and iteration operator

- *t is a subject-defined atomic task.*
- *t_i and t_f are unique initial and final tasks, respectively.*
- *$\otimes$ denotes the sequence operator. $W_1 \otimes W_2$ specifies W_2 is executed after W_1 completes.*
- *# denotes the exclusive choice operator. $W_1 \# W_2$ specifies that either W_1 executes or W_2 executes but not both.*
- *& denotes the and operator. $W_1 \& W_2$ specifies that both W_1 and W_2 must finish executing before the next task can start.*
- *$if\{C\}\, W_1\, else\, W_2$ denotes the conditioning operator. C is a Boolean-valued expression. Either W_1 or W_2 execute based on the result of evaluating C but not both.*
- *$while\{C\}\{W_1\}$ denotes iteration operator. If C evaluates to true, W_1 executes repeatedly until the expression C evaluates to false.*

A mission is defined as one or many subjects performing tasks to accomplish objectives.

Definition 2 (Mission). *A mission is defined as 5-tuple $\mathcal{M} = (W, S, \mathcal{ST}, I, O)$ where W is the workflow corresponding to the mission, S is a set of subjects, $\mathcal{ST} \subseteq S \times Tasks\,(W)$ is the set of subject to task assignments, I is the set initial conditions, and O is the set of mission objectives. The conditions and objectives are expressed in predicate logic.*

The SoS mission is divided into three sub-workflows. *Deployment* sub-workflow comprises check status and deploy tasks. *Moving-to-POI* sub-workflow consists of drive the vehicle, fly drones, keep 5-miles distance ahead, and scan routes tasks. *At-POI* sub-workflow includes evaluate situations, report to the base, and engaging tasks. The *Deployment*, *Moving-to-POI*, and *At-POI* sub-workflows respectively execute sequentially. These three sub-workflows connect to the main workflow that starts with the initial and final tasks. Each sub-workflow forms a set of tasks managed by operators. The operators are also used to connect sub-workflows to the main workflow. The complete workflow of the SoS mission is described as follows:

$$W = init \otimes \{check\_status\,(vehicle) \otimes check\_status\,([drones]) \otimes deploy \otimes$$
$$\{if\,(daytime)\,\{turn\_on\,([cameras])\,else\{turn\_on\,([IR\ sensors])\}\} \otimes fly\,([drones]) \otimes drive\_vehicle\} \otimes$$

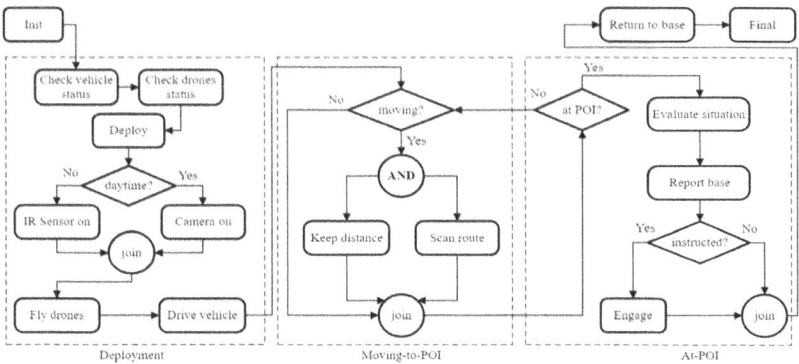

Fig. 2. SoS Mission Workflow

$while\,(moving)\,\{keep\_distance\ \&\ scan\_route\}\} \otimes if\,(locatio = POI)\,\{evaluate\_situation \otimes report\_base \otimes$
$if\,(instruction)\,\{engage\}\} \otimes return\_to\_base \otimes final$

Figure 2 illustrates the workflow diagram of the SoS mission. The dotted boxes represents the sub-workflows, *Deployment*, *Moving-to-POI*, and *At-POI*. The *return to base* task can be part of the main workflow since it does not represent any sub-workflow. Visualizing the workflow helps ensure the correctness of the workflow transformation to the CPN; hence, it is practical to draw the workflow.

Now, we instantiate the workflow. We initialize each subject and its attributes, assign each subject to a set of tasks, and set initial conditions for the subjects and objectives. There are two subject types *Vehicle* and *Drone*, and are instantiated as $S = \{vehicle : Vehicle; \, drones\,[] : Drone\}$. The subject type *Vehicle* has a set of attributes as *Vehicle.attributes* = $\{location : string; \, GPS : coordinate; \, speed \in \{0..240\}; \, fuel \in \{0..80\}; \, LOS : bool; BLOS : bool; \, evaluation\_reported : bool\}$. The subject type *Drone* has a set of attributes as *Drone.attributes* = $\{location : string; \, GPS : coordinate; \, fly : bool; \, acceleration \in \{1..10\}; \, battery \in \{1..100\}; \, camera : bool; \, IR\_sensor : bool; \, LOS : bool; \, data\_collected : bool\}$. The subjects *vehicle* and *[drones]* are assigned to tasks as $ST = \{((vehicle, [drones]), t) \mid t \in Tasks\,(W)\}$. Let I be a predicate logic formula giving the initialization conditions as:

$I = \exists\, s : S \mid (type = Vehicle) \wedge (location = \text{"station"}) \wedge (GPS = [34.0489, -111.0937]) \wedge (speed = 0) \wedge$
$(fuel = 80) \wedge (LOS = true) \wedge (BLOS = true) \wedge (evaluation\_reported = false) \wedge (type = Drone) \wedge$
$(location = \text{"station"}) \wedge (GPS = [34.0489, -111.0937]) \wedge (fly = false) \wedge (acceleration = 0) \wedge$
$(battery = 100) \wedge (camera = false) \wedge (IR\_sensor = false) \wedge (data\_collected = false).$

Let O be a predicate logic formula that defines the mission objectives as follows:
$O = \exists\, s : S \mid (type = Vehicle) \wedge (type = Drone) \wedge (location = \text{"station"}) \wedge (evaluation\_reported = true) \wedge$
$(data\_collected = true).$ We positioned the initial GPS coordinates for demonstration as [34.0489, -111.0937] and ranged the acceleration from 1 to 10. The GPS coordinate (i.e., Decimal Degrees (DD format) as [latitude, longitude]) is used to calculate the distance between the vehicle and the drones, and the acceleration is used to slow and speed up the drones to keep the 5-mile distance. As a result, the SoS mission specification is described as: $\mathcal{M}_{SoS} = (W, S, ST, I, O)$

4.3   Converting Formal Representation to SoS Mission Formal Specification

We consider CPN as a formal model that is practical to express mission state transitions due to CPN providing color sets, a flexible way to define data types to represent various types of subjects [9, 10]. The CPN is formally defined as:

Definition 3 (Coloured Petri Nets (CPN)). *CPN is defined as 9-tuple*
$CPN = (P, T, A, \Sigma, V, C, G, E, I)$, *where P, T, A, Σ, and V, are sets of places, transitions, arcs, colors, and variables, respectively. C, G, E, and I are functions that assign colors to places, guard expressions to transitions, arc expressions to arcs, and tokens at initialization expression, respectively.*

We use transformation rules defined in [1] to convert the mission workflow tuple into a CPN tuple as:

$$\mathcal{M} = (W, \mathcal{S}, \mathcal{ST}, I, O) \xrightarrow[Rules]{Trans.} CPN = (P, T, A, \Sigma, V, C, G, E, I)$$

The transformation rules are briefly explained through the SoS mission workflow:

Rule 1: Convert each task into a CPN transition.

Rule 2: Create input and output places for each transition, and create input and output arcs to pair the input place to the transition and the transition to the output place.

Rule 3: Produce a color set for each subject. The color set is a CPN datatype structure such as Product, Record, List, and Union [3]. The CPN color set combines primitive types such as *Bool*, *Int*, *Real*, and *String*. It also can be a compound of primitive types and color sets. The color sets represent subject attributes. The color set that represents the Vehicle is declared as:
$color : Vehicle = $ **record** $\{location : String; GPS : [Real, Real]; speed : Int \in \{1..240\}; fuel : Int \in \{1..80\}; LOS : Bool; BLOS : Bool; evaluation\_reported : Bool\}$
The color set that expresses the Drone is declared as:
$color : Drone = $ **record** $\{location : String; GPS : [Real, Real]; fly : Bool; acceleration : Int \in \{1..10\}; battery : Int \in \{1..100\}; camera : Bool; IR\_sensor : Bool; LOS : Bool; data\_collected : Bool\}$

Rule 4: Assign the color sets to the CPN places and arcs to declare their datatypes and declares a set of variables such that there is a variable for each color set. For instance, the declaration *var vehicle : Vehicle* declares a variable *vehicle* of type *Vehicle*. The expression *colset droneList = list Drone with* 2; declares a list of datatype *Drone* with maximum two elements, and the declaration *var drones : droneList* declares a variable *drones* of type *dronelist*.

Rule 5: Assign a guard expression for each CPN transition. The guard expression must be evaluated true in order for a transition to be executed (i.e., fired). For instance, $fly(drone_1) = true$ is a guard expression assigned to the deployment transition. It ensures that the deployment transition fires only if the $drone_1.fly$ attribute is true.

Rule 6: Assign an arc expression for each input and output arc of a transition. The arc expressions evaluate tokens to pass in and out from transitions (i.e., bindings). When a token is evaluated true by an arc expression, it enables the transition that connects to be fired. For instance, *if* (*at_POI* (*vehicle*) = *true*) {*vehicle*} *else* {*empty*} is an arc expression that ensures the *vehicle* token reaches the POI place so the token passes; otherwise, the POI place does not bind the next transition, and the next transition will not be executed.

Rule 7: Initialize variables with values and assign these variables to the CPN places as tokens. The initialization forms the initial conditions of the mission. The tokens are then manipulated in each CPN transition, generating the state transitions of the CPN. For instance, the initial conditions of the SoS mission are described as:

$vehicle = init\{(location = \text{"station"}) \land (GPS = [34.0489, -111.0937]) \land (speed = 0) \land (fuel = 80) \land$

$(LOS = true) \land (BLOS = true) \land (evaluation\_reported = false)\}$

$drone_1.drone_2 = init\{(location = \text{"station"}) \land (GPS = [34.0489, -111.0937]) \land (fly = false) \land$

$(acceleration = 0) \land (battery = 10) \land (camera = false) \land (IR\_sensor = false) \land (data\_collected = false)\}.$

Figure 3 illustrates the *CPN* model resulting from applying the conversion rules 1–7 to the SoS mission $\mathcal{M}_{SoS}$. We have explicitly omitted writing each arc expression into the *CPN* model to maintain readability. It shows the primary CPN mission, the SoS Mission main model, and three sub-CPN models: Deployment, Moving-to-POI, and At-POI. These three sub-CPNs are connected to the main CPN model through input and output ports, allowing the token to cross through the sub-CPN model and back to the main CPN. The small sub-CPNs, GPS attack, Battery, Camera, Sensor, LOS, and BLOS attack, depict CPN attack models, which are discussed next.

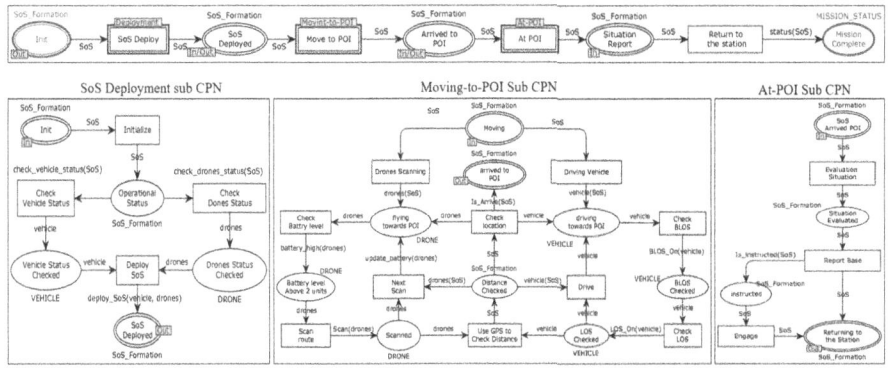

Fig. 3. CPN of the SoS Mission

5   Threat Model and Attacks Representation

This section describes the threat models and attack representations for the SoS mission.

5.1   SoS Threat Model

The threat model is derived directly from the mission description (Step 1). The vehicle and drones have mutable attributes that are attackable. We assume the drone can fly and accelerate are immutable since they are physical attributes (i.e., mechanical attributes), thus excluding them from the threat model. Other subjects with entirely immutable attributes, such as human elements and satellites, are also omitted from the threat model. We consider *GPS*, *LOS*, and *BLOS* to be mutable attributes of the vehicle. We also consider *GPS*, *battery*, *camera*, *IR_sensor*, *LOS*, are mutable attributes of the drones. Note that we focus on the communication and sensing capabilities of the military vehicle and drones since they are the best attackable attributes.

Table 1. Threat Model of the SoS Mission

Subject	Mutable Attribute	Attack Process	Impact on the Mission
Vehicle	GPS	Send out false signals to change the vehicle coordinates	The vehicle drives in the wrong direction, unable to reach the POI, leading to the mission to be *fail*. It may also cause the drones to hover away, unable to return, and the mission *fails*
	LOS	Disable the LOS communication	The vehicle loses communication with the drones, losing the route vision ahead; it may cause the mission to be *incomplete*
	BLOS	Disable the BLOS communication	The vehicle loses communication with the base, unable to receive instruction and unable to report the situation in real-time, leading the mission to be *failed*
Drone	GPS	Send out false signals to change the drones coordinates	The drones surveil incorrect route and collect useless data leading to the mission to be *incomplete*. They may also be unable to return to the vehicle and the mission *fails*
	battery	Consume one battery unit	It degrades the drones' ability to collect data and reduces flying time, leading to the mission to be *incomplete*; it might lead to mission *fails* if the drones do not have enough battery to return to the vehicle
	camera	Disable the camera	The drones are incapable of collecting data and cannot provide daytime vision to the vehicle, leading to the mission to be *incomplete*
	IR_sensor	Disable the camera	The drone is incapable of collecting data and cannot provide nighttime vision to the vehicle, leading to the mission to be *incomplete*
	LOS	Disable the LOS communication	The drone loses communication with the vehicle, unable to send the surveillance data to the vehicle, leading the mission to be *incomplete*

Table 1 illustrates the threat model for the SoS mission. The GPS attack changes the appearance of where the vehicle is located. This leads the vehicle to drive in a different direction, unable to reach the POI, leading the mission to *fail*. It also dislocates the drones because the GPS calculates the distance between the vehicle and the drones. The drone's battery attack is considered to be consuming battery units. The battery consumption attack degrades the drone's capability for surveillance and reconnaissance, which requires more battery. It also reduces the flying time of the drones (e.g., instead of flying 4 h in a regular situation, they will fly 3 h when an attack occurs). If the battery consumption attack occurs, we assume the drones should abort the mission and fly back to the vehicle, and the mission is incomplete. Another possibility is that the drones do not abort, keep flying, lose their battery, and then cannot return to the vehicle. In this case, the mission can fail because the data is not collected, and the vehicle cannot estimate the dangerous situations ahead. If the attacker turns off the camera and infrared sensor, the drones cannot collect data but can fly back to the vehicle, and the mission is incomplete. Besides mission *success*, two mission statuses are defined: mission is *incomplete* and mission *fails*. Mission success indicates that the drones collect data and fly back to the vehicle, and the vehicle reaches the POI and reports to the military base. The incomplete mission indicates that the drones cannot collect data but can fly back to the vehicle, and the vehicle reaches the POI but cannot report to the military base. Mission failure demonstrates that the drones cannot return to the vehicle, the vehicle cannot reach the POI, or the vehicle or the drones do not return to the station. Analyzing the mission status from the mission specification is essential to translate the attacks into formal specifications.

5.2   Formal Specification of Attacks

Each attack described in the threat model is formally specified as a CPN attack (i.e., a sub-CPN model). The CPN attack is plugged into the proper place of the CPN mission based on precondition and postcondition. From the SoS threat model in Table 1, eight CPN attacks are formalized as three CPN attacks related to the vehicle, and five CPN attacks demonstrate the drone attacks. The three CPN attacks of the vehicle are presented in Fig. 4 as vehicle GPS attack, vehicle LOS attack, and BLOS attack. These CPN attacks have the same precondition and postcondition as the *Vehicle* datatype. They must receive a variable of type *Vehicle*, change its values, and return a new state of this variable. These CPN attacks can be attached to any place in the CPN mission if that place has a precondition and postcondition of *Vehicle* datatype.

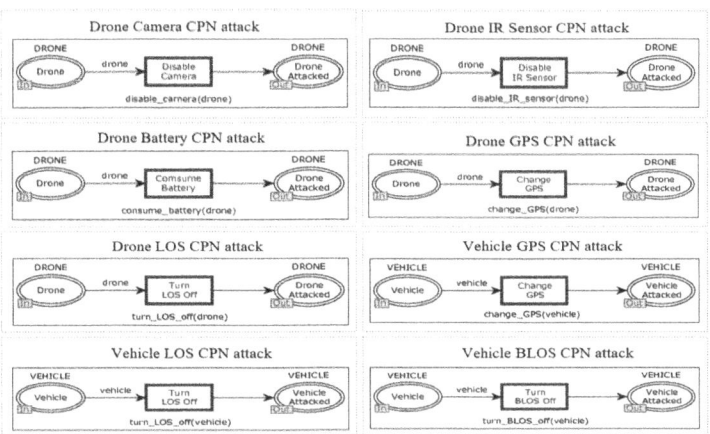

Fig. 4. CPN Attacks for the SoS Mission

Similarly, the five CPN attacks are formalized for the drones: drone GPS attack, battery attack, camera attack, infrared sensor attack, and drone LOS attack. These CPN attacks have precondition and postcondition of datatype *Drone* and must receive a variable of type *Drone*, change its values, and return the new state of this variable. For instance, the CPN battery attack decreases the value of the drone variable's attribute $drone_1.battery$ by 10 units and returns the drone variable with a new state of values. The CPN camera attack changes the value of the drone variable's attribute $drone_1.camera$ to false, and the CPN infrared sensor attack changes the value of the drone variable's attribute $drone_1.IR\_sensor$ to false, and return the drone with new values to the CPN mission. For these changes in the drone's attributes, resiliency analysis is executed to verify where the CPN mission can succeed, be incomplete, or fail.

6 SoS Mission Resiliency Analysis

This section applies Steps 6 and 7 of the methodology to the SoS mission to analyze its resiliency. State space and transitions of the CPN mission attached to CPN attacks are attached are verified using CPNTools [16].

6.1 SoS Mission Formal Specification Analysis

We investigate the decremental resiliency of a mission in which a mission-critical SoS is degraded during execution, and no previously unavailable capability may recover. We start with one attack that degrades a subject's capability. We then decrease subjects' capabilities by different attacks and degrade them to demonstrate a decremental resiliency analysis. Table 2 examines 4 attack scenarios, demonstrating decremental resiliency.

Table 2. Decremental Attack Scenarios

Tasks	Location	Attack	Mission Status
Two drones scan the route and transmit to the military vehicle	Moving to POI	Disable cameras and IR sensors, and consume drones' battery	Incomplete
Two drones scan the route and transmit to the military vehicle	Moving to POI	Change GPS of the drones	Fail
The vehicle and drones are mapping the route and collecting data	Moving to POI	Disable LOS of the drones and the vehicle's BLOS	Fail
Report the situation to the base and wait for engaging instruction	At POI	Disable LOS, cameras and IR sensors of the drones and the vehicle's BLOS	Incomplete

The first attack scenario expresses that the two drones were attacked many times, turning off their cameras and IR sensors and consuming their batteries with 10 units each. These attacks occurred while the SoS was moving toward the POI. The drones could not collect data but had enough battery to safely return to the military vehicle using GPS. The mission status is incomplete because the data was never collected. The second attack scenario shows the drones' GPS being attacked while scanning the route, and they could not maintain the 5-mile distance. They may hover away from the military vehicle and lost. The mission failed because the drones did not return to the military vehicle. The third attack scenario also occurred while the SoS moving toward the POI. The attack disabled the vehicle's BLOS and the drones' LOS, rendering them unable to communicate. The drones could not transmit their GPS and data to the vehicle, nor did the vehicle communicate with the base. The mission failed because the drones could not reach the POI nor return to the vehicle. The fourth attack scenario happened at the POI location, where all communication and sensing capabilities of the vehicle and the drones were disabled. The mission was incomplete because the objectives of collecting data and reporting the base were unmet, but the SoS returned safely to the station.

State Space Analysis. The analysis examines the state space of the CPN model with each attack scenario. Table 3 illustrates the number of nodes and arcs of the state space model (i.e., reachability graph) generated for each attack scenario. It also shows the number of nodes and arcs of the strongly connected components (SCC) graphs generated from the state space models. Table 3 reports that all CPNs of attack scenarios are full SCC. This indicates that the state space models generated from these CPNs are acyclic; they do not have infinite loops. However, it reports dead markings and dead transitions for each attack scenario, which are considered for analysis. Dead markings are states where the CPN is terminated, and dead transitions are those transitions that are not executed (no token visits them). Checking the dead markings gives the values of the SoS attributes when the mission is terminated, whereas analyzing the dead transitions illustrates why the SoS did not execute specific tasks.

Table 3. State Space Report

Location & Attack Scenario	State Space		SCC Graph		Status
	#Node42	#Arcs 61	#Node42	#Arcs 61	Full
Moving to POI, disable cameras and sensors, and consume battery	Dead Markings [30,34,37,42]		#Dead Transition 5		Live Transition None
	#Node31	#Arcs 38	#Node31	#Arcs 38	Full
Moving to POI, change GPS of the drones	Dead Markings [25,26,31]		#Dead Transition 2		Live Transition None
	#Node37	#Arcs 54	#Node37	#Arcs 54	Full
Moving to POI, disable LOS of the drones, and the vehicle's BLOS	Dead Markings [29,33,34,37]		#Dead Transition 5		Live Transition None
	#Node63	#Arcs 85	#Node63	#Arcs 85	Full
At POI, disable drones cameras, sensors, LOS, and vehicle's BLOS	Dead Markings [47,54,55,58,63]		#Dead Transition 7		Live Transition None

We use CPN-ML programming language [8] to verify dead marking and dead transition for each attack scenario. We also use built-in functions provided by the CPNTools [16], such as $Reachable(x, y)$, $AllReachable()$, $NodesInPath(x, y)$, $DeadMarking(x)$, $ListDeadMarkings()$, and $ListDeadTransitions()$, for the verification process. These functions return an execution sequence for each dead marking where the mission is succeeded, incomplete, or failed. We backtrack through the execution sequence and locate the state where the attack occurs. While backtracking, we verify the change of states in the sequence to inspect the reasons that lead to dead transitions and what state values the sequence terminates with at dead markings.

The first attack scenario, moving to POI and disabling cameras and sensors and consuming battery, reports 4 dead markings ([30,34,37,42]) and 5 dead transitions. These dead markings show that the CPN mission terminates where the drone's tokens loop over the battery check transition, which is attacked and ends with an empty mission status. State 42 expresses that the CPN mission terminates with mission status "Incomplete" because the *data_collected* value is false as part of the mission objectives is unsatisfied.

The state space of the second attack scenario, moving to POI and changing the drones' GPS, reports 3 dead markings ([25,26,31]) and 2 dead transitions. States 25 and 26 show that the drones' GPS was attacked in the moving-to-POI location, causing the CPN mission to terminate with an empty mission status. In State 31, the CPN mission completed the mission with mission status "Fail". Verifying this state shows that the drone's GPS differs from the vehicle's GPS, indicating that the drone is not returning to the station. It also shows that the drone *data_collected* attribute is false, and the mission objectives are unmet. The two dead transitions are "$SoS'UseGPStoCheckDistance$ 1" and "$SoS'Checklocation$ 2", indicating that the GPS attack prevented the drones' token from passing through these transitions.

The third attack scenario, moving to POI and disabling the drones' LOS and the vehicle's BLOS, reports 4 dead markings ([29,33,34,37]) and 5 dead transitions. States 29, 33, and 34 show that the CPN mission cannot continue and terminates with an empty mission status. State 37 shows that the CPN mission ended with the mission state "Fail" because the drone's GPS differs from the vehicle's GPS due to disconnection, and the vehicle disconnected from the base. Verifying the 5 dead transitions shows that the vehicle BLOS checks transition leads to check vehicle LOS, uses GPS to check the distance, and checks the location to be dead transitions. It is because these transitions are related to one another; failing one causes failing another. The impact of attacking LOS communication cascades to the GPS capability.

The fourth attack scenario, at POI disabling drones cameras, sensors, LOS, and vehicle's BLOS, reports 5 dead markings ([47,54,55,58,63]) and 7 dead transitions. States 47, 54, 55, and 58 show the CPN mission terminated with an empty mission status. These are because of the number of CPN attacks disabling cameras, sensors, LOS, and BLOS. State 63 shows that the CPN mission ended with the mission state "Incomplete". Even though the number of attacks in the fourth scenario is more than the number of attacks in the third scenario, the CPN mission status ends with "Incomplete" compared to the third scenario that ended with "Fail". In the fourth attack scenario, the drone's GPS is the same as the vehicle's. The drones can collect data because the attacks occurred at the POI location, not the route. The CPN mission status in the fourth attack scenario is incomplete because the SoS cannot report the base; the BLOS is disconnected.

Even though the fourth attack scenario includes more attacks than the third attack scenario, the fourth attack scenario terminates with an incomplete mission status compared to the third attack scenario, which terminates with a failed mission status. Interestingly, the state space of these attack scenarios reports an exact number of dead markings, only one node, and an exact number of dead transitions, 8 dead transitions. The location where the attack scenario occurs significantly varies the mission outcomes.

The verification shows where the SoS mission fails in dispute of attacks. It also points out that attacking one SoS capability affects others, and the mission fails. Lastly, the impact of attacks is different from one location to another (geographically); an attack is critical at one location and ineffective at another. SOS engineering must take into account these facts when describing resilient SoS missions.

6.2   SoS Mission Resiliency Analysis

The verification indicates that the SoS must immediately abort the mission in some instances, especially when LOS and BLOS communication is attacked. Continuous surveilling and moving toward the POI without communication minimizes the resilience level of the mission. To enhance the resiliency of SoS critical missions, suitable restrictions must be counted on the SOS missions, and improvements for SoS capabilities must be considered. Table 4 expresses constraints and improvements to increase the resiliency level of the SoS critical-missions. That is, if we add these constraints and improvements to the SoS mission described in Sect. 2, this mission becomes resilient and resistant to various types of attacks.

Table 4. Mission Restrictions & Improvements

Task	Location	Restriction & Improvement
Drones scan the route and collect data	Moving to POI	Abort the mission if the drones' LOS are disabled. Provide recoverable LOS to pursue the mission
Drones scan the route and collect data	Moving to POI	Abort the mission if the drones' LOS are attacked. Or, provide recoverable LOS to pursue the mission
Drones scan the route and collect data	Moving to POI	Provide recoverable cameras and IR sensors for the drones as well as backup battery
Military vehicle moves toward POI	Moving to POI	Abort the mission if the vehicle' BLOS is attacked. Or, provide recoverable BLOS to communicate with the base
SoS reports POI situation to the base	At POI	Use backup communication channels when BLOS is attacked and provide recoverable BLOS for reporting

7   Related Work

The literature on resiliency issues discusses the unavailability [12–14, 18, 20]. We propose that the resiliency problem can be considered as degradation.

Wang *et al.* [18] introduce three types of resilience, static, decremental, and dynamic. They use the term a user to refer to a subject, an active entity doing tasks. Static resilience refers to a situation in which subjects become unavailable before the workflow executes, and no subjects may become available during the execution. Decremental resiliency expresses a condition where subjects become unavailable before or during the execution of the workflow, and no previously unavailable subjects may become available during execution, while dynamic resilience describes the situation where a subject may become unavailable at any time; a previously unavailable subject may become available at any time. The different types of resilience formulations capture various types of attack scenarios.

Yang *et al.* [20] address the workflow satisfiability problem that ascertains whether a set of subjects can complete a workflow. The authors investigate the computational complexity of the workflow satisfiability problem in two aspects. One aspect considers either one path or all paths of a workflow, and the other focuses on the possible patterns in a workflow. They present a set of algorithms for solving various types of problems of workflow satisfiability. They show that many existential and universal workflow satisfiability problems are NP-complete and NP-hard. Thus, they conclude that restrictions on workflow patterns induce such problems to be solvable in polynomial time.

Mace *et al.* [12–14] propose a quantitative measure of workflow resiliency. They use a Markov Decision Process (MDP) to model workflow to provide a quantitative measure of resilience. They refer to binary classification, such as returning an execution sequence if one exists and declaring the workflow resilient or returning false and declaring the workflow not resilient. The authors show that the MDP models give a termination rate and an expected termination step.

8   Conclusion

This paper introduces a methodology that addresses the mission resiliency analysis for mission-critical System of Systems (SoS). The methodology systematically analyzes various decremental resiliency of SoS missions. The methodology uses workflow for the formal representation of the SoS mission. It also constructs threat models and converts them into formal representation of attacks. The methodology then utilizes Coloured Petri Nets (CPN) to specify the SoS missions with attacks attached.

We applied the methodology to a SoS formation demonstrating a military vehicle and two drones working together to monitor a national border. The results show that the resiliency of the SoS mission is a degradation issue. When attacks occur, a mission-critical SoS cannot continue, and the mission fails or is incomplete. The methodology indicates which state the SoS mission succeeds, fails, and is incomplete. It also highlights restrictions that must be added to the SoS mission to improve resiliency.

Future work will investigate the methodology's application to dynamic resiliency, where a system is able to recover its components after an attack. The model state explosion will be investigated to verify such scenarios. Future work will also use threat modeling frameworks such as STRIDE and PASTA to cover various types of attacks that violate the confidentiality, integrity, and availability in the context of a mission.

Acknowledgements. This work was supported in part by funding from NIST under Award Number 60NANB23D152 and NSF under Award Numbers CNS 2335687, DMS 2123761, CNS 1822118, NIST, ARL, Statnett, AMI, NewPush, and Cyber Risk Research.

References

1. Abdelgawad, M., Ray, I., Vasquez, T.: Workflow Resilience for Mission Critical Systems. In: Dolev, S., Schieber, B. (eds.) Stabilization, Safety, and Security of Distributed Systems. SSS 2023. LNCS, vol. 14310. Springer, Cham (2023). https://doi.org/10.1007/978-3-031-44274-2_37
2. Cook, S., Pratt, J.: Advances in systems of systems engineering foundations and methodologies. Aust. J. Multi-Disciplinary Eng. **17**(1), 9–22 (2021)
3. CPN Tools Administration: Help topics for verification (2018). http://cpntools.org/2018/01/15/verification/. Accessed 12 Jan 2024
4. Dahmann, J., Roedler, G.: Systems of systems engineering standards. Insight **19**(3), 23–26 (2016)
5. Figueira, N., Pochmann, P., Oliveira, A., de Freitas, E.P.: A C4ISR application on the swarm drones context in a low infrastructure scenario. In: the proceedings of the International Conference on Electrical, Computer and Energy Technologies (ICECET), pp. 1–7. IEEE, Prague, Czech Republic (2022)
6. Houliotis, K., Oikonomidis, P., Charchalakis, P., Stipidis, E.: Mission-critical systems design framework. Adv. Sci. Technol. Eng. Syst. J. **3**(2), 128–137 (2018)
7. ISO/IEC/IEEE: International standard – systems and software engineering – system of systems (SoS) considerations in life cycle stages of a system. ISO/IEC/IEEE 21839:2019(E) **1**(8767114), 1–40 (2019)
8. Jensen, K., Kristensen, L.M.: CPN ML programming. In: Coloured Petri Nets, pp. 43–77. Springer, Heidelberg (2009). https://doi.org/10.1007/b95112_3
9. Jensen, K., Kristensen, L.M.: Coloured Petri Nets. Springer, Heidelberg (2009). https://doi.org/10.1007/b95112
10. Jensen, K., Kristensen, L.M., Wells, L.: Coloured Petri Nets and CPN tools for modelling and validation of concurrent systems. Int. J. Softw. Tools Technol. Transfer **9**(3–4), 213–254 (2007)
11. Lappas, V., et al.: Autonomous unmanned heterogeneous vehicles for persistent monitoring. Drones **6**(4), 94 (2022)
12. Mace, J.C., Morisset, C., van Moorsel, A.: Quantitative workflow resiliency. In: Kutyłowski, M., Vaidya, J. (eds.) ESORICS 2014. LNCS, vol. 8712, pp. 344–361. Springer, Cham (2014). https://doi.org/10.1007/978-3-319-11203-9_20
13. Mace, J., Morisset, C., van Moorsel, A.: Modelling user availability in workflow resiliency analysis. In: the proceedings of the 2015 Symposium and Bootcamp on the Science of Security (HotSoS), pp. 1–10. ACM, Urbana Illinois, USA (2015)
14. Mace, J.C., Morisset, C., van Moorsel, A.: Resiliency variance in workflows with choice. In: Fantechi, A., Pelliccione, P. (eds.) SERENE 2015. LNCS, vol. 9274, pp. 128–143. Springer, Cham (2015). https://doi.org/10.1007/978-3-319-23129-7_10
15. Ponsard, C., Massonet, P., Molderez, J.F., Rifaut, A., Lamsweerde, A.V., Van, H.T.: Early verification and validation of mission critical systems. Formal Methods Syst. Des. **30**(3), 233–247 (2007)
16. Ratzer, A.V., et al.: CPN tools for editing, simulating, and analysing coloured petri nets. In: van der Aalst, W.M.P., Best, E. (eds.) ICATPN 2003. LNCS, vol. 2679, pp. 450–462. Springer, Heidelberg (2003). https://doi.org/10.1007/3-540-44919-1_28
17. Ross, R., Pillitteri, V., Graubart, R., Bodeau, D., McQuaid, R.: NIST special publication 800–160 volume 2: developing cyber resilient systems. NIST Spec. Publ. **2**, 224 (2019)
18. Wang, Q., Li, N.: Satisfiability and resiliency in workflow authorization systems. ACM Trans. Inf. Syst. Secur. **13**(4), 1–35 (2010)

19. Weisman, M., et al.: Quantitative measurement of cyber resilience: Modeling and experimentation. Computing Research Repository (2023). (CoRR) ABS/2303.16307
20. Yang, P., Xie, X., Ray, I., Lu, S.: Satisfiability analysis of workflows with control-flow patterns and authorization constraints. IEEE Trans. Serv. Comput. **7**(2), 237–251 (2014)

Enhancing EV Charging Station Security Using a Multi-dimensional Dataset: CICEVSE2024

Emmanuel Dana Buedi$^{(\boxtimes)}$ ⓘ, Ali A. Ghorbani ⓘ, Sajjad Dadkhah ⓘ, and Raphael Lionel Ferreira

University of New Brunswick, 3 Bailey, Fredericton, NB, Canada
{dana.buedi,ghorbani,sdadkhah,raphael.ferreira}@unb.ca

Abstract. The rapid adoption of electric vehicles (EVs) is fundamentally transforming the automotive industry, prompting a surge in the installation of charging stations to accommodate the growing number of EVs and enhance overall mobility and user experience. Efforts to conduct machine learning-based cybersecurity research and develop solutions to address the growing threats and vulnerabilities in EV charging station infrastructure face challenges stemming from the unavailability of suitable datasets. The primary contribution of this study is addressing these challenges by publishing a multi-dimensional dataset that comprises power consumption data, network traffic and host activities of the EVSE in both benign and attack conditions. The experimental testbed uses a real Electric Vehicle Supply Equipment (EVSE), Raspberry Pi and standard industry communication protocols for EV charging infrastructure, with the scenarios observing the EVSE in both idle and charging states. The results of statistical analysis and machine learning classification tasks demonstrate the suitability of this dataset for baseline behavioral profiling, classification and anomaly detection tasks.

Keywords: Electric Vehicle Charging Station · EVSE · Multi-dimensional Dataset · Cybersecurity

1 Introduction

The automotive landscape is undergoing a profound transformation with the rapid proliferation of electric vehicles. It is anticipated that electric cars will account for more than 60% of all vehicles sold globally by the year 2030 [7], indicating a substantial shift from the existing internal combustion engine (ICE) paradigm. This surge is fueled by a confluence of factors, including advances in battery technology, government incentives and the ever-increasing demand for sustainable mobility solutions focused on environmental consciousness [16,29]. As the number of electric vehicles on the road increases, more charging stations are being installed to facilitate mobility and improve the user experience.

A. L. Ferrara and R. Krishnan (Eds.): DBSec 2024, LNCS 14901, pp. 171–190, 2024.
https://doi.org/10.1007/978-3-031-65172-4_11

The EVSE, which refers to the physical hardware supplying electric energy to recharge EVs, integrates both hardware and software elements. It is designed to facilitate seamless and safe operations, catering to the diverse needs of electric vehicle users and charging station operators [31]. In order to match or surpass the user experience and convenience provided by conventional fuel stations, the electric vehicle charging infrastructure has been improved through the integration of communication and software features. The hardware implementation of the EVSE as described in [15] and presented in Fig. 1, reflects the level of sophistication and supported services. The Secondary Equipment Communication Controller (SECC) or the main board functions as the primary intelligence hub, hosting all implemented logic, Open Charge Point Protocol (OCPP) [2] client, and ISO15118 server applications. The HomePlug Green PHY modem facilitates V2G communication via the Control Pilot (CP) pin. For remote communication with the Charging Station Monitoring System (CSMS), WiFi, cellular, or Ethernet connections can be used. Communication with Energy Management Systems (EMS) is achievable through RS232 or RS485 interfaces typically using the Modbus protocol. Payment services can be conducted locally using RFID, NFC cards, Mobile applications or by scanning a QR code. The power electronics components are used for power conditioning and, in the case of DC fast chargers, for converting AC to DC.

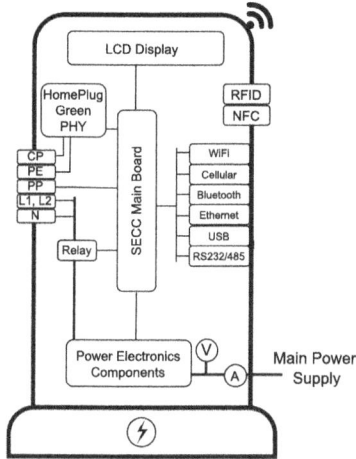

Fig. 1. EVSE Hardware Components

While improving user experience and convenience through enhanced services is desirable, the associated hardware and software features may be susceptible to cybersecurity threats [14]. Researchers are actively addressing the vulnerabilities and threats in electric vehicle charging infrastructure [3,14,30,32,34]. Machine learning models have emerged as promising tools for cybersecurity applications, particularly in intrusion and anomaly detection. However, these models rely on datasets that comprise the charging station's behavior under both normal and anomalous conditions. Currently, the application of machine learning models in cybersecurity research for electric vehicle charging stations faces several open issues relating to availability of datasets. The existing datasets primarily concentrate on EV charging sessions and revolve around modeling and planning power system load profiles. Datasets sourced from other domains, such as the Internet of Things (IoT) and Operational Technology (OT), fail to offer a comprehensive representation of the operational behavior of a charging station, rendering them unsuitable as alternatives. To the best of our knowledge, there are no publicly available datasets that capture the internal behavior and communication network activities of the EVSE host under both normal and anomalous conditions. The main contribution of this work are:

- Generating and publishing a multi-dimensional dataset constituting power consumption data, network traffic captures and host activities of the EVSE under normal and anomalous conditions.
- Performing statistical analysis on the dataset, comparing some features to assess their adequacy for behavioral profiling and anomaly detection tasks.
- Evaluating the suitability of the dataset for binary and multi-class classification of benign and attack scenarios using machine learning algorithms.

The remainder of this paper is organized as follows: Sect. 2 compares the contribution of this work to other related research. Section 3 presents the details of the EV charging station lab setup and the various scenarios implemented. Section 4 presents the dataset and statistical analysis of some key features. Section 5 presents a machine learning evaluation of the dataset for the classification of benign and attack scenarios. Finally, Sect. 6 presents the conclusion and some future directions of this research work.

2   Related Work

Within cybersecurity research, this work can be positioned in the domain of studying the baseline behaviour of EV charging stations for intrusion or anomaly detection purposes. In this regard, the most closely related work, [20] focuses on EV authentication within the charging station infrastructure. In their work, the authors set up a simulation environment using a Linux-based general-purpose computer. Three main entity categories, Electric vehicles, charging stations and a grid server were simulated on the host computer. The Perf tool is then used to record specific kernel events from the host computer for each entity referenced by the process ID. They considered four main EV authentication DDoS attack scenarios involving a combination of full, random, gaussian and non-gaussian attacks. On the other hand, our work generates network, host and power consumption data from a testbed setup consisting of an actual EV charging station and Raspberry Pi devices. The attack scenarios in our work include both host and network based attacks (Table 1).

Table 1. Existing Datasets for Electric Vehicle Charging Stations

Dataset	Year	Hardware Device Platform	Data	EVSE Communication Protocols	Attack Scenario(s)	Accessibility
Pecan Street Dataport [17]	2023	EVSE	Energy Consumption	None	None	Commercial
CICEV2023 [20]	2023	General Purpose Computer	Host Events	None	Network Attacks	Public
Asensio [5]	2020	EVSE	Charging Sessions	None	None	Public
ACN-data [21]	2019	EVSE, Mobile App	Charging Sessions, EV User	None	None	Public
E. Xydas et al. [40]	2016	EVSE	EV Charging Demands	None	None	Private
iMove [10]	2015	EVSE	Charging Sessions	None	None	Public
ELaadNL [10]	2015	EVSE	Charging Sessions	None	None	Public
This Work	2024	EVSE, Raspberry Pi	Network, Host, SECC Power Consumption	OCPP, ISO15118 (V2G)	Network and Host Attacks	Public

In the area of characterizing EV charging behaviour based on charging session data, a lot of work has been done to generate benchmark datasets to support ongoing research. Typically, charging session features includes EV power consumption, EV-to-EVSE connection and disconnection time, EV arrival and departure time, charging current and status of some pilot signals. The Adaptive Charging Network, ACN-Data [21], provides a comprehensive, up-to-date and publicly accessible dataset on EV charging sessions. The network started with over 100 EVSEs located in two sites within the Caltech and JPL campuses. The authors collect data directly from the charging stations and the EV user mobile application. In their work, the authors of [40] monitored the EV charging demands for 255 charging stations and weather data from 2012 to 2013. Similar to the ACN-Data, the Asensio [5], Pecan Street Dataport [17], iMove and ELaadNL [10] all focus on charging session data. The focus of these datasets has been on the overall impact of EV charging station demand on the electrical grid. Even though these works could be used for profiling EV charging sessions and to some extent EV user behaviours, they however unlike our work, are not cybersecurity focused and hence cannot be used for detecting security attacks launched against the EVSE.

Another associated area of study explores the application of hardware events for cybersecurity solutions, including device behavioral fingerprinting for anomaly detection [26,33,39] and malware classification [1,27,35]. However, the datasets used for most of these works are unavailable for further experiments and validation. Distinguishing itself from previous studies, this research introduces additional data from sources such as power consumption and network traffic, providing researchers with a publicly available dataset with a broader range of options for detecting malicious activities in resource constrained devices.

3   EV Charging Station Lab Setup

Our testbed configuration setup presented in Fig. 2 comprises an operational Level 2 charging station [36], EVSE-A and a Raspberry Pi and communication setup. The EVSE-A is configured to communicate to the remote CSMS platform using OCPP protocol and could be installed for use in a public, home or shared facility [4]. The EVSE-B is represented by a Raspberry Pi, it communicates with the EVCC and CSMS using ISO15118 and OCPP respectively. The power consumption of the EVSE-B is also monitored by another Raspberry Pi using a wattmeter and Inter-Integrated Circuit (I2C) protocol [22].

3.1   Electric Vehicle Supply Equipment

The EVSE-A employed in our laboratory setup is the Grizzl-E smart EV charger[1], a Level 2 AC charging unit supporting a 40A/9.6kW charging output. The ESP32 controller powers the intelligent functionalities of this EV charger, providing WiFi capabilities for local and remote communication. A detailed description of EVSE-A features is provided in Table 2.

Table 2. EVSE-A Specifications

Feature	Specification
Model	Grizzl-E Smart Connect - GRS-6-24-PB
Type	AC Level 2 (208-240 VAC)
Maximum Output Ratings	40A (9.6 kW)
Charging cable Connector	J1772
Connectivity	2.4Ghz Wifi network
Mobile App	Yes
Protocols	OCPP 1.6
Main board controller	ESP32
Status Indicators	LED, Buzzer

The EVSE-B is represented by a Raspberry Pi 4 model B, featuring a Quad-core Cortex-A72 (ARM v8) 64-bit processor, 2GB RAM, running Ubuntu 22.04.3 LTS (Jammy), and using the Linux 5.15.0-1044-raspi kernel. The deliberate choice of Raspberry Pi hardware is motivated by several key factors that highlight its suitability for this project:

---

[1] https://grizzl-e.com/products-specs/grizzl-e-smart/.

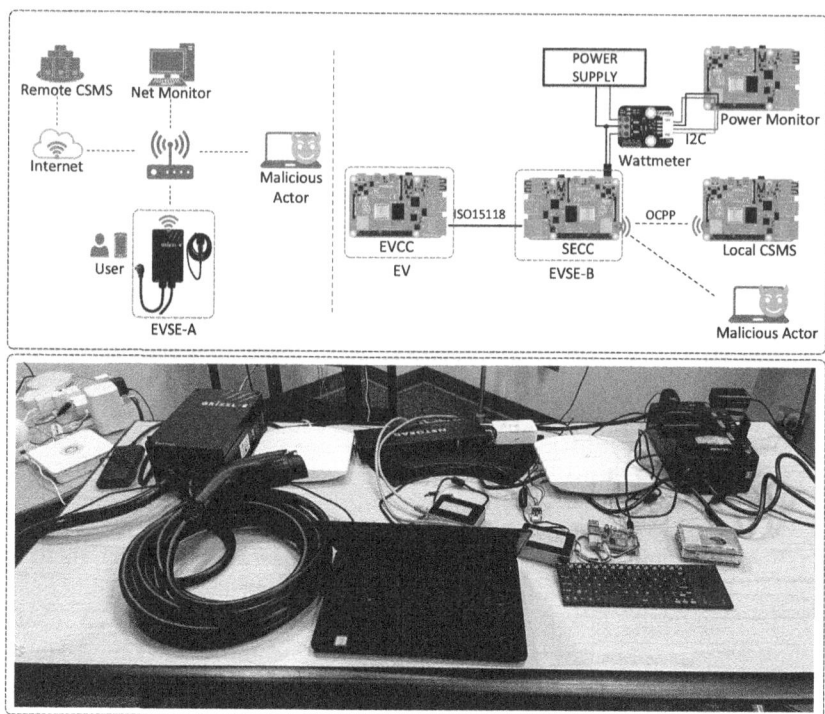

Fig. 2. EV Charging Station Lab Setups

– **Alignment with Typical EVSE Main Board Hardware Requirements:** the features available on Raspberry Pi closely align with the essential requirements of typical EVSE hardware, as illustrated in Fig. 1. The input/output, memory, storage, compute and communication interface resources make the Raspberry Pi a suitable hardware platform for this task. This strategic alignment ensures that our lab setup accurately mirrors the characteristics of the main board in real-world EVSE, fostering a more representative and effective testing environment.
– **Proven Implementation in Existing EV Charging Stations:** Raspberry Pi hardware has a track record of successful implementation in various commercial EVSEs[2]
– **Accessibility and Replicability:** the widespread availability and ease of acquisition of Raspberry Pis enhance the accessibility of our research methodology. This deliberate choice ensures that other cybersecurity researchers can readily replicate our setup for their investigations.

---

[2] https://wallbox.com/en_ca/wallbox-delivers-a-secure-ev-charging-experience, https://www.eocharging.com/support/home-charging/eo-mini-pro-2, https://support.hypervolt.co.uk/knowledge-base/quick-start-guide-and-operation-manual.

3.2   Power Consumption Circuit

The power consumed by the SECC or main board of EVSE-B is monitored and recorded to serve as additional data used for baseline profiling of the EVSE. To achieve this, we used the I2C Digital Wattmeter [11] sensor board. This board supports load within 26VDC, 8A ratings. The input of the module is connected in series between the power supply adapter and the EVSE-B. The output of the module is connected to the power monitoring Raspberry Pi through the SDA and SCL pins using the I2C protocol. We used a Python script to log the data from the Wattmenter. The script is composed of two parts:

- **Initialization:** this part of the script initializes the I2C interface of the Raspberry Pi, loads the calibrations for the Wattmeter and creates a CSV file to log the data.
- **Logging Loop:** this section of the script iteratively queries the I2C bus, employing functions such as getBusVoltage_V(), getShuntVoltage_mV(), getCurrent_mA(), and getPower_mW(). The purpose is to retrieve and record data from the Wattmeter, executing this process at a regular sampling interval of 1 s.

3.3   Network Topology

The network packets are captured from the ISO15118 interface of EVSE-B and the OCPP interfaces of EVSE-A and EVSE-B using Wireshark and TCPdump [6] on a Linux computer. The Ethernet connection of the two Access Points, AP1 and AP2 is mirrored [41] to the monitoring and capturing port of the switch as presented in Fig. 3. This allows the network capture tool to gain access to all the network communication traffic traversing the switch, facilitating the capture of the specific traffic of interest. Attack traffic generated by the Attacker and benign traffic communication between any of the EVSE and the CSMS or EVCC can be captured with this topology.

3.4   Experiment Scenarios

Our experiments include scenarios that depict both the normal operational patterns and attack conditions of the EVSE. Under normal operations, the EVSE is allowed to carry out its routine functions in both idle and charging states. For EVSE-B, the status of V2G communication with the EVCC determines its operational state. The benign dataset is captured during normal operational modes in scenarios SNI1 and SNC1, providing valuable information for establishing baseline behavioral profiles of the EVSE. A summary of the various scenarios explored in this experiment is presented in Table 3.

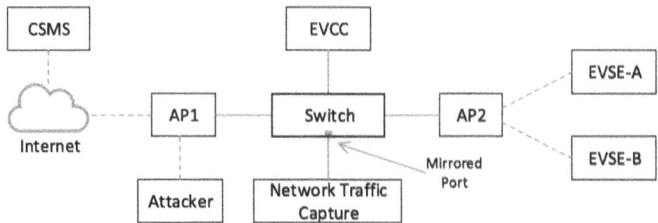

Fig. 3. Topology used for Capturing Network Traffic pcap files. The computer hosting the Wireshark and TCPdump monitoring and capture tools is connected to the mirrored port of the Ethernet switch.

Table 3. Experiment Scenarios

Scenario	EVSE State	Behavior	Description
SNI1	Idle	Benign	
			– No connection to EVCC – OCPP heartbeat communication between EVSE and CSMS
SNC1	Charging	Benign	
			– EVCC Connected to EVSE-B with ISO 15118 V2G Communication – OCPP charging data transmission between EVSE and CSMS
SMI1	Idle	Network Attacks	
			– No conecttion to EVCC – OCPP heartbeat communication between EVSE-A / EVSE-B and CSMS – Scenario for each network attack type describe in Table 4
SMC1	Charging	Network Attacks	
			– EVCC Connected to EVSE-B with ISO 15118 V2G Communication – OCPP charging data transmission between EVSE-B and CSMS – Scenario for each network attack type describe in Table 4
SMI2	Idle	Host Attacks	
			– No connection to EVCC – OCPP heartbeat communication between EVSE-B and CSMS – Scenario for each host-based attack type describe in Table 4
SMC2	Charging	Host Attacks	
			– EVCC Connected to EVSE-B with ISO 15118 V2G Communication – OCPP charging data transmission between EVSE-B and CSMS – Scenario for each host-based attack type describe in Table 4

The attack scenarios execute various host and network based attacks on the EVSE in both the idle and charging states. The details of the executed attacks are presented in Table 4. For network attacks, the communication interfaces of the EVSE are exposed to reconnaissance and Denial of Service (DoS) attacks from a malicious actor with access to the wireless interface of the EVSE. We also implement the case where an already compromised EV launches an attack against the V2G interface of the EVSE. A malicious device equipped with a HomePlug Green PHY modem, legitimate MAC address or digital certificate could act as an EV and pose a threat to any V2G communication enabled EVSE. [13]. Executing various scans during the reconnaissance stage provides useful information on open ports, available services and system information. The malicious actor can proceed to exploit any discovered vulnerabilities to gain privileged access to the EVSE. For the case of EVSE host based attacks, we assume that the worst-case scenario where the EVSE main board has already been compromised through a discovered vulnerability or through privileged access to the diagnosis interface ports of the EVSE [18]. The attacker performs various malicious activities through an established backdoor connection to a Command and Control (C2) server. The backdoor enables the attacker to execute malicious scripts to perform activities such as

- Payload download
- File access permission changes
- File creation, deletion and checking present working directory
- File encryption and decryption

Another host attack scenario is cryptojacking, which refers to the case where an attacker hijacks the computing resource of a victim to mine cryptocurrency [37]. Resource-constrained devices such as IoT are prone to this attack where attackers hijack these devices to download crypto mining malware. Attacker could deploy crypto malware payloads such as Mal/Miner-c [28] to victims devices in a botnet to harvest computing resources into a pool to mine cryptocurrencies. The main controller board or SECC of the EVSE presents a potential target for such an attack due to its constant connection to a power source. Moreover, the auxiliary energy usage by the internal components of the EVSE is often inadequately monitored. In our case, we configure EVSE-B to join a legitimate pool server to mine the Monero[3] cryptocurrency.

4   The CICEVSE2024 Dataset

4.1   Power Consumption Data

The energy consumption data provided by the CSMS and mobile application only indicates the amount of power delivered to the EV during charging [15,24,25]. However, the energy consumed by the internal components of the EVSE, such as the main-board, is usually not reported. As observed in Fig. 4, this data provides a good source of information and, to some extent, an indication of the workload of the main-board of the EVSE. Notably, unlike backdoor and Recon, the current drawn during DoS and cryptojacking attack scenarios are significantly higher, reaching almost double the levels observed in benign conditions in certain cases. Depending on the attack, the percentage deviations of current consumed by the main board of the EVSE present a good Indication of Compromise (IOC) useful for building anomaly detection systems for EV charging stations. By analysing the power consumption data we can determine if the workload deviates from the expected baseline behavior.

---

[3] https://www.getmonero.org/downloads/.

Table 4. Host and Network Attacks

Attack Type	Attack Tool	Interface / Target	Description	MITRE ATT&CK Tactics and Techniques	C	I	A
			Attack			Impact	
TCP Port Scan	nmap	EVSE-A: OCPP EVSE-B: OCPP & ISO15118	Identify open TCP ports on a target system.	Discovery - Service Discovery (T1007) Network Service Discovery (T1046)	●	○	○
Service Version Detection	nmap	EVSE-A: OCPP EVSE-B: OCPP & ISO15118	Determine the software and version running on open ports.	Discovery - Service Discovery (T1007), Software Discovery (T1518)	●	○	○
OS Fingerprinting	nmap	EVSE-A: OCPP EVSE-B: OCPP & ISO15118	Attempt to identify the operating system.	Reconnaissance - Gather Victim Host Information (T1595) Discovery - System Information Discovery (T1082)	●	○	○
Aggressive Scan	nmap	EVSE-A: OCPP EVSE-B: OCPP & ISO15118	Combine various scan types for a comprehensive view of the target.	Reconnaissance - Gather Victim Host Information (T1595) Discovery - Service Discovery (T1007), Collection (T1119)	●	○	○
SYN Stealth Scan	nmap	EVSE-A: OCPP EVSE-B: OCPP & ISO15118	Use SYN packets to identify open ports.	Discovery - Service Discovery (T1007)	●	○	○
Vulnerability Scan	nmap	EVSE-A: OCPP EVSE-B: OCPP & ISO15118	Systematically scanning network interface to identify potentially exploitable security flaws.	Reconnaissance - Gather Victim Host Information (T1595) Discovery - System Information Discovery (T1082) Network Service Discovery (T1046)	●	○	○
Slowloris Scan	nmap	EVSE-A: OCPP	Exploit local webserver on EVSE by keeping multiple connections, exploit limitations on concurrent connections to exhaust its resources and render it unresponsive.	Impact - Endpoint Denial of Service (T1499)	○	○	●
UDP Flood	hping3	EVSE-A: OCPP EVSE-B: OCPP & ISO15118	Sends large volumes of UDP packets to overwhelm targeted network interfaces.	Impact - Network Denial of Service (T1498), Endpoint Denial of Service (T1499).	○	○	●
ICMP Flood	hping3	EVSE-A: OCPP EVSE-B: OCPP & ISO15118	Sends large volumes of ICMP packets to overwhelm targeted network interfaces.	Impact - Network Denial of Service (T1498), Endpoint Denial of Service (T1499)	○	○	●
PSHACK Flood	hping3	EVSE-A: OCPP EVSE-B: OCPP & ISO15118	Flood attack exploiting the TCP PSH flag to overwhelm targeted network interfaces.	Impact - Network Denial of Service (T1498), Endpoint Denial of Service (T1499)	○	○	●

Table 4. continued

Attack Type	Attack Tool	Interface / Target	Description	MITRE ATT&CK Tactics and Techniques	Impact C	I	A
ICMP Fragmentation	hping3	EVSE-A: OCPP EVSE-B: OCPP & ISO15118	Sending fragmented ICMP packets to exploit vulnerabilities in handling fragmented traffic on the target network interface.	Impact - Network Denial of Service (T1498), Endpoint Denial of Service (T1499)	○	○	●
TCP Flood	hping3	EVSE-A: OCPP EVSE-B: OCPP & ISO15118	Overloading target's network interface with a large volume of TCP packets to disrupt services and cause unresponsiveness.	Impact - Network Denial of Service (T1498), Endpoint Denial of Service (T1499)	○	○	●
SYN Flood	hping3	EVSE-A: OCPP EVSE-B: OCPP & ISO15118	Flooding a target's network interface with a high volume of TCP SYN packets, exploiting the three-way handshake process and causing service disruption.	Impact - Network Denial of Service (T1498), Endpoint Denial of Service (T1499)	○	○	●
SynonymousIP Flood	hping3	EVSE-A: OCPP EVSE-B: OCPP & ISO15118	Sending a flood of packets from synonymous IP addresses, potentially evading detection and overwhelming target network interface.	Impact - Network Denial of Service (T1498), Endpoint Denial of Service (T1499)	○	○	●
Cryptojacking	Monero	EVSE-B	Illegitimate use of victim's computing resources to mine cryptocurrency without consent	Impact - Resource Hijacking (T1496)	●	●	●
Backdoor	C2 Server	EVSE-B	Unauthorized remote access to a victim. For our experiment, the C2 server drops and executes malicious scripts on the victim EVSE. Malicious Activities: Payload download, File access permission changes, File encryption and decryption, File creation, deletion and checking PWD.	Execution - Command and Scripting Interpreter (T1059) Command and Control - Content Injection (T1659) Defense Evasion - File and Directory Permissions Modification (T1222), Indicator Remover (T1070) Impact - Data Encrypted for Impact (T1486) Discovery - File and Directory Discovery (T1083)	●	●	●

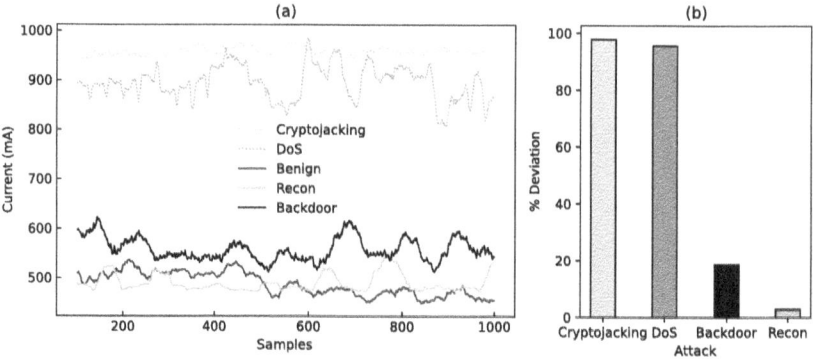

Fig. 4. Wattmeter Current Reading for EVSE-B. (a) presents current reading with MA window size = 20 while (b) shows the percentage deviation of attacks from benign samples.

4.2   HPC and Kernel Events

The EVSE employs resource-constrained or embedded computing hardware as its main board or SECC to execute essential intelligent functions necessary for charging station operation [15]. Conventional host security solutions, such as Anti-Virus software may not be applicable for such resource constrained devices. As an alternative, a viable solution leverages the computing events of the system hardware to establish robust security measures [27]. Modern processors from manufactures such as intel [38], ARM [23] and AMD [12] are equipped with Hardware Performance Counters (HPC), which record low-level microarchitecture events [9].

Table 5. Labelling of the Host Dataset

Column ID	Entry
State	Idle, Charging
Scenario	Recon, DoS, Cryptojacking, Backdoor, Benign
Attack	Cryptojacking, Backdoor, None, TCP-port-scan, service-version-detection, os-fingerpriting, aggressive-scan, syn-stealth-scan, vulnerability-scan, slowloris-scan, upd-flood, icmp-flood, pshack-flood, icmp-fragmentation, TCP-flood, syn-flood, synonymousIP-flood
Label	Attack, Benign
Interface	OCPP, ISO15118

These hardware events prove instrumental in baseline behavioral fingerprinting of devices, classifying malware, and employing machine learning techniques for anomaly detection [8]. We use Perf[4] to gather approximately 900 kernel and HPC events from the Raspberry Pi (EVSE-B) during the experimental scenarios.

Table 5 provides a summary of the host dataset labels for each scenario. Figure 5 illustrates a T-SNE visualization of the samples, presenting an unsupervised classification of samples based on the charging station's state and the various attack scenarios. Notably, T-SNE exhibits its capability to effectively represent this high-dimensional dataset in two dimensions, revealing a clear separation between attack and benign scenarios.

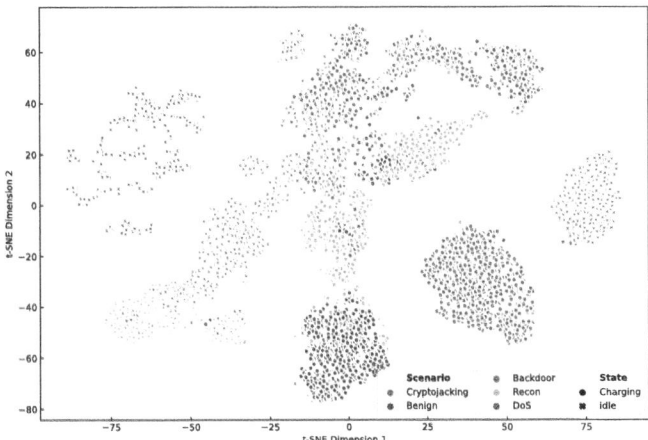

Fig. 5. T-SNE visualisation of Hardware Events Samples Grouped by Scenarios and Charging State of EVSE

Figure 6 depicts a plot illustrating two HPC events and two kernel events recorded for the experiment scenarios. In the case of host-based attacks such as Cryptojacking, there is a notable increase in instruction execution and memory access, reaching up to 800% of the benign scenario. A comparable pattern is observed for network-based attacks. However, reconnaissance attacks exhibit a comparatively lower deviation from the benign scenario, making them potentially challenging to detect using HPC or kernel events.

In Table 6, the mean deviations of some HPC and kernel events have been computed. Generally, the malicious scenarios are observed to exhibit a substantial deviation from benign samples. These events offer valuable insights for crafting lightweight anomaly detection security solutions tailored to resource-constrained devices, such as the EVSE in this context. Rule-based algorithms can be leveraged to formulate IF-THEN rules, providing a foundation for identifying malicious system behavior.

---

[4] https://perf.wiki.kernel.org/index.php/Main_Page.

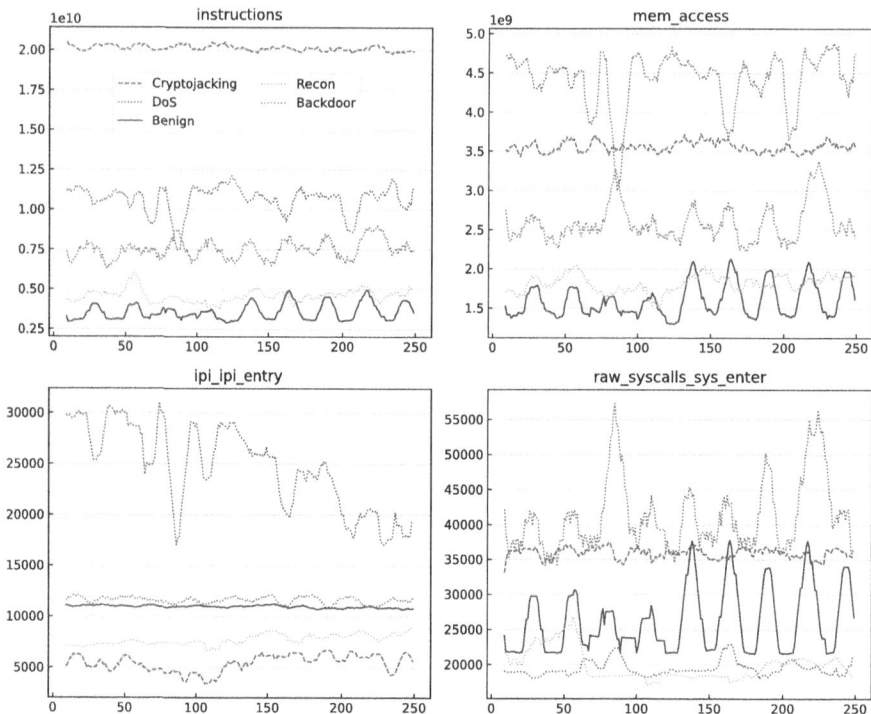

Fig. 6. HPC and Kernel Events under Benign and Attack Conditions. MA window size = 10

5   Machine Learning Experiments

We assess the dataset's suitability for machine learning applications by conducting multi-class and binary classification tasks using eight machine learning algorithms. The network traffic, HPC and kernel event datasets are evaluated independently to facilitate a comparison of algorithm performance on each of these data sources. To mitigate the curse of dimensionality, we employ Principal Component Analysis (PCA) [19] for feature extraction during the pre-processing stage. Subsequently, we use 10-fold cross-validation during the model training and testing stages in an attempt at generalization and to mitigate the impact of model over-fitting. In the binary classification task, the models are trained to classify samples as either benign or attack using these data sources separately. In the multi-class classification task, we employ *Scenario* labels with five classes - Cryptojacking, Backdoor, Recon, DoS, and Benign as presented in Table 5.

The results of the machine learning classification experiments are presented in Fig. 7. Across both binary and multi-class classification tasks, the HPC data source consistently outperforms kernel events in nearly all experiment instances. Notably, the models exhibit superior performance in the binary classification task compared to the multi-class classification task. In the binary classification

Table 6. Selected HPC and Kernel Events: Computed Average Percentage Deviations of Attack Sampled from Benign Samples

Event	Description	Backdoor	Crypto	Dos	Recon
instructions	Number of instructions executed	214.19	943.54	316.65	130.69
cache-misses	Number of cache misses	39.92	397.06	112.83	64.70
exc_taken	Exception taken	86.82	61.42	169.62	44.55
cpu-migrations	CPU Migrations	842.43	−100.00	1425.70	110.05
dTLB-store-misses	Data TLB - Write Misses	39.56	1068.57	101.48	72.97
l1d_cache_wr	Level 1 data cache access - Write	143.87	121.56	321.24	113.14
L1-icache-loads	Level 1 instruction cache access - Read	163.10	538.63	283.92	107.18
l2d_cache_rd	Level 2 data cache - Read	51.47	422.95	151.50	70.75
mem_access_rd	Data memory access - read	118.19	387.56	253.17	91.42
mem_access_wr	Data memory access - write	142.94	119.96	322.17	114.86
kmem_kfree	Kernel memory freeing event	149.06	67.66	2087.07	187.25
net_dev_xmit	Network device transmission event	27.42	33.79	172.88	59.64
qdisc_dequeue	Dequeue event	31.47	42.66	261.16	88.42
raw_syscalls_sys_enter	System call entry (raw) event	107.66	79.70	2.35	44.13
irq_softirq_raise	Software interrupt - Raised	90.29	65.72	1573.65	109.45
sched_migrate_task	Task migration event in the scheduler	670.49	-100.00	1091.74	78.18
sched_switch	Task switch event in the scheduler	42.10	39.22	332.82	24.59
syscalls_sys_enter_close	System call entry for close syscall	421.15	273.72	15.58	231.63
syscalls_sys_enter_read	System call entry for read syscall	44.32	30.19	37.09	56.21
syscalls_sys_enter_write	System call entry for write syscall	119.30	35.73	57.34	76.92

Table 7. Binary Classification Task: Weighted Average Results for HPC and Kernel Events Employing Various Machine Learning Models

Model	HPC				Kernel Events			
	Accuracy	F1-score	Precision	Recall	Accuracy	F1-score	Precision	Recall
Decision Tree	0.901	0.901	0.902	0.901	0.900	0.900	0.901	0.900
KNN	**0.913**	**0.913**	**0.913**	**0.913**	0.915	0.915	0.915	0.915
Adaboost	0.906	0.904	0.904	0.906	0.893	0.891	0.891	0.893
MLP	0.801	0.804	0.811	0.801	0.720	0.723	0.726	0.720
Naive Bayes	0.798	0.807	0.846	0.798	0.760	0.773	0.832	0.760
Logistic Regression	0.721	0.737	0.831	0.721	0.607	0.618	0.825	0.607
Random Forest	0.910	0.910	0.910	0.910	**0.918**	**0.917**	**0.917**	**0.918**
SVM	0.833	0.833	0.834	0.833	0.813	0.793	0.808	0.813

scenario, Decision Tree, KNN, Adaboost, Random Forest, and Support Vector Machine achieved weighted accuracy exceeding 80%. Tables 7 and 8 provide a comprehensive report on the experiments, presenting details such as weighted accuracy, F1-score, Precision, and Recall for all the models. For binary classification, KNN and Random Forest reported the highest accuracy of more than 91% for HPC and kernel events respectively. For multi-class classification, Random Forest reported the highest accuracy of 78.87% and 71.99% for HPC and

Table 8. Multi-class Classification Task: Weighted Average Results for HPC and Kernel Events Employing Various Machine Learning Models

Model	HPC				Kernel Events			
	Accuracy	F1-score	Precision	Recall	Accuracy	F1-score	Precision	Recall
Decision Tree	0.749	0.750	0.750	0.749	0.682	0.681	0.681	0.682
KNN	0.785	0.777	0.773	0.785	0.716	0.711	0.714	0.716
Adaboost	0.583	0.569	0.568	0.583	0.432	0.442	0.502	0.432
MLP	0.580	0.556	0.552	0.580	0.511	0.492	0.490	0.511
Naive Bayes	0.654	0.608	0.573	0.654	0.617	0.577	0.554	0.617
Logistic Regression	0.620	0.573	0.539	0.620	0.534	0.406	0.343	0.534
Random Forest	**0.788**	**0.782**	**0.778**	**0.788**	**0.719**	**0.717**	**0.718**	**0.719**
SVM	0.769	0.714	0.819	0.769	0.642	0.598	0.654	0.642

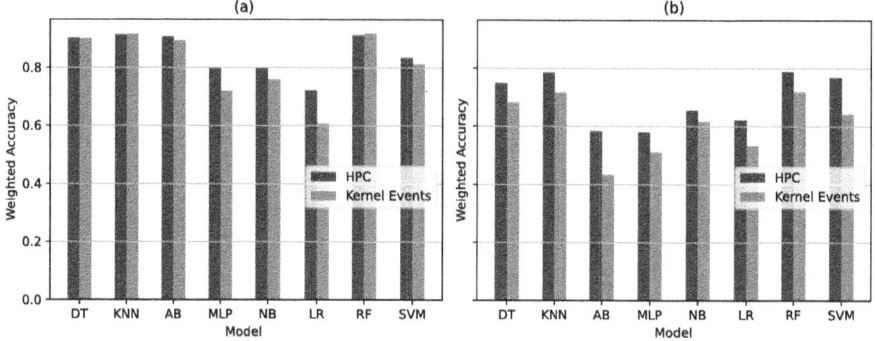

Fig. 7. Weighted Accuracy for HPC and Kernel Events using Different Machine Learning models. The performance of the models for Binary Classification Task is presented in (a) whilst the results for multi-class classification task is presented in (b)

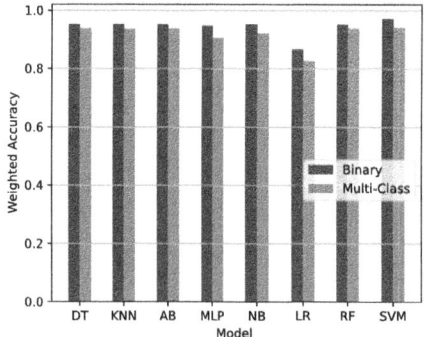

Multi-class scores			
Model	Accuracy	F1-score	Precision	Recall
Decision Tree	0.937	0.947	0.969	0.937
KNN	0.936	0.947	0.968	0.936
Adaboost	0.937	0.948	0.969	0.937
MLP	0.905	0.914	0.932	0.905
Naive Bayes	0.921	0.930	0.950	0.921
Logistic Regression	0.826	0.872	0.942	0.826
Random Forest	0.937	**0.947**	**0.969**	0.937
SVM	**0.941**	0.927	0.915	**0.941**

Fig. 8. Weighted Scores for Network Traffic Dataset Binary and Multi-class Classification using Different Machine Learning models.

kernel events respectively. The classification results for network traffic data is presented in Fig. 8. For Multi-class classification results, the Support Vector Machine (SVM) reported the highest accuracy and recall, with both values reaching 94.13%. On the other hand, Random Forest achieves the highest F1-score at 94.78% and Precision at 96.94%.

6   Conclusion

Investigating cybersecurity solutions for EV charging stations is hindered by the lack of a suitable dataset. This work tackles this pivotal challenge by generating a multi-dimensional dataset that captures network traffic, hardware events, and power consumption data of the EV charging station in both normal and anomalous conditions. Our testbed incorporates a real EV charger, Raspberry Pi, and standard industry communication protocols, enabling us to investigate a variety of attack scenarios. We perform 15 network-based attack scenarios, which include variants of reconnaissance and Denial of Service (DoS), as well as host-based attacks featuring backdoor and cryptojacking. To assess the dataset's suitability for cybersecurity research, we conduct statistical analysis and also, evaluate the use of various machine learning models for classification tasks. There is a notable difference in the power consumption of the main board of EVSE under attack conditions, with the least variation occurring during reconnaissance attacks and the most significant variation seen during cryptojacking attacks. The outcomes of machine learning classification tasks applied to network traffic and host data demonstrate the capability to distinguish between benign activities and various attack scenarios. The dataset is available publicly for download on the CIC Dataset website[5]. Future efforts could focus on expanding this dataset to include scenarios aimed at exploitation of weaknesses in protocols such as ISO15118 and OCPP, along with attacks on underlying technologies like Power Line Communication (PLC). This study lays the groundwork for research endeavors aimed at developing robust and effective security solutions for electric vehicle charging infrastructure.

References

1. Alam, M., Sinha, S., Bhattacharya, S., Dutta, S., Mukhopadhyay, D., Chattopadhyay, A.: Rapper: ransomware prevention via performance counters. arXiv preprint arXiv:2004.01712 (2020)
2. Alcaraz, C., Lopez, J., Wolthusen, S.: OCPP protocol: security threats and challenges. IEEE Trans. Smart Grid **8**(5), 2452–2459 (2017)
3. Antoun, J., Kabir, M.E., Moussa, B., Atallah, R., Assi, C.: A detailed security assessment of the EV charging ecosystem. IEEE Network **34**(3), 200–207 (2020)
4. Antoun, J., Kabir, M.E., Moussa, B., Atallah, R., Assi, C.: Impact analysis of level 2 EV chargers on residential power distribution grids. In: 2020 IEEE 14th International Conference on Compatibility, Power Electronics and Power Engineering (CPE-POWERENG), vol. 1, pp. 523–529. IEEE (2020)

---

[5] https://www.unb.ca/cic/datasets/evse-dataset-2024.html.

5. Asensio, O.I., Apablaza, C.Z., Lawson, M.C., Walsh, S.E.: Replication Data for: A Field Experiment on Workplace Norms and Electric Vehicle Charging Etiquette (2020). https://doi.org/10.7910/DVN/NFPQLW

6. Banerjee, U., Vashishtha, A., Saxena, M.: Evaluation of the capabilities of wireshark as a tool for intrusion detection. Int. J. Comput. Appl. **6**, 1–5 (2010). https://doi.org/10.5120/1092-1427

7. Boudina, R., Wang, J., Benbouzid, M., Yao, G., Zhou, L.: A review on stochastic approach for PHEV integration control in a distribution system with an optimized battery power demand model. Electronics **9**, 139 (2020). https://doi.org/10.3390/electronics9010139

8. Celdrán, A.H., Sánchez, P.M.S., Castillo, M.A., Bovet, G., Pérez, G.M., Stiller, B.: Intelligent and behavioral-based detection of malware in IoT spectrum sensors. Int. J. Inf. Secur. **22**(3), 541–561 (2023)

9. Chenet, C.P., Savino, A., Di Carlo, S.: A survey of hardware-based malware detection approach. arXiv preprint arXiv:2303.12525 (2023)

10. Develder, C., Sadeghianpourhamami, N., Strobbe, M., Refa, N.: Quantifying flexibility in EV charging as DR potential: analysis of two real-world data sets. In: 2016 IEEE International Conference on Smart Grid Communications (SmartGridComm), pp. 600–605 (2016). https://doi.org/10.1109/SmartGridComm.2016.7778827

11. DFRobot. https://wiki.dfrobot.com/gravity_i2c_digital_wattmeter_sku_sen0291. Accessed 2 Dec 2023

12. Dongarra, J., London, K., Moore, S., Mucci, P., Terpstra, D.: Using PAPI for hardware performance monitoring on Linux systems. In: Conference on Linux Clusters: The HPC Revolution, vol. 5. Linux Clusters Institute (2001)

13. EEPOWER. https://eepower.com/market-insights/new-tool-searches-for-ev-charging-security-vulnerabilities. Accessed 27 Nov 2023

14. Hamdare, S., et al.: Cybersecurity risk analysis of electric vehicles charging stations. Sensors **23**(15), 6716 (2023)

15. Harper, J.D., Jun, M., Bohn, T., Meintz, A., Bennett, J.: EVS@ scale deep dive-SCM/VGI (day 2: SCM/VGI demonstration). Technical report, National Renewable Energy Lab (NREL), Golden, CO (United States) (2022)

16. Hawkins, T., Gausen, O., Strømman, A.: Environmental impacts of hybrid and electric vehicles-a review. Int. J. Life Cycle Assess. **17**, 997–1014 (2012). https://doi.org/10.1007/s11367-012-0440-9

17. Pecan Street Inc. https://www.pecanstreet.org/dataport. Accessed 30 Nov 2023

18. Johnson, J., et al.: Cybersecurity for electric vehicle charging infrastructure. Technical report, Sandia National Lab (SNL-NM), Albuquerque, NM (United States) (2022)

19. Jolliffe, I.T.: Principal Component Analysis. Springer Series in Statistics. Springer, New York (1986). https://doi.org/10.1007/978-1-4757-1904-8

20. Kim, Y., Hakak, S., Ghorbani, A.: DDoS attack dataset (CICEV2023) against EV authentication in charging infrastructure. In: 2023 20th Annual International Conference on Privacy, Security and Trust (PST), pp. 1–9 (2023). https://doi.org/10.1109/PST58708.2023.10320202

21. Lee, Z.J., Li, T., Low, S.H.: ACN-data: analysis and applications of an open EV charging dataset. In: Proceedings of the Tenth International Conference on Future Energy Systems. e-Energy 2019 (2019)

22. Mankar, J., Darode, C., Trivedi, K., Kanoje, M., Shahare, P.: Review of I2C protocol. Int. J. Res. Advent Technol. **2**(1) (2014)

23. Mathew, D., Jose, B.A., Mathew, J., Patra, P.: Enabling hardware performance counters for microkernel-based virtualization on embedded systems. IEEE Access **8**, 110550–110564 (2020)

24. Nguyen, Q.H.: Evaluation and development of the bridging application between ISO 15118 and OCPP 2.0.1 protocols. masterthesis, Rheinland-Pfälzische Technische Universität Kaiserslautern-Landau (2023). https://nbn-resolving.de/urn:nbn:de:hbz:386-kluedo-73253

25. Orcioni, S., Conti, M.: EV smart charging with advance reservation extension to the OCPP standard. Energies **13**(12) (2020). https://doi.org/10.3390/en13123263. https://www.mdpi.com/1996-1073/13/12/3263

26. Ott, K., Mahapatra, R.: Hardware performance counters for embedded software anomaly detection. In: 2018 IEEE 16th International Conference on Dependable, Autonomic and Secure Computing, 16th International Conference on Pervasive Intelligence and Computing, 4th International Conference on Big Data Intelligence and Computing and Cyber Science and Technology Congress (DASC/PiCom/DataCom/CyberSciTech), pp. 528–535. IEEE (2018)

27. Patel, N., Sasan, A., Homayoun, H.: Analyzing hardware based malware detectors. In: Proceedings of the 54th Annual Design Automation Conference 2017, pp. 1–6 (2017)

28. Perez, J.: Analysis and detection of the silent thieves. Ph.D. thesis, Utica College (2018)

29. Pevec, D., Babic, J., Carvalho, A., Ghiassi-Farrokhfal, Y., Ketter, W., Podobnik, V.: A survey-based assessment of how existing and potential electric vehicle owners perceive range anxiety. J. Clean. Prod. **276**, 122779 (2020). https://doi.org/10.1016/j.jclepro.2020.122779

30. Pourmirza, Z., Walker, S.: Electric vehicle charging station: cyber security challenges and perspective. In: 2021 IEEE 9th International Conference on Smart Energy Grid Engineering (SEGE), pp. 111–116. IEEE (2021)

31. Saadat, S., Maingot, S., Bahizad, S.: Electric vehicle charging station security enhancement measures. In: 2020 5th IEEE Workshop on the Electronic Grid (eGRID), pp. 1–8 (2020). https://doi.org/10.1109/eGRID48559.2020.9330666

32. Saadat, S., Maingot, S., Bahizad, S.: Electric vehicle charging station security enhancement measures. In: 2020 5th IEEE Workshop on the Electronic Grid (eGRID), pp. 1–8. IEEE (2020)

33. Sánchez, P.M.S., Valero, J.M.J., Celdrán, A.H., Bovet, G., Pérez, M.G., Pérez, G.M.: A survey on device behavior fingerprinting: data sources, techniques, application scenarios, and datasets. IEEE Commun. Surv. Tutor. **23**(2), 1048–1077 (2021)

34. Sarieddine, K., Sayed, M.A., Torabi, S., Atallah, R., Assi, C.: Investigating the security of EV charging mobile applications as an attack surface. ACM Trans. Cyber-Phys. Syst. **7**(4), 1–28 (2023)

35. Sayadi, H., et al.: Towards accurate run-time hardware-assisted stealthy malware detection: a lightweight, yet effective time series CNN-based approach. Cryptography **5**(4), 28 (2021)

36. Schwarzer, V., Ghorbani, R.: Current state-of-the-art of EV chargers. EVTC Electric Vehicle Transportation Centre, p. 169 (2015)

37. Varlioglu, S., Elsayed, N., ElSayed, Z., Ozer, M.: The dangerous combo: fileless malware and cryptojacking. In: SoutheastCon 2022, pp. 125–132 (2022). https://doi.org/10.1109/SoutheastCon48659.2022.9764043

38. Vogl, S., Eckert, C.: Using hardware performance events for instruction-level monitoring on the X86 architecture. In: Proceedings of the 2012 European Workshop on System Security EuroSec, vol. 12. Citeseer (2012)
39. Woo, L.L., Zwolinski, M., Halak, B.: Early detection of system-level anomalous behaviour using hardware performance counters. In: 2018 Design, Automation & Test in Europe Conference & Exhibition (DATE), pp. 485–490. IEEE (2018)
40. Xydas, E., Marmaras, C., Cipcigan, L., Jenkins, N., Carroll, S., Barker, M.: A data-driven approach for characterising the charging demand of electric vehicles: a UK case study. Appl. Energy **162**, 763–771 (2016). https://doi.org/10.1016/j.apenergy.2015.10.151
41. Zhang, J., Moore, A.W.: Traffic trace artifacts due to monitoring via port mirroring. In: 2007 Workshop on End-to-End Monitoring Techniques and Services (2007). https://doi.org/10.1109/e2emon.2007.375317

Optimal Automated Generation
of Playbooks

Kéren A. Saint-Hilaire[1,2(✉)] ⓘ, Christopher Neal[1,2] ⓘ, Frédéric Cuppens[1] ⓘ,
Nora Boulahia-Cuppens[1] ⓘ, and Makhlouf Hadji[2] ⓘ

[1] Polytechnique Montréal, Montréal, Canada
{christopher.neal,frederic.cuppens,nora.boulahia-cuppens}@polymtl.ca,
christopher.neal@irt-systemx.fr
[2] IRT SystemX, Palaiseau, France
keren-a.saint-hilaire@polymtl.ca,
{keren.saint-hilaire,makhlouf.hadji}@irt-systemx.fr

Abstract. Cyberattacks have become more complex and analysts need
help managing all alerts promptly. Many organizations implement Security, Orchestration, Automation, and Response (SOAR) tools to automate Incident Response (IR). However, it is challenging to integrate
these tools, often delaying expected Return on Investment (ROI). Our
approach aims to automatically generate optimal playbooks using the
Pareto front, which balances impact, loss, and complexity. These playbooks are populated in an ontology that aims to be integrated with a
SOAR to overcome the SOAR limitations. Using the Pareto Front, we
aim to reduce the generated playbooks by an average of over 75%.

Keywords: Network security · Playbooks · Security management

1 Introduction

The global increase in cyberattacks concerns all organizations using IT components. Organizations monitor network traffic and information systems to identify
potential attacks and trigger security alerts for investigation. However, the Computer Security Incident Response Teams (CSRIT) are overcome by the volume
of incidents they must analyze manually. Analysts face too many alerts, leading
to delayed response or desensitization to the alerts.

To ease analysts' work, Security, Orchestration, Automation, and Response
(SOAR) systems help by ingesting data from sensors and Security Information and Event Management (SIEM) systems to automate responses. However,
SOAR integration is complex and time-consuming, requiring the integration of
all security tools. It also demands severe financial investment, such as buying
tools and engaging specialized personnel. Therefore, organizations often see a
delayed Return on Investment (ROI) and only automate small tasks.

There is a lack of research concerning the SOAR, while no standards exist for
its interoperability. Playbooks, which define how to execute automated actions,

ⓒ IFIP International Federation for Information Processing 2024
Published by Springer Nature Switzerland AG 2024
A. L. Ferrara and R. Krishnan (Eds.): DBSec 2024, LNCS 14901, pp. 191–199, 2024.
https://doi.org/10.1007/978-3-031-65172-4_12

vary in format and lack formalization, so responses are not uniform for the analyst. We propose using an ontology to formalize Incident Response (IR) playbooks to address SOAR limitations. To enhance interoperability, we propose automatically generating optimal playbooks, balancing conflicting objectives, and integrating them into an IR playbook ontology for use in any SOAR system.

We base our approach on countermeasures selected from matching the MITRE D3FEND KG[1] and the Vulnerability Description Ontology (VDO) KG [1]. The playbooks are automatically generated, matching the selected countermeasures with the IR actions from the RE&CT framework[2].

We identify only the optimal playbooks thanks to the Pareto front [12] and choose from any of the optimal playbooks. We populate the IR playbook ontology with instances for the selected playbook, then validate the approach through a use-case scenario. We evaluate the approach by comparing the gap of playbooks generated before and after applying Pareto and based on the time cost.

2   Related Work

The National Institute of Standards and Technology (NIST) describes IR as an action plan an organization can follow to contain an attack and recover from it [2]. According to NIST, the lifecycle of an IR includes the steps of Preparation, Detection & Analysis, Containment, Eradication & Recovery, and Post-incident activities.

The RE&CT framework inspired by MITRE ATT&CK[3] is a knowledge base of attack techniques and tactics that allows for categorizing IR techniques and actions. However, dependencies and automation amongst actions are not explained within RE&CT. Therefore, we define playbooks using our IR ontology and provide more information for executing actions. Oasis's Collaborative Automated Course of Action Operations (CACAO) playbook schema offers structured and standardized playbooks with logical steps and workflow definitions, enhancing automation potential [7] that we use in our approach.

Automated playbook generation varies. Empl et al. [3] extract CACAO playbooks from CSAF documents, but limitations arise when documentation lacks asset information. We propose a knowledge-based approach to fill this gap. Sarda et al. [8] use Large Language Models (LLMs) for anomaly detection and remediation. Islam et al. [6] automate security tool integration with NLP models and an ontology. Mern et al. [9] propose an Automated Cyber Security Orchestrator (ACSO) using Reinforcement Learning (RL), which can be time-consuming. Our approach generates playbooks by aligning knowledge bases to populate a playbook ontology for SOAR integration.

Moreira et al. [10] advocate using an ontology for security incident management, defining critical concepts like action and event, also found in the Unified

---

[1] https://d3fend.mitre.org/.
[2] https://atc-project.github.io/atc-react/.
[3] https://attack.mitre.org/.

Cybersecurity Ontology (UCO) [11]. Hutschenreuter et al. [5] apply ontologies to port infrastructure resilience, introducing concepts like threat and security measures. While these ontologies offer relevant concepts, adaptation is needed. We employ an ontology to formalize IR playbooks for SOAR integration. We reuse some concepts proposed in [5,10] in our ontology.

Ganin et al. [4] introduce a multicriteria framework for risk management. Our approach similarly considers multiple criteria, utilizing the Pareto front to select dominant playbooks based on impact, complexity, and loss. We choose these parameters based on knowledge from the literature review. We also rationally fix their values based on state-of-the-art knowledge. The Pareto front prioritizes dominant playbooks regardless of specific criteria values, ensuring fairness in selection.

3   Our Approach

3.1   The Proposed Approach

Figure 1 illustrates our optimal IR playbook generation process. Countermeasures are selected through KG matching, obtaining a list categorized by the RE&CT framework's IR phases. Our IR playbook ontology provides attributes for the IR actions, such as required tools, complexity, loss, and impact values. The list of actions is pruned, matching these attributes with system and attack details. All possible playbooks are generated from the refined list, ensuring each playbook exceeds a set impact threshold. The Pareto front allows the identification of the optimal playbooks and the selection of one to create ontological individuals for our playbook ontology.

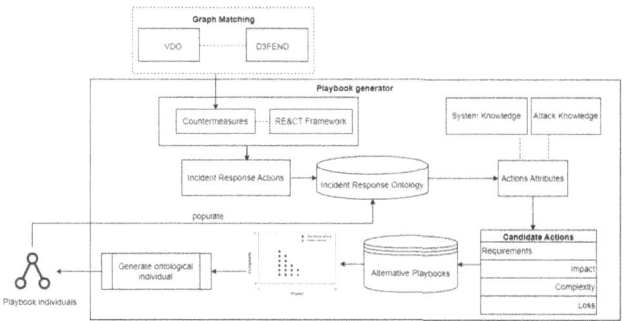

Fig. 1. Our Optimal Playbook Generation Process

3.2 IR Playbook Ontology

This section describes our IR ontology[4] represented by Fig. 2. We reuse some classes from existing ontologies. However, they are insufficient for modeling the playbook concept. We create new classes to represent better and formalize the domain. The class **Playbook** is a super-class of **Playbook Step**, identifying the successive steps corresponding to IR actions; each individual of the playbook step class defines the action's position in the playbook. We take the class **Action** from the UCO [11] ontology. The classes **Tool** and **Security Tool**, taken from the existing ontologies UCO and SecOrp [6], represent assets involved in action completion. The class **Requirement**, created based on the requirements linked to each action from RE&CT, is equivalent with **Security Tool**, which is a subclass of **Tool**. Data represented by the class **Artifact** can also release actions. These actions released allow us to respond to attack tactics and techniques used by an attacker as described in the MITRE ATT&CK matrix; we represent these concepts by the class **ATTACK Thing**, which has the subclasses **Offensive Tactic** and **Offensive Technique**. The execution of an action requires releasing commands. Each **Command** individual is linked to a security tool and an action.

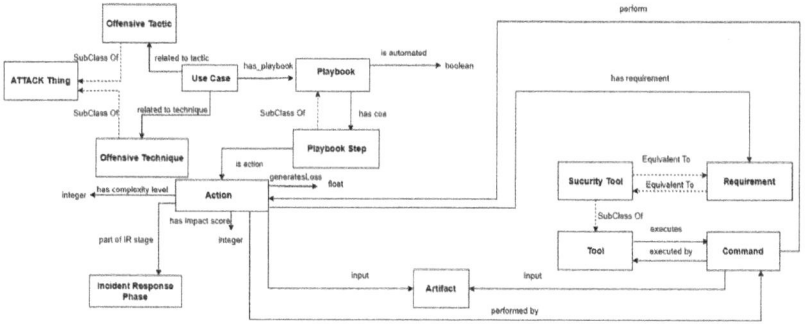

Fig. 2. Our IR Playbook Ontology.

The class **Use Case** concerns a specific use case monitored by an Intrusion Detection System (IDS). We also create our ontology properties representing the relationship between the individuals of the different classes. Each containment action is linked to a recovery action thanks to the property **linked to recovery action**. There is no concept in an ontology that allows different action sequences. We then create the following properties. The property **has coa** is directed to the first step of a playbook (*coa* signifies *Course of Action*). The property **has next** determines which playbook step should follow in the playbook workflow. Properties **has next if false** and **has next if true** allow representing the order of the actions when a condition exists. The property **parallel** expresses the fact that two actions can be executed at the same time.

---

[4] https://github.com/phDimplKS/playbook_ontology.

We populate our ontology with individuals created from two existing security knowledge bases: 1) the MITRE ATT&CK framework constitutes a base for representing attack concepts, and 2) the RE&CT framework is the base for expressing IR. We instantiate the sub-classes **Offensive Technique** and **Offensive Tactic** with knowledge from MITRE ATT&CK. We do this population process semi-automatically. We instantiate the class **Incident Response Phase** with the phases defined in RE&CT; the RE&CT actions allow instantiating the class **Action**. For the **Requirement** class, RE&CT proposes requirements for some actions, so we populate the ontology with them. However, no requirement is proposed for some actions, so we should consult security blogs to find their requirements and populate the ontology. We populate the classes **Playbook Step** and **Playbook** with the individuals created for the optimal playbooks selected.

3.3   Countermeasures Selection

This section details the automated process of selecting countermeasures for an exploited vulnerability using KG matching. The dashed part in Fig. 1 illustrates the graph-matching process, with VDO KG and D3FEND KG as inputs. After parsing the KGs and conducting text processing to create corpora, Word2Vec models are trained, and KGs are embedded. Cosine similarity calculation matches attack impacts and attacker methods entities with their most similar offensive techniques in D3FEND. A query on D3FEND retrieves artifact entities linked to the matched offensive techniques.

When no artifact is linked to the offensive techniques, cosine similarity helps find the most similar artifact entity of D3FEND to its corresponding entity in VDO. Finally, our solution automatically queries the defensive techniques linked to this artifact from D3FEND, the countermeasures against the vulnerability exploited. Each countermeasure is automatically linked with an IR phase and a category of the RE&CT framework. The candidate countermeasures selected serve as input to our optimal playbook generation solution.

3.4   Optimal Playbooks Generation

As Fig. 1 shows, the playbook generator first generates a list of IR actions, correlating the countermeasures with the RE&CT framework. Our solution queries the action attributes for its release. The playbook generator verifies that the system contains at least one required security tool for the action execution. In addition, it verifies the needed attacker's position to release the attack. For a remote attack, it discards the action *block internal IP address*.

The playbook generator generates different combinations of actions considering the following conditions. We fix a threshold value that the sum of all actions' impact values must exceed. A playbook should contain at least 1 action from two different phases: Identification, Preparation, Containment, and Eradication phases, considering that an action of the identification phase is necessary to detect and analyze an anomaly, and then at least one action from the containment or eradication phase should be taken to mitigate the threat detected.

We fix a maximum number of actions in a playbook to reduce the execution time required to generate a playbook. Based on our experiments presented in Sect. 4.1, we notice that we get an average of more than 20 alternative playbooks for a vulnerability. In order to choose the optimal one, we use the concept of Pareto optimality in our process. The optimality objectives are: **a)** maximize the impact, **b)** minimize the complexity, and **c)** minimize the loss. All the playbooks that have no other playbooks dominating them are Pareto optimal.

The recovery action for each containment action in the optimal playbooks and the lessons learned phase's actions are added automatically to these playbooks. One of the optimal playbooks can be selected randomly, since they are all equally optimal. Our approach automatically generates individuals for the chosen playbook to populate our playbook ontology. An instance for the playbook class is created, linked with the first playbook step through the **has coa** property. Playbook steps are associated using the **has next** and **parallel** properties, with each step linked to a playbook action.

4   Results and Evaluation

4.1   Illustrative Example

The chosen scenario to explain the application of our approach implementation[5] includes the exploitation of CVE-2021-21277. The remote exploitation of this vulnerability may lead to sensitive information discovery and privilege escalation, allowing the adversary to exploit other vulnerabilities in the system. Figure 3 sketches the network environment. The system includes a vulnerable web server, a terminal with antivirus and firewall software, a SIEM, and an IDS.

Our solution gets the candidate IR actions, matching RE&CT with selected countermeasures. A query retrieves their attributes from the playbook ontology. Considering system software and attacker position, only 14 actions are used in playbook generation. Table 1 lists some of these actions and their parameters.

Playbooks are generated based on the constraints from Sect. 3.1 for 2 conditions. For the first condition, because of an organization's limited resources, each phase can only be repeated once; this condition results in 28 playbooks. For the second condition, considering that an organization may have more resources available, phases can be repeated multiple times; however, each category can be repeated once per phase. This condition leads to 214 playbooks.

We implement a Pareto front algorithm to select one optimal playbook. The playbooks dominating others and not dominated by any playbook constitute the Pareto front; they are optimal. For the first condition, the algorithm allows us to get 2 optimal playbooks. For the second condition, we get 6 optimal playbooks. In both cases, the solution randomly selects 1 of the optimal playbooks.

---

[5] https://github.com/phDimplKS/play.

Figure 4 represents the optimal playbook selected for the second. In this playbook, 2 identification and 2 containment actions can be executed in parallel; their execution requires different tools. However, the action *Patch vulnerability* has higher priority than the action *Block process by executable format* because it has a higher impact value. Executing actions follows a logical workflow; containment actions follow identification actions. Then, the recovery action *Unblock blocked process* follows. Afterward, the playbook ontology is populated with instances of the selected playbook.

Table 1. List of actions with their parameters' values for CVE-2021-21277

Defensive Technique	Action	Impact	Complexity	Loss	Phase	Category
Process Segment Execution Prevention	find process by executable format	5	3	0.3	Identification	Process
Asset Vulnerability Enumeration	list victims of security alert	4	4	0.2	Identification	General
Process Segment Execution Prevention	list processes executed	4	3	0.2	Identification	Process
Asset Vulnerability Enumeration	put compromised acounts on monitoring	3	7	0.7	Identification	General
Software Update	patch vulnerability	4	3	0.0	Containment	General
Segment Address Offset Randomization	block process by executable format	4	3	0.0	Containment	Process

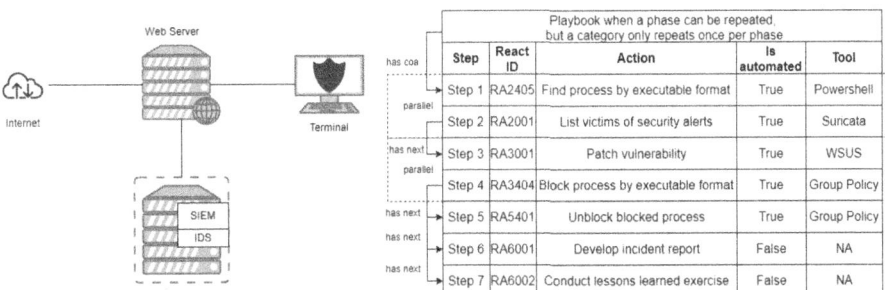

Fig. 3. Example System   **Fig. 4.** Optimal Playbook Generated

4.2   Evaluation

We evaluate our approach over 40 CVEs for the 2 different conditions: **First condition** considering that each phase can only be repeated once in a playbook and **Second condition** considering that a phase can be repeated, but a category repeats only once per phase. We base our evaluation on the following metrics:

- The percentage gap between the number of playbooks generated before n and after m applying Pareto optimality using the formula: $(n - m)/n * 100$.
- The execution time of our optimal playbook generation process in seconds.

Table 2 compares the average metrics for both conditions. The Pareto front reduces generated playbooks by over 75%. Despite higher generation time in the second condition, model performance remains unaffected mainly due to initially higher playbook numbers.

Table 2. Evaluation of our approach

	No Playbooks before Pareto	No Playbooks after Pareto	First metric: Gap in %	Second metric: Time in s
First condition	26.5	2.1	78.59	7.07
Second condition	8417.5	10.175	82.22	13.01

5 Conclusion

Our proposed approach tackles the SOAR integration problem with a scalable and reusable ontology. It is easier to populate an ontology communicating with a SOAR than to manage multiple playbook formats. Thanks to all the knowledge bases involved in our approach, the actions of a playbook compose a logical workflow ensuring optimal responses. An optimal playbook is generated each time we launch our system for the same attack scenario on an unchanged system. In the future, we will focus on instantiating the playbook actions on an AG.

References

1. Booth, H., Turner, C.: Vulnerability description ontology (VDO): a framework for characterizing vulnerabilities. Technical report, NIST
2. Cichonski, P., Millar, T., Grance, T., Scarfone, K., et al.: Computer security incident handling guide. NIST Spec. Publ. **800**(61), 1–147 (2012)
3. Empl, P., Schlette, D., Stöger, L., Pernul, G.: Generating ICS vulnerability playbooks with open standards. Int. J. Inf. Secur. **23**, 1215–1230 (2023)
4. Ganin, A.A., Quach, P., Panwar, M., Collier, Z.A., Keisler, J.M., Marchese, D., Linkov, I.: Multicriteria decision framework for cybersecurity risk assessment and management. Risk Anal. **40**(1), 183–199 (2020)
5. Hutschenreuter, H., Çakmakçi, S.D., Maeder, C., Kemmerich, T.: Ontology-based cybersecurity and resilience framework. In: ICISSP, pp. 458–466 (2021)
6. Islam, C., Babar, M.A., Nepal, S.: Automated interpretation and integration of security tools using semantic knowledge. In: Giorgini, P., Weber, B. (eds.) CAiSE 2019. LNCS, vol. 11483, pp. 513–528. Springer, Cham (2019). https://doi.org/10.1007/978-3-030-21290-2_32

7. Jordan, B., Thomson, A.: Cacao security playbooks version 1.0. OASIS. Committee Specificat. **2** (2021)
8. Komal, S., et al.: Adarma auto-detection and auto-remediation of microservice anomalies by leveraging large language models. In: Proceedings of the 33rd Annual International Conference on Computer Science and Software Engineering, pp. 200–205 (2023)
9. Mern, J., Hatch, K., Silva, R., Hickert, C., Sookoor, T., Kochenderfer, M.J.: Autonomous attack mitigation for industrial control systems. In: 2022 52nd Annual IEEE/IFIP International Conference on Dependable Systems and Networks Workshops (DSN-W), pp. 28–36. IEEE (2022)
10. Moreira, G.B., Calegario, V.M., Duarte, J.C., dos Santos, A.F.P.: Csiho: an ontology for computer security incident handling. In: Anais do XVIII Simpósio Brasileiro de Segurança da Informação e de Sistemas Computacionais, pp. 1–14. SBC (2018)
11. Syed, Z., Padia, A., Finin, T., Mathews, L., Joshi, A.: UCO: a unified cybersecurity ontology. UMBC Student Collection (2016)
12. Van Veldhuizen, D.A., Lamont, G.B., et al.: Evolutionary computation and convergence to a pareto front. In: Late Breaking Papers at the Genetic Programming 1998 Conference, pp. 221–228. Citeseer (1998)

ML Attack, Vulnerability

ALERT: A Framework for Efficient Extraction of Attack Techniques from Cyber Threat Intelligence Reports Using Active Learning

Fariha Ishrat Rahman[1(⊠)], Sadaf Md Halim[1], Anoop Singhal[2], and Latifur Khan[1]

[1] The University of Texas at Dallas, Richardson, USA
{farihaishrat.rahman,sadafmd.halim,lkhan}@utdallas.edu
[2] National Institute of Standards and Technology, Gaithersburg, USA
anoop.singhal@nist.gov

Abstract. In the dynamic landscape of cybersecurity, curated knowledge plays a pivotal role in empowering security analysts to respond effectively to cyber threats. Cyber Threat Intelligence (CTI) reports offer valuable insights into adversary behavior, but their length, complexity, and inconsistent structure pose challenges for extracting actionable information. To address this, our research focuses on automating the extraction of attack techniques from CTI reports and mapping them to the MITRE ATT&CK framework. For this task, fine-tuning Large Language Models (LLMs) for downstream sequence classification shows promise due to their ability to comprehend complex natural language. However, fine-tuning LLMs requires vast amounts of annotated domain-specific data, which is costly and time-intensive, relying on the expertise of security professionals. To meet these challenges, we propose ALERT, a novel cybersecurity framework which leverages active learning strategies in conjunction with an LLM. This approach dynamically selects the most informative instances for annotation, thereby achieving comparable performance with a significantly smaller dataset. By prioritizing the annotation of samples that contribute the most to the model's learning, our methodology optimizes the allocation of resources. As a result, our framework achieves comparable performance with a dataset that is 77% smaller, making it more efficient for extracting and mapping attack techniques from CTI reports to the ATT&CK framework.

Keywords: Active Learning · LLM · ATT&CK · CTI

1 Introduction

Cybersecurity is a major challenge in today's world, as cyber criminals use advanced and diverse TTPs (Tactics, Techniques, and Procedures) to evade detection and cause damage. One of the most notorious examples of this is

© IFIP International Federation for Information Processing 2024
Published by Springer Nature Switzerland AG 2024
A. L. Ferrara and R. Krishnan (Eds.): DBSec 2024, LNCS 14901, pp. 203–220, 2024.
https://doi.org/10.1007/978-3-031-65172-4_13

the WannaCry ransomware attack [2], which encrypted the data and systems of around 200,000 computers across over 150 countries and demanded ransom for their release. To help combat these growing threats, security analysts rely on Cyber Threat Intelligence (CTI) reports [28], which provide insights into the behavior of cyber adversaries. These reports contain useful information about CTI artifacts (e.g., IoCs) and attacker behavior (e.g., TTPs). However, applying threat information contained within these resources effectively is still a challenge. The main reason is that CTI reports are often lengthy, complex, and inconsistent in their structure and terminology, which makes it difficult for analysts to extract and use the relevant information.

Various methods have been developed for extracting pertinent information from Cyber Threat Intelligence (CTI) reports. Techniques such as ChainSmith [34] and iACE [17] primarily focus on the identification of Indicators of Compromise (IoCs) using natural language processing (NLP). However, there is a noticeable shift towards higher-level extraction of Tactics, Techniques, and Procedures (TTPs) as they offer greater resilience against changes by adversaries [6], exemplified by methods like TTPDrill [13] and AttacKG [16]. However, despite the progress demonstrated by these approaches, rule-based methods exhibit limitations when dealing with the complexity of CTI reports [3]. Consequently, the field is increasingly turning towards more sophisticated models, such as Large Language Models (LLMs).

Recent research has explored leveraging pre-trained LLMs to extract attack techniques from CTI reports and map them to the ATT&CK framework [3,25]. ATT&CK [22] offers a comprehensive repository of TTPs utilized by real-world attackers. These TTPs are presented in a matrix of tactics and techniques, providing valuable insights for security analysts. Identifying attack behavior is the first step in defense against adversaries. Once ATT&CK techniques are identified, appropriate countermeasures can be taken using tools like D3FEND [20].

For this task, pre-trained LLMs were further fine-tuned by training on domain-specific datasets such as texts from CTI reports. While fine-tuned LLMs demonstrate good performance, their training demands substantial annotated data, which is scarce in the cybersecurity domain. Given the ever-evolving threat landscape, annotating large datasets becomes impractical. Hence, there is a pressing need for more efficient annotation methods in this domain-an area that remains largely unexplored.

Our proposed solution involves integrating active learning strategies with an LLM. By dynamically selecting informative instances for annotation, we achieve comparable performance while significantly reducing the annotation burden. The guiding intuition here is that not all data instances provide equal value to a learner at any given point in time. For instance, a lot of data points may involve repetition or similar data – i.e., they discuss similar content and are therefore labeled similarly. Therefore, if a learner has already encountered similar samples and learned from them, it would be wasteful to utilize annotation resources in order to annotate yet another similar unlabeled data instance. With large volumes of unlabeled threat data arriving continuously, and with annotation requiring a security expert's valuable time, it is crucial that we perform annotation in the most effective way possible. Active Learning helps identify the utility of each potential unlabeled sample with respect to the current learner, using various heuristics. Thus, this approach optimizes the allocation of resources, ensuring timely and accurate extraction of attack techniques from CTI reports.

1.1 Challenges

In this section, we discuss the primary challenges associated with constructing an automated tool for extracting and mapping attack techniques from Cyber Threat Intelligence (CTI) reports to ATT&CK framework.

- **Extracting Attack Techniques from Noisy, Unstructured CTI Reports.** Structured Cyber Threat Intelligence (CTI) reports are essential for efficient automation and information aggregation across diverse sources. However, creating these reports is labor-intensive, leading many security firms like Avast, McAfee, and Kaspersky to present their public threat reports in unstructured formats. These reports, while informative, are often lengthy, written in natural language, and relevant information often becomes obscured by extraneous text. Additionally, inherent ambiguity in the text further complicates the extraction of relevant information. Traditional rule-based systems such as regular expressions, heuristics, and dependency parsing [13,16] are inadequate for this task. Hence, despite the widespread availability of these reports, the extraction of pertinent information remains a challenging task, necessitating the adoption of more sophisticated machine-learning-based approaches, such as Large Language Models (LLMs) [3].
- **Annotation is Expensive and Time-Consuming.** A major hurdle in utilizing LLMs for real-world applications is the need for vast amounts of labeled data specific to the task at hand. Unfortunately, in our domain, creating a comprehensive dataset mapping CTI reports to all ATT&CK techniques is a significant challenge. While MITRE's annotated dataset used in TRAM [23] offers a valuable starting point, it only covers a limited portion – 50 out of the 625 existing techniques. Even for these 50 techniques, acquiring the 11,000 instance used in fine-tuning an LLM demanded extensive human effort. Scaling this annotation process to encompass the remaining 575 techniques is simply impractical. Security experts' time is precious, and the ever-evolving threat landscape constantly adds new attack techniques, rendering large-scale manual annotation inefficient and unsustainable. This is why we propose a method that achieves performance comparable to TRAM, but with a significantly reduced requirement for annotated training data.

1.2 Contribution

We demonstrate the effectiveness of Active Learning as an intelligent approach that optimizes the annotation process for mapping CTI (Cyber Threat Intelligence) reports to the ATT&CK framework. Our approach strategically selects the most informative instances from CTI reports for human annotation, significantly reducing annotation efforts while achieving performance comparable to models trained on large datasets. This translates to **reduced costs** for organizations, **improved efficiency** for security teams, and **greater scalability** as the ATT&CK framework and threat landscape evolve. We provide extensive analysis and provide empirical projections on the annotation-savings that will be made possible for analysts when using the ALERT framework. We also explore additional data augmentation techniques to evaluate its effectiveness. Our contributions include:

1. Development of a novel framework, **ALERT** (**A**ctive **L**earning **E**nhanced **R**obust **T**hreat-Mapper) that leverages Active Learning in conjunction with a Large Language Model for resource-efficient extraction of ATT&CK techniques from CTI reports.
2. Conducting an extensive empirical study to compare the performance of different active learning strategies and evaluating the effectiveness of data augmentation to determine the optimal strategy for our task.
3. Providing projections on how the ALERT framework can reduce analysts' effort by intelligently annotating unlabeled ATT&CK techniques in a real-world dataset.

Furthermore, we are open-sourcing our code on Github (https://github.com/space-urchin/ALERT/tree/main) for further research efforts in this direction and for the benefit of the security community.

2   Preliminaries

2.1   Cyber Threat Intelligence Reports

Combating the ever-evolving threat landscape demands a collaborative approach. Public Cyber Threat Intelligence (CTI) reports [28] serve as an effective tool to counter cyber threats. CTI reports equip security professionals with actionable information, which may include Indicators of Compromise (IoCs) such as malicious IP addresses and domain names. Additionally, they may offer in-depth explorations that provide invaluable insights into the motivations, tactics, techniques, and procedures (TTPs) employed by cyber adversaries, and outline mitigation strategies and best practices for defense. Regularly published by security vendors like Trustwave, SonicWall, and CrowdStrike, CTI reports empower organizations to proactively identify and defend against emerging threats. Figure 1 is an excerpt from a CTI report, "NotPetya Technical Analysis," [7] published by CrowdStrike which describes an attack technique from the ATT&CK framework.

2.2   ATT&CK Framework

ATT&CK [22] (Adversarial Tactics, Techniques, and Common Knowledge) is a framework designed to organize and categorize adversary behavior and strategies

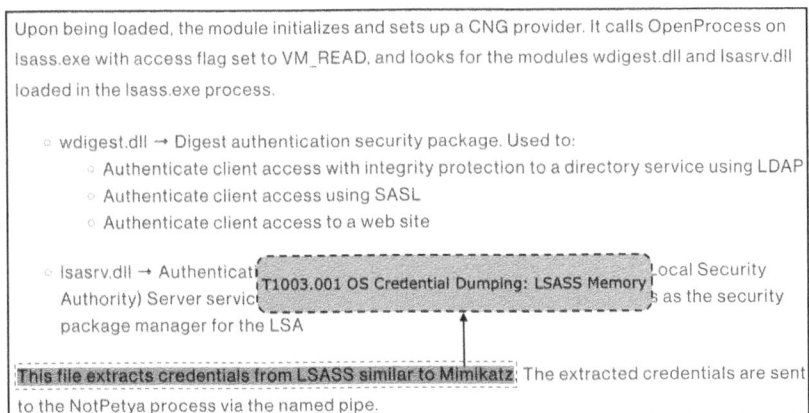

Upon being loaded, the module initializes and sets up a CNG provider. It calls OpenProcess on lsass.exe with access flag set to VM_READ, and looks for the modules wdigest.dll and lsasrv.dll loaded in the lsass.exe process.

- wdigest.dll → Digest authentication security package. Used to:
  - Authenticate client access with integrity protection to a directory service using LDAP
  - Authenticate client access using SASL
  - Authenticate client access to a web site

- lsasrv.dll → Authenticat... T1003.001 OS Credential Dumping: LSASS Memory ...ocal Security Authority) Server servic... ...s as the security package manager for the LSA

This file extracts credentials from LSASS similar to Mimikatz. The extracted credentials are sent to the NotPetya process via the named pipe.

Fig. 1. Excerpt from: NotPetya Technical Analysis - A Triple Threat: File Encryption, MFT Encryption, Credential Theft. The annotation illustrates the mapping of the highlighted line to an attack technique (T1003.001) in the MITRE ATT&CK framework.

observed in real-world cyber attacks. It classifies various tactics, techniques, and sub-techniques employed by threat actors, offering insights into their methodologies. Tactics denote overarching strategies utilized by threat actors to achieve their objective, while techniques are the specific methods or tools they employ to execute these tactics. Techniques can further be broken down into sub-techniques. The ATT&CK matrix organizes these elements, encompassing 14 tactics, 201 techniques, and 424 sub-techniques. In this work, when we refer to techniques, we collectively refer to all 625 techniques and sub-techniques. In Fig. 2, we see the hierarchical view of the attack technique referenced in Fig. 1. Specifically, the sub-technique T1003.001 is categorized under the technique T1003: Credential Dumping, which falls within the broader tactic TA0006: Credential Access.

2.3 Active Learning

We will now outline the problem setting for pool-based Active Learning (AL) [30] described in Algorithm 1. Let $D^{\mathcal{U}}$ represent the pool of unlabeled data, and $D^{labeled}$ denote the set of initially labeled instances. The objective is to iteratively choose a subset $D^{query} \subset D^{\mathcal{U}}$ for annotation by an oracle (in our context, a security expert). The learner is typically allocated a budget $\mathcal{B}$ which limits the maximum number of labels it can query. Samples are typically selected by some measure of the informativeness of the unlabeled points with respect to the current model, utilizing an acquisition function that employs an active learning strategy. The query strategy can therefore be defined as follows:

$$D^{query} = \arg\max_{D \subset D^{\mathcal{U}}} \mathcal{I}(D, f) \qquad (1)$$

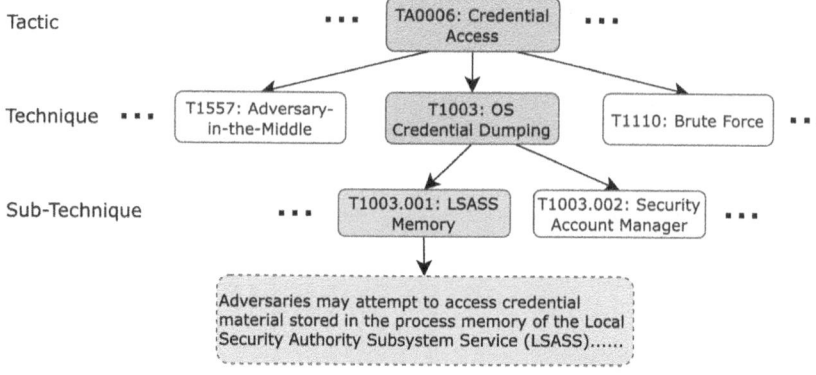

Fig. 2. Hierarchical view of the attack technique T1003.001 from the ATT&CK matrix

where D represents a subset of $D^{\mathcal{U}}$, f is the model trained on the currently labeled data $D^{labeled}$, and $\mathcal{I}(D, f)$ is a measure of informativeness.

Algorithm 1. Pool-based Active Learning

1: **while** $|D^{labeled}| < \mathcal{B}$ **do**
2:     Train the model on $D^{labeled}$.
3:     Evaluate informativeness of each data point in $D^{\mathcal{U}}$ using an acquisition function.
4:     Select D^{query} based on the most informative instances.
5:     Request labels for D^{query} from the oracle.
6:     Update $D^{labeled}$ with the newly acquired labels.
7: **end while**

We explore strategies spanning 3 state-of-the-art areas in Active Learning: Uncertainty Sampling, Expected Model Change, and Diversity Sampling.

Uncertainty Sampling selects instances for which the model is least certain according to a decision rule [14]. In multiclass scenarios, uncertainty can be defined in various ways. We investigate 4 uncertainty sampling based strategies as described below:

- **Top Confidence**: Let $P(y|x)$ represent the model's predicted probability of label y given the instance x. The instance with the smallest probability for its top predicted label is chosen. This can be expressed as:

$$\underset{x}{\operatorname{argmin}} \max_{y}(P(y|x)) \qquad (2)$$

- **Margin Sampling**: The margin is defined as the difference between the highest and second-highest predicted probabilities. In this strategy, we select the instance with the smallest margin [27], defined as:

$$\arg\min_{x} P(y_1|x) - P(y_2|x) \tag{3}$$

where where y_1 and y_2 are the first and second most probable class labels under the model.

- **Maximum Entropy**: It considers the entropy [31] of the label distribution, which reflects the uncertainty of the model's prediction for an instance. Higher values of entropy indicate greater uncertainty in the probability distribution. It is defined as:

$$\underset{x}{argmax} - \sum_{y \in Y} P(y|x) \log(P(y|x)) \tag{4}$$

- **Monte Carlo Dropout**: While dropout is typically applied during training, Gal et al. [10] proposed a method for uncertainty estimation where dropout is applied at test time. For each inference cycle, dropout is applied with a different dropout mask. Each forward pass produces a prediction, so over a number of inference cycles, MC (Monte Carlo) Dropout provides a distribution of predictions rather than a single point estimate. Instance uncertainty is calculated using the average of these predictions using an acquisition function such as max entropy.

Expected Model Change is an approach that aims to select examples expected to induce the most significant changes in the model. For models where gradients can be computed (such as neural networks), one potential method is to select the instance with the highest **expected gradient length (EGL)**, where the expectation is calculated over the probabilities assigned by the model to the labels. This strategy selects instances with the largest expected gradient norm, as they are expected to have a large influence on the model [12,30]. The decision rule is defined as:

$$\underset{x}{argmax} \sum_{y \in Y} P(y|x) \|\nabla l_\theta(\{x, y\})\| \tag{5}$$

where ∇l_θ is the gradient of the objective function l with respect to the model parameters θ, and $\|\nabla l_\theta(\{x, y\})\|$ is the Euclidean norm of the gradient vector for instance $\{x, y\}$.

Diversity Sampling. The goal of this approach is to improve the model's understanding by actively seeking out diverse instances that can provide new information independently of other labeled samples. One way of doing this is to define active learning as a **Core-Set** selection problem [29]. This problem involves selecting a subset from a fully labeled dataset such that a model trained on this subset performs as closely as possible to the model trained on the entire dataset. This is equivalent to a k-Center problem which is solved by a greedy approximation, called **Greedy Core-Set**, as shown in Algorithm 2.

Algorithm 2. Greedy Core-Set

1: Assign each point in $D^{labeled}$ as a cluster center.
2: **while** $|D^{labeled}| < \mathcal{B}$ **do**
3:     Calculate pairwise distance between each data point in $D^{\mathcal{U}}$ and its closest cluster center.
4:     Select the point c^{new} in $D^{\mathcal{U}}$ that is farthest from its cluster center.
5:     Assign c^{new} as a new cluster center.
6:     Add c^{new} to D^{query}.
7: **end while**

---

3 Proposed Approach

We propose ALERT, which utilizes an active learning pipeline for fine-tuning our learner for mapping text to ATT&CK. This is illustrated in Fig. 3. In this section, we discuss the components in detail.

Learner. Mapping an instance from a CTI report to an attack technique in ATT&CK is a sequence classification task suited for Large Language Models, particularly encoder-only architectures such as BERT [8]. Variants of BERT, further fine-tuned on additional datasets [5,18] have also demonstrated strong performance in this task. For our learner, we opt for SciBERT [5], a pre-trained BERT-based model specifically trained on scientific data. SciBERT has shown robust performance in computer science tasks, making it a suitable choice for our purpose. Notably, it is also the model of choice for MITRE's TRAM [25] for their mapping task, facilitating direct comparisons.

Pipeline. We employ pool-based active learning with a batch mode for our experiments. To warm-start our fine-tuning process, we randomly select 1% of the training pool to create our initial labeled set, $D^{labeled}$. In each annotation cycle, we fine-tune our SciBERT model on the labeled dataset, $D^{labeled}$, for 10 epochs. Subsequently, we evaluate the remaining unlabeled set, $D^{\mathcal{U}}$, using the fine-tuned model. To select an informative set of unlabeled samples D^{query}, we apply one of the acquisition functions described in Sect. 2.3.

Since we use deep learning based transformer models like SciBERT, training requires significant computational resources, and thus retraining the model after every new data point addition is highly impractical. To mitigate this, a batch-mode approach is often adopted [9,11], where the model queries for a set of points, instead of a single one, at each iteration. We thus adopt the same batch-mode setting.

In each cycle, we select 10 unlabeled instances for annotation (i.e., $|D^{query}|$ = 10), a popular and effective choice in batch-mode AL [19,26]. These unlabeled samples are then labeled and added to $D^{labeled}$, and the process repeats until the budget, $\mathcal{B}$ is exhausted. Lastly, we evaluate the final model on a test set.

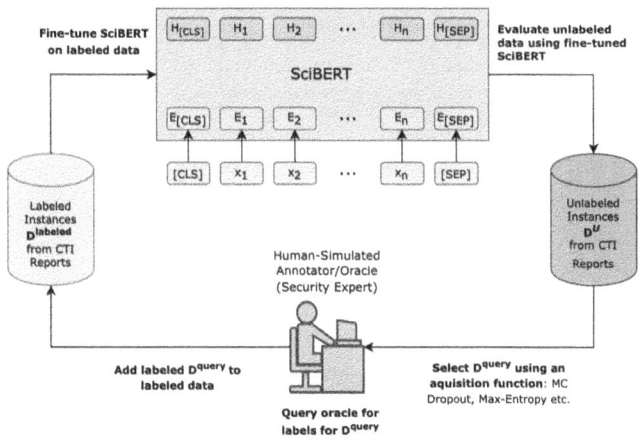

Fig. 3. Pool-based Active Learning Pipeline in ALERT

3.1   Active Learning Strategies

We compared 6 different AL strategies - Top Confidence, Maximum Entropy, Margin Sampling, Monte Carlo Dropout, Approximated Gradient Length and Core-Set.

In **Top Confidence**, **Maximum Entropy**, and **Margin Sampling**, we utilize the logit outputs from the SciBERT model. The logits scores in SciBERT are generated by applying a classifier to the [CLS] (classification) output. The [CLS] vector represents the aggregated sequence representation for classification tasks, and the classifier computes the logits scores based on this representation. By applying softmax to the logit scores, we obtain the probability distribution $P(y|x)$. We then apply the corresponding formula defined in Sect. 2.3 to select the instances.

In **Monte Carlo Dropout**, we employ a method similar to the one described in [9]. We introduce dropout via a perceptron, where the input to the perceptron is the CLS vectors from the fine-tuned SciBERT model. The dropout rate is set to 0.9. Instance uncertainty is then calculated by averaging the softmax probabilities over 10 inference cycles, using the max-entropy acquisition function. This is illustrated in Fig. 4.

In **Approximated Gradient Length**, we adopt an idea similar to that of EGL defined in Sect. 2.3. However, computing the expected gradient over all 50 class labels is computationally expensive. Instead, in our approach, for faster computation, we compute the gradient as if the model's prediction on the unlabeled instance were the true label, similar to the idea presented in [4].

In **Core-Set**, we apply the greedy approach outlined in Algorithm 2. We choose the Euclidean distance as the pairwise distance metric and it is calculated in the CLS vector space.

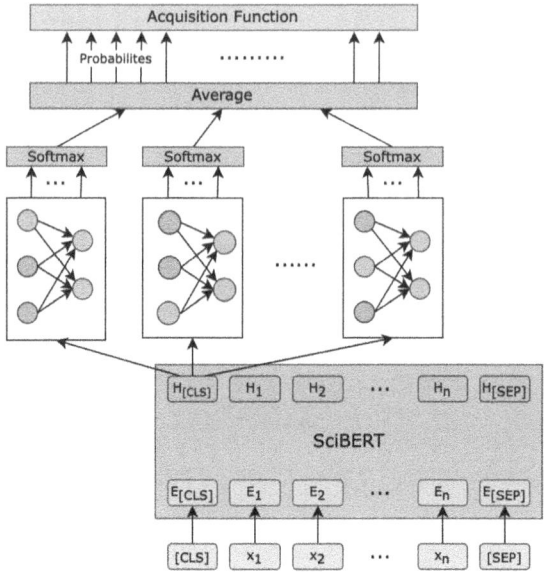

Fig. 4. Pipeline for calculating instance uncertainty using Monte Carlo Dropout

4   Experiments and Results

4.1   Dataset

To evaluate our approach, we selected MITRE's corpus [23], which to our knowledge is the only large scale dataset mapping CTI reports to ATT&CK techniques. The corpus consists of 11,130 sentences and phrases extracted from CTI reports describing ATT&CK techniques. Although ATT&CK comprises of 625 ATT&CK techniques, MITRE focused their efforts on annotating a subset of the 50 most common techniques [21] for the dataset. We split the dataset into training and test sets using an 80/20 ratio. Specifically, we allocate 8,904 instances to the training set, while the remaining 2,226 instances form our test set.

We evaluate our model on 4 different budgets, based on annotation cycles. As an example, if we have 50 annotation cycles with $|D^{query}| = 10$, this gives 500 samples that are actively acquired. With an initial warm start of 1% data (89 samples), this gives a total of 589 samples when using 50 annotation cycles. The budgets are as follows:

1. $\mathcal{B}_1$: 50 annotation cycles, corresponding to 589 samples, approximately 7% of the training pool
2. $\mathcal{B}_2$: 100 annotation cycles, corresponding to 1089 samples, approximately 12% of the training pool
3. $\mathcal{B}_3$: 150 annotation cycles, corresponding to 1589 samples, approximately 18% of the training pool
4. $\mathcal{B}_4$: 200 annotation cycles, corresponding to 2089 samples, approximately 23% of the training pool.

4.2   Comparative Study of AL Strategies

Our findings in Table 1 confirm the significant advantage of AL strategies compared to random selection. Random Selection serves as a control experiment for Active Learning (AL). During each annotation cycle, instead of selecting instances using an AL strategy, we randomly select 10 instances. This allows us to assess the effectiveness of AL in improving the overall performance. All AL methods outperform the random baseline, highlighting the importance of selecting informative data points for effective model training.

When we examine Fig. 5 which is a plot of the F1 scores (y-axis) of different AL strategies across different annotation budgets (x-axis), three AL strategies stand out: Top Confidence, Monte Carlo Dropout, and Core-Set. Top Confidence offers the most efficient choice for smaller budgets ($\mathcal{B}_1$ and $\mathcal{B}_2$) - it achieves the best performance while boasting the lowest computational cost as seen in Table 2. This efficiency stems from its reliance on a simple minimum operation on the model's confidence scores. For the largest budget ($\mathcal{B}_4$), Core-Set delivers the highest accuracy, closely mirroring the performance of fine-tuning on the entire training pool. However, this enhanced accuracy comes at the cost of computational speed, as it involves calculating pairwise Euclidean distances between all labeled and unlabeled data points. Finally, while Monte Carlo Dropout demonstrates effectiveness for budget $\mathcal{B}_3$, it's the most computationally expensive option. This method requires multiple model inferences per annotation cycle, leading to significantly longer runtimes.

In conclusion, when computational resources are limited, Top Confidence stands out for its efficiency. For the strategy that most closely approaches the best performance at our annotation budget limit, Core-Set offers the best results.

4.3   Performance Against Baselines

In our work, we compare our proposed approach against three baseline methods that leverage Large Language Models (LLMs) for ATT&CK mapping: TRAM, TTPClassifier and SMET.

TRAM (Threat ATT&CK Mapper) [25], developed by MITRE, utilizes SciBERT for sequence classification. We fine-tune SciBERT on the entire training pool and evaluate its performance on the test set.

TTPCLassifier [3] uses pre-trained Sentence-BERT (SBERT) [24] to generate embeddings and uses cosine similarity between embeddings to map text to ATT&CK. SMET [1], on the other hand, employs SBERT fine-tuned on 38,396 pairs of attack vectors extracted from ATT&CK technique descriptions and procedure examples to generate embeddings used to train a multinomial logistic regression model. It's important to note that both these approaches were originally designed to map only to attack techniques, not sub-techniques. Since our test set includes both, we consider the prediction of these methods correct during evaluation even if it maps to the higher-level attack technique instead of the sub-technique (e.g. for Fig. 1, mapping to T1003 instead of T1003.001 is considered a correct prediction).

Table 1. Performance Metrics for different Active Learning Strategies

Budget	Model	Precision	Recall	F1
B_1 **7%** of data 589 samples	**ALERT w/Top Confidence**	**0.79**	**0.75**	**0.75**
	ALERT w/Maximum Entropy	0.73	0.70	0.70
	ALERT w/Margin Sampling	0.74	0.73	0.71
	ALERT w/Monte Carlo Dropout	0.76	0.74	0.74
	ALERT w/Approximated Gradient Length	0.74	0.72	0.71
	ALERT w/Core-Set	0.75	0.73	0.72
	ALERT w/Random Selection	0.16	0.15	0.14
B_2 **12%** of data 1089 samples	**ALERT w/Top Confidence**	**0.81**	**0.80**	**0.80**
	ALERT w/Maximum Entropy	0.78	0.77	0.77
	ALERT w/Margin Sampling	0.79	0.77	0.76
	ALERT w/Monte Carlo Dropout	0.80	0.79	0.79
	ALERT w/Approximated Gradient Length	0.79	0.77	0.77
	ALERT w/Core-Set	0.80	0.79	0.79
	ALERT w/Random Selection	0.10	0.09	0.09
B_3 **18%** of data 1589 samples	ALERT w/Top Confidence	0.84	0.82	0.83
	ALERT w/Maximum Entropy	0.84	0.83	0.83
	ALERT w/Margin Sampling	0.79	0.78	0.78
	ALERT w/Monte Carlo Dropout	**0.85**	**0.84**	**0.84**
	ALERT w/Approximated Gradient Length	0.82	0.81	0.81
	ALERT w/Core-Set	0.84	0.83	0.83
	ALERT w/Random Selection	0.06	0.06	0.06
B_4 **23%** of data 2089 samples	ALERT w/Top Confidence	0.85	0.84	0.84
	ALERT w/Maximum Entropy	0.85	0.84	0.84
	ALERT w/Margin Sampling	0.79	0.79	0.79
	ALERT w/Monte Carlo Dropout	0.85	0.84	0.84
	ALERT w/Approximated Gradient Length	0.82	0.81	0.81
	ALERT w/Core-Set	**0.86**	**0.85**	**0.85**
	ALERT w/Random Selection	0.07	0.06	0.06

Table 2. Runtime of three top performing AL strategies for 50 annotation cycles - Top Confidence, Monte Carlo Dropout, Core-Set

	ALERT w/Top Confidence	ALERT w/Monte Carlo Dropout	ALERT w/Core-Set
Time(s)	11098	13210	11815

The results are seen in Table 3. We observe that simply using a pre-trained SBERT as in the case of TTPClassifier achieved poor performance - an F1 score of only 0.46. However, fine-tuning SBERT on a domain-specific dataset such as in the case of SMET achieved a better performance. Nonetheless, our approach still significantly outperformed SMET eventhough it was trained on a larger dataset, achieving an F1 score of 0.85 compared to SMET's score of 0.69. Moreover, our approach was fine-tuned on a subset of dataset which was 77% smaller than that of TRAM yet achieved comparable performance to TRAM (F1 score of 0.88).

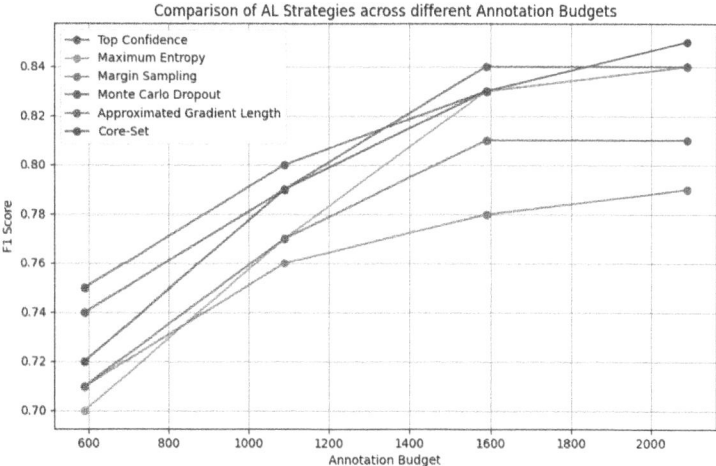

Fig. 5. Comparison of F1 scores obtained by different AL Strategies at different annotation budgets

4.4   Efficacy of Augmentation

In scenarios where training data is scarce, augmentation helps by artificially expanding the dataset. Easy Data Augmentation (EDA) [32] provides a simple yet effective way to augment training data, improving performance on low-resource tasks. EDA involves four simple processes: synonym replacement, random insertion, random deletion, and random swap. In our experiments, we further investigate whether augmentation techniques like EDA enhances model performance in conjunction with active learning strategies.

The data augmentation experiment, as reported in Table 4, further emphasizes the importance of selecting informative data for annotation. We conducted

Table 3. Performance Metrics against Baselines

Model	Fine-tuning Corpus	Precision	Recall	F1
ALERT w/Core-Set	**2089 instances** of labeled text from CTI Reports	0.86	0.85	0.85
TRAM (SciBERT)	**8904 instances** of labeled text from CTI Reports	0.88	0.88	0.88
SMET (Fine-tuned SBERT)	**38,396 instances** of attack vectors extracted from ATT&CK technique description and procedure examples	0.78	0.63	0.69
TTPClassifier (Pre-trained SBERT)	Zero-shot	0.72	0.42	0.46

Table 4. Performance of AL strategies with Augmentation

AL Strategy	Fine-tuning Corpus (total instances: real/synthetic)	Precision	Recall	F1
Top Confidence + EDA	**2945 instances:** 589 / 2356	0.75	0.73	0.73
Top Confidence + EDA	**5445 instances:** 1089 / 4356	0.82	0.80	0.80
Monte Carlo Dropout + EDA	**7945 inst ances:** 1589 / 6356	0.82	0.81	0.81
Core-set + EDA	**10445 instances:** 2089 / 8356	0.83	0.82	0.82

this experiment with the top-performing AL strategies for each annotation budget. After selecting samples using the respective AL strategy, for each sample in the labeled set, we generated four new synthetic samples using the four processes. The models were then fine-tuned on significantly larger augmented datasets. However, we observed no improvement compared to the results in Table 1. Since the new synthetic data were derived from the original samples selected by the AL strategy, they provided no new information to the model. This finding highlights that augmentation on top of AL strategies to is not useful and instead efforts should be directed toward curating a diverse dataset to improve generalization.

We can therefore conclude that for our task, simply increasing the size of the fine-tuning corpus is not as effective as selecting diverse samples that provide more informative data for the learner.

4.5   Case Study on ALERT's Impact

We perform a case study to highlight the impact of ALERT's efficiency. These experiments are conducted using the ALERT w/Core-set model. MITRE's dataset containing CTI reports currently contains annotations of 50 different ATT&CK techniques. However, in total, we know that there are 625 techniques as outlined in Sect. 2.2. Thus, the remaining 575 techniques currently remain unlabeled, and we would like to provide approximations for the amount of annotation effort we can save by using ALERT.

To do so, we create subsets of MITRE's labeled dataset (with 50 techniques), by removing all data instances belonging to randomly chosen techniques. In this way, we create subsets with reduced numbers of techniques. Specifically, we create 4 subsets containing 2, 5, 10, and 25 techniques respectively. Along with the original dataset with 50 techniques, this gives 5 *scenarios* ranging from datasets with 2 techniques (binary classification), to 50. Now, for each of these scenarios, we use TRAM (a baseline model without AL optimizations), to calculate the F-1 scores it achieves. Next, we use ALERT to identify the minimum number of annotations necessary to achieve an F-1 score which is within $\triangle f$ of TRAM's F-1 scores, where $\triangle f$ is a very small value. From Table 3, we noted that ALERT's F1 score is only 0.03 short of TRAM while using 6815 samples fewer (over 77% fewer) samples. We thus set $\triangle f = 0.03$ as our target, and check how many training samples ALERT requires to achieve an F-1 score within $\triangle f$ of TRAM.

After identifying the required number of samples in the 5 different scenarios, we plot these points, shown in Fig. 6(a). Finally, we use a simple curve-fitting approach based on least-squares-optimization [33], to extrapolate the number of training samples that experts would need to annotate to classify larger numbers of ATT&CK techniques, all the way up to 625 (the total number of techniques found in the ATT&CK framework). Figure 6(b) visualizes these projected values. Thus, for 625 techniques, we project that the ALERT framework would require 27,418 samples instead of TRAM's 115,682. In other words, ALERT would require a dataset over 4 times smaller to achieve comparable performance, making this much more feasible for annotators and security analysts. Since only 50 techniques are currently labeled in the MITRE dataset, the practical utility is clear.

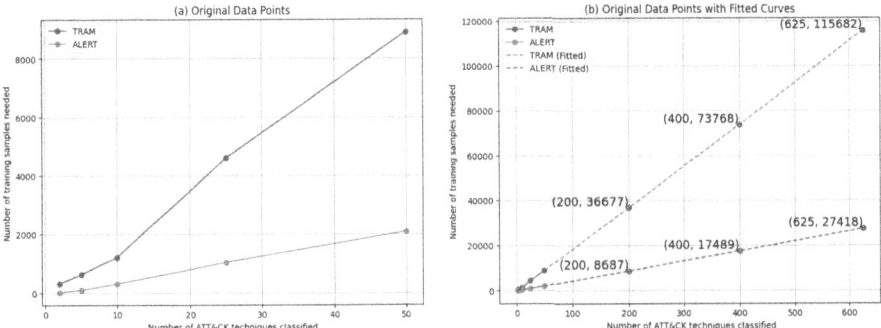

Fig. 6. (a) Relationship between the number of different ATT&CK techniques classified against the number of training samples (b) An extrapolation from our real data points to project the number of of samples needed for all ATT&CK techniques

5   Related Work

To tackle the complexity of CTI reports, recent approaches have used Large Language Models (LLMs) [1,3,25] to extract and map text to ATT&CK techniques.

TTPClassifier [3] computes embeddings for extracted text from CTI reports using a pre-trained Sentence BERT (SBERT) [24]. It also computes embeddings for the title and description of each ATT&CK technique. It then uses a similarity metric (weighted cosine similarity) to match the text with the title and description to map to the most relevant technique. However, this zero-shot approach is not very effective, and their method exhibits high false positive rates.

SMET [1] fine-tunes the SBERT model on 38,396 pairs of attack vectors extracted from each ATT&CK technique description and procedure example to generate a semantically meaningful embedding of attack vectors. Finally, SMET uses a logistic regression model trained on these embeddings to estimate the probability of an attack vector belonging to each ATT&CK technique and rank techniques based on the estimated probability.

MITRE developed Threat Report ATT&CK Mapper (TRAM) [23] to automate the mapping of CTI reports to ATT&CK. They built a training corpus of
11,300 sentences and phrases extracted from CTI reports that map to the 50 most
common ATT&CK techniques. This corpus was then used to fine-tune SciBERT
[5] for sequence classification, to map the sentences/phrases to ATT&CK.

While previous work focused on classification metrics, to our knowledge, ours
is the first to address the *efficiency* of the process, by utilizing Active Learning
to attain comparable and consistent performance while using a much smaller
yet highly informative data subset. Efficiency is key in this domain. In fact, the
main challenge of MITRE's work with TRAM has been the annotation process,
and hence, they were only able to annotate 50 out of the total 625 techniques.

Active learning has been a widely adopted solution for data-driven machine
learning [30], especially when annotation is expensive and time-consuming. This
idea has been proven effective in various domains, including image classification
[29], speech recognition [12], and natural language processing [9]. There has been
limited work using active learning in the cybersecurity domain, some focusing
on LSTMs [15]. However, leveraging active learning with LLMs in cybersecurity,
particularly for the extraction of ATT&CK techniques, has yet to be explored.

Our work bridges this gap by conducting an extensive study on using active
learning strategies with an LLM in this domain. We propose a solution that
reduces the annotation burden for efficient extraction and mapping of ATT&CK
techniques from CTI reports.

6   Conclusion and Future Work

Our evaluations demonstrate that ALERT significantly improves the efficiency of
ATT&CK technique extraction from CTI reports by reducing annotation costs.
Thus, ALERT will be a valuable tool for analysts going forward, as they set out
to efficiently annotate unlabeled threat reports in the future. For future work, we
aim to go beyond extracting attack techniques. We aim to identify *relationships*
between different techniques in CTI reports. For instance, if reports describe a
chronological order of techniques in an APT kill chain, detecting a technique
at a specific stage may enable proactive defenses against later attacks. Finally,
we aim to leverage out-of-distribution detection methods to find novel attack
techniques not yet present in the ATT&CK framework.

Acknowledgements. The research reported herein was supported in part by NIST
Award # 60NANB23D007, NSF awards DMS-1737978, DGE-2039542, OAC-1828467,
OAC-1931541, and DGE-1906630, ONR awards N00014-17-1-2995 and N00014-20-
1-2738, and the National Center for Transportation Cybersecurity and Resiliency
(TraCR).

Disclaimer. Certain equipment, instruments, software, or materials are identified in
this paper in order to specify the experimental procedure adequately. Such identification is not intended to imply recommendation or endorsement of any product or
service by NIST, nor is it intended to imply that the materials or equipment identified
are necessarily the best available for the purpose.

References

1. Abdeen, B., Al-Shaer, E., Singhal, A., Khan, L., Hamlen, K.: SMET: semantic mapping of CVE to ATT&CK and its application to cybersecurity. In: Atluri, V., Ferrara, A.L. (eds.) IFIP Annual Conference on Data and Applications Security and Privacy. pp. 243–260. Springer, Heidelberg (2023). https://doi.org/10.1007/978-3-031-37586-6_15

2. Akbanov, M., Vassilakis, V.G., Logothetis, M.D.: Wannacry ransomware: analysis of infection, persistence, recovery prevention and propagation mechanisms. J. Telecommun. Inf. Technol. 1, 113–124 (2019)

3. Alam, M.T., Bhusal, D., Park, Y., Rastogi, N.: Looking beyond iocs: automatically extracting attack patterns from external CTI. In: Proceedings of the 26th International Symposium on Research in Attacks, Intrusions and Defenses, pp. 92–108 (2023)

4. Ash, J.T., Zhang, C., Krishnamurthy, A., Langford, J., Agarwal, A.: Deep batch active learning by diverse, uncertain gradient lower bounds. arXiv preprint arXiv:1906.03671 (2019)

5. Beltagy, I., Lo, K., Cohan, A.: Scibert: a pretrained language model for scientific text. arXiv preprint arXiv:1903.10676 (2019)

6. Bianco, D.: The pyramid of pain (2013). https://detect-respond.blogspot.com/2013/03/the-pyramid-of-pain.html

7. CrowdStrike: Notpetya technical analysis - a triple threat: File encryption, mft encryption, credential theft. https://www.crowdstrike.com/blog/petrwrap-ransomware-technical-analysis-triple-threat-file-encryption-mft-encryption-credential-theft/

8. Devlin, J., Chang, M.W., Lee, K., Toutanova, K.: Bert: pre-training of deep bidirectional transformers for language understanding. arXiv preprint arXiv:1810.04805 (2018)

9. Dor, L.E., et al.: Active learning for bert: an empirical study. In: Proceedings of the 2020 Conference on Empirical Methods in Natural Language Processing (EMNLP), pp. 7949–7962 (2020)

10. Gal, Y., Ghahramani, Z.: Dropout as a bayesian approximation: representing model uncertainty in deep learning. In: International Conference on Machine Learning, pp. 1050–1059. PMLR (2016)

11. Gentile, C., Wang, Z., Zhang, T.: Achieving minimax rates in pool-based batch active learning. In: Chaudhuri, K., Jegelka, S., Song, L., Szepesvari, C., Niu, G., Sabato, S. (eds.) Proceedings of the 39th International Conference on Machine Learning. Proceedings of Machine Learning Research, vol. 162, pp. 7339–7367. PMLR (2022). https://proceedings.mlr.press/v162/gentile22a.html

12. Huang, J., Child, R., Rao, V., Liu, H., Satheesh, S., Coates, A.: Active learning for speech recognition: the power of gradients. arXiv preprint arXiv:1612.03226 (2016)

13. Husari, G., Al-Shaer, E., Ahmed, M., Chu, B., Niu, X.: TTPDRILL: automatic and accurate extraction of threat actions from unstructured text of CTI sources. In: Proceedings of the 33rd Annual Computer Security Applications Conference, pp. 103–115 (2017)

14. Lewis, D.D.: A sequential algorithm for training text classifiers: corrigendum and additional data. In: ACM SIGIR Forum, vol. 29, pp. 13–19. ACM, New York (1995)

15. Li, T., Hu, Y., Ju, A., Hu, Z.: Adversarial active learning for named entity recognition in cybersecurity. Comput. Mater. Continua 66(1) (2021)

16. Li, Z., Zeng, J., Chen, Y., Liang, Z.: AttacKG: constructing technique knowledge graph from cyber threat intelligence reports. In: Atluri, V., Di Pietro, R., Jensen, C.D., Meng, W. (eds.) ESORICS 2022. LNCS, vol. 13554, pp. 589–609. Springer, Heidelberg (2022). https://doi.org/10.1007/978-3-031-17140-6_29

17. Liao, X., Yuan, K., Wang, X., Li, Z., Xing, L., Beyah, R.: Acing the ioc game: toward automatic discovery and analysis of open-source cyber threat intelligence. In: Proceedings of the 2016 ACM SIGSAC Conference on Computer and Communications Security, pp. 755–766 (2016)

18. Liu, Y., et al.: Roberta: a robustly optimized bert pretraining approach. arXiv preprint arXiv:1907.11692 (2019)

19. Lourentzou, I., Gruhl, D., Welch, S.: Exploring the efficiency of batch active learning for human-in-the-loop relation extraction. In: Companion Proceedings of the the Web Conference 2018, pp. 1131–1138 (2018)

20. MITRE: D3fend (2023). https://d3fend.mitre.org/

21. MITRE: Large language models: Architecture (2023). https://github.com/center-for-threat-informed-defense/tram/wiki/Large-Language-Models#architecture

22. MITRE: Mitre att&ck framework (2023). https://attack.mitre.org

23. MITRE: Threat report att&ck mapper (tram) (2023). https://github.com/center-for-threat-informed-defense/tram/

24. Reimers, N., Gurevych, I.: Sentence-bert: sentence embeddings using siamese bert-networks. arXiv preprint arXiv:1908.10084 (2019)

25. Ross, J., Lasky, J.: Our tram large language model automates TTP identification in CTI reports. MITRE-Engenuity (2023). https://medium.com/mitre-engenuity/our-tram-large-language-model-automates-ttp-identification-in-cti-reports-5bc0a30d4567

26. Sahan, M., Smidl, V., Marik, R.: Batch active learning for text classification and sentiment analysis. In: Proceedings of the 2022 3rd International Conference on Control, Robotics and Intelligent System, pp. 111–116 (2022)

27. Scheffer, T., Decomain, C., Wrobel, S.: Active hidden Markov models for information extraction. In: Hoffmann, F., Hand, D.J., Adams, N., Fisher, D., Guimaraes, G. (eds.) IDA 2001. LNCS, vol. 2189, pp. 309–318. Springer, Heidelberg (2001). https://doi.org/10.1007/3-540-44816-0_31

28. Schlette, D., Caselli, M., Pernul, G.: A comparative study on cyber threat intelligence: the security incident response perspective. IEEE Commun. Surv. Tutor. **23**(4), 2525–2556 (2021)

29. Sener, O., Savarese, S.: Active learning for convolutional neural networks: a core-set approach. arXiv preprint arXiv:1708.00489 (2017)

30. Settles, B.: Active learning literature survey (computer sciences technical report 1648) University of Wisconsin-Madison, Madison, WI, USA (2009)

31. Shannon, C.E.: A mathematical theory of communication. Bell Syst. Tech. J. **27**(3), 379–423 (1948)

32. Wei, J., Zou, K.: EDA: easy data augmentation techniques for boosting performance on text classification tasks. arXiv preprint arXiv:1901.11196 (2019)

33. Weisstein, E.W.: Least squares fitting (2002). https://mathworld.wolfram.com/

34. Zhu, Z., Dumitras, T.: Chainsmith: automatically learning the semantics of malicious campaigns by mining threat intelligence reports. In: 2018 IEEE European Symposium on Security and Privacy (EuroS&P), pp. 458–472. IEEE (2018)

VulPrompt: Prompt-Based Vulnerability Detection Using Few-Shot Graph Learning

Saquib Irtiza$^{(\boxtimes)}$ ⓘ, Xiaodi Li ⓘ, Mahmoud Zamani ⓘ, Latifur Khan ⓘ,
and Kevin W. Hamlen ⓘ

The University of Texas at Dallas, Richardson, TX 75080, USA
{saquib.irtiza,xiaodi.li,mahmoud.zamani,lkhan,hamlen}@utdallas.edu

Abstract. VulPrompt is a new approach for detecting software vulnerabilities from source code by employing a prompt-based graph learning technique within a few-shot learning framework. Rather than adopting the *Pretrain-Finetune* paradigm typical of prior works, it is the first to adopt the more recent *Pretrain-Prompt* paradigm in this domain, which affords the creation of a smaller, lightweight model that outperforms larger models within other baseline methods. Evaluations conducted in a few-shot setting reflect the scarcity of large, high-quality labeled datasets for vulnerability detection in large software products—a prevalent issue in cybersecurity. Results show that the reduced number of trainable parameters for prompt-based learning models make them well-suited for this learning scenario, requiring only n instances to train efficiently. The learnable prompt reduces the gap between the pretrain and downstream objectives for a particular task by adjusting the input data for the downstream task to fit the pretrained model. Comparative analyses between VulPrompt and other baseline methods demonstrate the model's robust performance across all datasets tested, consistently achieving notable results. This success showcases the efficacy and adaptability of VulPrompt for detecting software vulnerabilities across different datasets, highlighting its potential as an impactful tool in the cybersecurity domain.

Keywords: Prompt · Graph Neural Network · Few-shot Learning · Vulnerability Detection

1 Introduction

The prevalence of open source software within the developer community has led to a notable increase in reported vulnerabilities annually. As of 2023, an estimated 96% of codebases contain open source components, 84% of which exhibit at least one security vulnerability, and 48% of those vulnerabilities are considered high-risk [46]. These concerning statistics underscore the escalating issue of software vulnerabilities, demanding urgent attention. However, addressing this challenge is complicated by the rising complexity and scale of modern-day codebases. Hence, there is a critical need for an automatic and intelligent method to identify vulnerabilities that minimizes the burden of human code auditing.

© IFIP International Federation for Information Processing 2024
Published by Springer Nature Switzerland AG 2024
A. L. Ferrara and R. Krishnan (Eds.): DBSec 2024, LNCS 14901, pp. 221–240, 2024.
https://doi.org/10.1007/978-3-031-65172-4_14

To address this challenge, *Deep Learning* (DL) methods have emerged as a potential solution, aiming to automatically extract features from program source codes. Prior works [10,11,25,26,28,55] have demonstrated the efficiency of DL models in detecting various software security issues, often achieving high accuracy. However, existing DL approaches encounter at least two significant limitations: First, many of these methods [13,49,54] rely on supervised learning that require a large volume of labeled data for their effective training. This poses significant challenges due to its complexity, and requires specialized expertise and time. Furthermore, the resulting data is prone to human error, which tends to be significant since vulnerabilities are coding errors missed by human code authors during code development, testing, and review. Second, many of these works [29,40] treat code as natural language inputs, overlooking crucial structural and semantic elements, such as control-flows, dataflows, and abstract syntax trees. This drawback limits their ability to capture essential aspects of code functionality and relationships, hindering their effectiveness in vulnerability detection.

These limitations motivate using graphical representations of programs in a few-shot learning setting [42,52], to facilitate the development of efficient learning models in scenarios where obtaining a high ratio of vulnerable to non-vulnerable instances is impractical. Few-shot learning ensures that we utilize a limited number of instances, matching the number of shots, and maintaining a balanced representation in both classes. This approach mitigates the necessity for a perfectly balanced labeled dataset with a large number of instances in both classes, effectively addressing data dependency challenges. Also, few-shot learning promotes lightweight models with fewer parameters, enabling rapid training due to the small training sample size. This is highly advantageous in a domain where new vulnerability types emerge regularly, potentially lacking an abundance of available instances for that type. Few-shot learning accommodates this dynamic nature of vulnerability discovery by efficiently training models with limited data.

Furthermore, it opens avenues for employing prompt tuning, a new paradigm in transfer learning that involves training a separate vector while maintaining the pretrained model's frozen weights. This strategy is feasible due to prompts containing only a limited number of trainable parameters, which can be effectively trained using the limited instances available in few-shot learning scenarios. Our specific emphasis lies in graph-based prompting techniques, such as Graphprompt [34], to finetune prompts based on the topologies of the graph extracted from the individual source codes. However, as discussed in Sect. 3, the current state-of-the-art graph prompting methods have inherent limitations that render them unsuitable for our intended downstream task. These limitations motivate the development of our domain-specific prompting method, VulPrompt, which leverages attention mechanisms to generate multiple prompts, one for each node in the graph. This enables more nuanced and tailored prompts in the few-shot learning setting within our domain. Our key contributions are as follows:

– We explore the problem of detecting software vulnerabilities in a few-shot setting to simulate a more realistic real-world scenario where vulnerable classes have few labeled instances.

– We develop VulPrompt, the first prompt-based vulnerability detection model that alleviates the need for a large and balanced dataset with labels to train an efficient detection model.
– We improve upon GraphPrompt to include attention-based prompting to account for the different graph structures in the domain of cybersecurity while avoiding overfitting issues.

The rest of this paper is arranged as follows: Sect. 2 describes some relevant background concepts, Sect. 3 discusses some related works in this domain, Sect. 4 contains detailed description of our approach along with the motivation behind our work, Sect. 5 reports the quantitative analysis of our method, Sect. 6 includes some of the limitations and potential future research directions and Sect. 7 contains the conclusion. All source code related to the project can be found at: https://github.com/Saquibirtiza/VulPrompt.git.

2 Background

To ensure that the small training data that we use for few-shot learning contains sufficient features for the models to train efficiently, we use graph representations of the source codes instead of their textual form, in contrast to existing works [29, 40]. This is because graphs offer superior representations of complex relationships among program components (e.g., control- and data-flows), and they enhance scalability—particularly when handling large codebases. This strategy enables us to use graph-based learning models such as *graph neural networks* (GNNs) [28, 55], which are capable of performing node- and graph-level classifications on specific tasks. These models have demonstrated superior performance compared to those relying solely on sequential textual data, often resulting in improved false positive and false negative rates. By leveraging graph representations, we can capture nuanced program structures effectively, empowering the models with richer features for enhanced vulnerability detection accuracy.

Typically, researchers tend to divide the training process of GNNs into two parts: *pretraining* and *finetuning*, to reduce their dependency on large, task-specific labeled data. Instead of training a model from scratch in a supervised fashion, this paradigm helps GNNs to rely on readily available, label-free data to learn intrinsic graph properties in a task-agnostic manner. Depending on the specific downstream task, a relatively smaller labeled dataset can then be used to finetune the pretrained model. But this paradigm often yields suboptimal performance due to the inconsistency between the objectives of the pretraining and downstream tasks [31]. For instance, the pretraining task might be related to learning connectivity patterns between nodes (associated with link prediction) [14,15], while the finetuning task might be related to node or graph classification tasks. This mismatch in objectives can lead to significant differences in how the model learns and adapts, often resulting in performance compromises when transitioning from pretraining to finetuning.

Figure 1 shows how this research gap can be bridged by bringing the downstream task closer to the pretrained model without altering the model's weights.

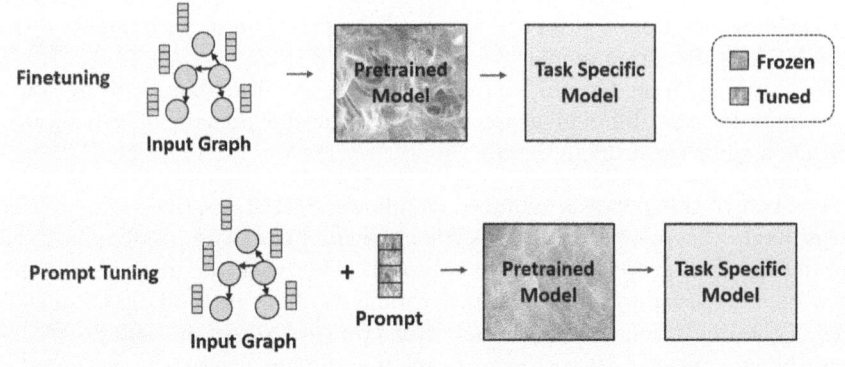

Fig. 1. Finetuning versus Prompt Tuning

This is achieved by introducing a learnable prompt embedding that transforms the input data space instead of fine-tuning the parameters of the pretrained model based on the downstream task. This creates an impression that the downstream task adapts to the pretrained model, rather than vice versa. Prompting was initially used in language models [2] and subsequently extended to graphs [45], but was initially confined to node classification due to the intricate nature of graphs. To address this limitation, GraphPrompt [34] unified node and graph classification tasks by framing them as variations of the link prediction task used in the pretrained model. This extension allowed prompting to be employed across a broader range of downstream tasks. However, to prevent overfitting of prompts due to limited data in few-shot tasks, GraphPrompt utilizes relatively simple prompts. These limitations are overcome by our method, VulPrompt.

3   Related Works

Software vulnerability detectors have employed various techniques, ranging from similarity-based methods such as text-based approaches [17,20,41], graph-based solutions [23,39], and combinations of the two [27]. However, many of these methods are primarily tailored for code clone detection and cannot effectively identify new vulnerabilities [29]. To address this, pattern-based techniques have emerged, prompting the use of semi-automatic methods [43,44,51]. These require human intervention to manually extract features, which can be utilized in traditional machine learning models, such as support vector machines and k-nearest neighbors, to identify vulnerabilities. Nevertheless, manual feature extraction remains a laborious and resource-intensive process, often posing cost challenges.

Automated methods [5,6,30] on the contrary, automatically extract the features. Devign [54] utilizes a GNN with a unique convolutional module capable of extracting features from node representations across various graph types; *abstract syntax trees* (ASTs), *control-flow graphs* (CFGs), *dataflow graphs* (DFGs), and *natural code sequence* (NCS). Although this method is generalizable across different graph structures, it is less effective with imbalanced data, prompting the development of alternative approaches. VulCNN [49] converts graphs into images through centrality scores derived from graph neighborhood information. This transformation allows the utilization of *convolutional neural networks* (CNNs), achieving competitive results. However, being a supervised model, it requires a large labeled dataset for training, which is often unavailable for data from real-world software repositories.

Vulchecker [36] addresses the data dependency challenge by developing an augmentation technique that merges *program dependency graphs* (PDGs) from real-world projects with vulnerable subgraphs from synthetic datasets. The aim is to enhance the model's performance by learning from extensive real-world programs rather than shorter synthetic ones. While this augmentation technique is beneficial for generating real-world vulnerabilities, it still requires a labeled synthetic dataset like Juliet [19] to augment the real-world projects. This implies that the effectiveness of this approach depends on having sufficiently large labeled synthetic datasets.

In contrast, our approach focuses on few-shot learning, enabling comparable performance with a minimal number of instances in each class, making it less reliant on large labeled datasets. It is different from the other few-shot work in this domain, Vul-Mirror [12], which exclusively relies on similarity-based methods for code clone detection. Consequently, Vul-Mirror encounters the same issue highlighted in prior works [29]-its inability to effectively identify novel vulnerabilities. VulPrompt overcomes this problem by performing vulnerability detection task instead of code clone detection which can find new unseen vulnerabilities.

To reduce our model's reliance on labeled data, we leverage graph prompting techniques, inspired by their success in text [22, 24, 31, 32] and image domains [1, 18]. These methods share the same strategy of keeping the pretrained model weights frozen while introducing some learnable components to the input to make them compatible with the pretrained model. However, applying these techniques to graphs is more intricate, with numerous open challenges. Initial efforts [34, 45] have employed prompts to align pretrained models trained on link prediction data [21] with downstream tasks. However, these methods show limited generalizability across different pretrained models. A recent study [8] addresses this by developing a prompting method applicable regardless of the specific pretrained model in use. The approach exhibits superior performance compared to finetuning in various scenarios, as highlighted through theoretical analyses. Our work integrates insights from all these techniques to devise a domain-specific prompting approach that outperforms existing vulnerability detection models.

4 Proposed Method

4.1 Graph Conversion and Vector Generation

Our approach begins by converting source codes into graphs and identifying high-quality representations for its nodes. We accomplish this through a series of static analysis steps, as outlined in prior studies [20,49], which transforms input source code into graphs while preserving the program's semantics. Instead of treating the entire file as a single input, we divide the program into smaller snippets containing code from individual functions. This decreases the analysis granularity, leading to enhanced model accuracy.

Figure 2 and lines 3–5 of Algorithm 1 outline the initial step, which consists of snippet normalization via a three-step process. Firstly, we eliminate any comments in the code since they do not impact the program's control- or data-flow. Following that, user-defined function names are substituted with generic names (e.g., "FUN1") to ensure that they do not influence the program's representation. Likewise, user-defined variables are replaced with generic names (e.g., "VAR1"). This process yields a cleaned and standardized code snippet, which serves as the basis for generating graphs.

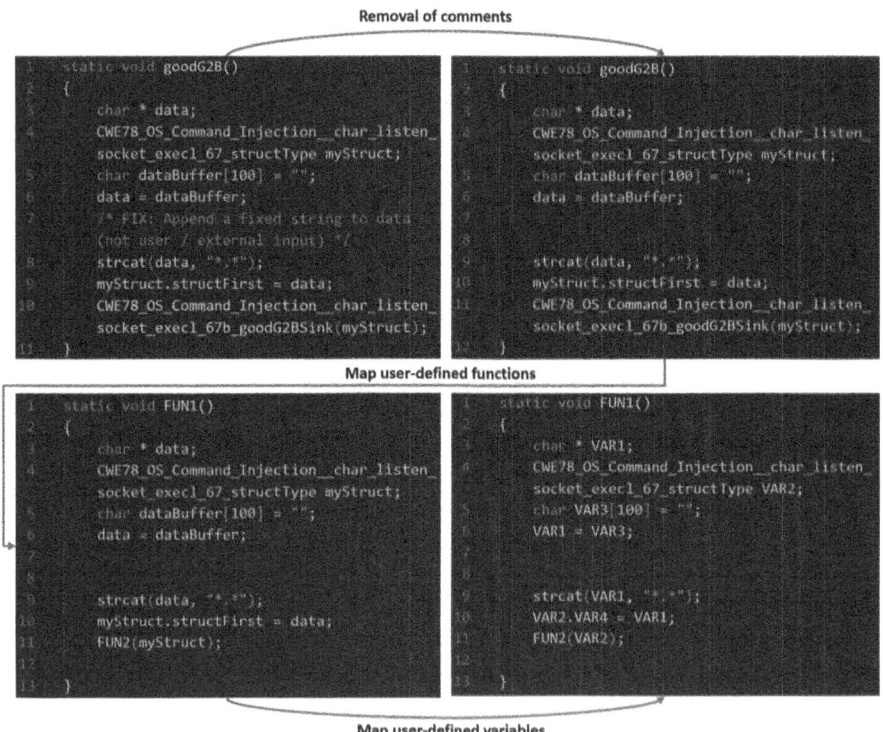

Fig. 2. Code normalization steps to format source code before graph conversion

Figure 3 visualizes the graph generation process, which is also described by lines 6–10 of Algorithm 1. Leveraging the program analysis tool *Joern* [50] (cf., [25,54]), we generate PDGs that encapsulate both control-flow and data-flow information. Within these graphs, each node corresponds to a line in the snippet,

---

Algorithm 1: Data preparation pipeline of VulPrompt

Data: functions from source code, S
Result: processed graph dataset, $\mathcal{T}_{processed}$

1   $\mathcal{T}_{processed} \leftarrow \emptyset$;
2   **for** *every s in S* **do**
3   remove comments from s
4   replace user-defined functions in s with generic names
5   replace user-defined variables in s with generic names
6      $pdg \leftarrow$ convert s to graph using *Joern*
7   convert nodes in pdg into embeddings $v_1, v_2, ..., v_m$ using sent2vec
8   append pdg to $\mathcal{T}_{processed}$
9   **end**
10   return $\mathcal{T}_{processed}$

Algorithm 2: Pretraining steps of VulPrompt

Data: processed graph dataset, $\mathcal{T}_{processed}$
Result: pretrained weights, Θ

1   $\mathcal{T}_{pre} \leftarrow \emptyset$
2   **for** *every t in $\mathcal{T}_{processed}$* **do**
3      $S_v, S_a, S_b \leftarrow$ use subgraph sampler to generate triplets from pdg of t
4   append the triplets to $\mathcal{T}_{pre}$
5   **end**
6   $\Theta \leftarrow$ optimize $\mathcal{L}_{pre}$ from Eq. 1 using $\mathcal{T}_{pre}$ and aggregator function, f_{ag}
7   return Θ

---

viewed as individual sentences to be transformed into embedding vectors. The edges in the graph depict the relationships of control- and data-flow between these nodes. To obtain node representation vectors, we utilize sent2vec [38] embeddings due to its popularity and capability to train in an unsupervised manner. All embeddings have a fixed length of 128 regardless of the size of the code.

4.2 Developing Pretrain Model

Algorithm 2 and the first row of Fig. 4 summarize the pretraining stage, which leverages readily available data [15,35] in the form of label-free graphs, $\mathcal{G}_{pre}$, to optimize a link prediction model after the data is processed by the steps in Sect. 4.1 giving $\mathcal{T}_{processed}$. We first process the data using a subgraph sampler, which identifies contextual subgraphs centered around various nodes within the graph. This extracts three subgraphs from each graph; two are connected and treated as positive candidates, while the third is disconnected and considered a negative candidate. Our objective is to enhance the similarity between positive subgraphs and diminish the similarity between negative ones through self-supervised techniques. This yields embeddings that naturally translate to

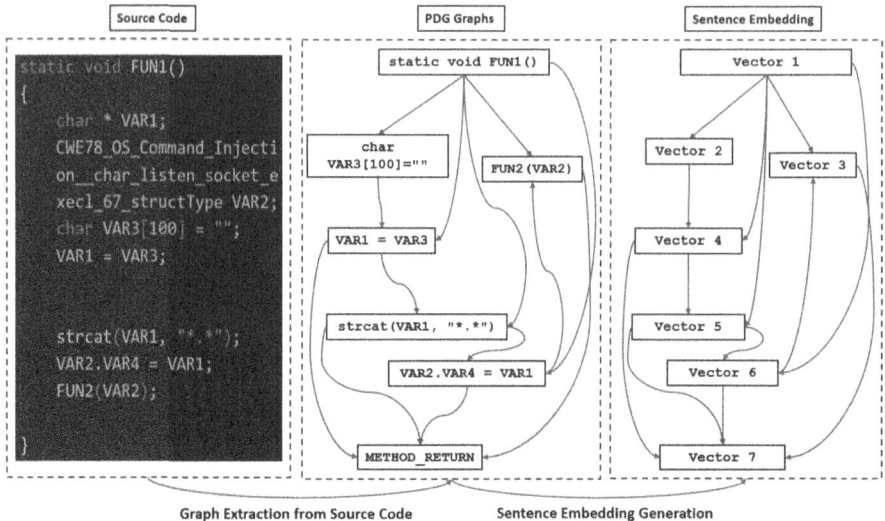

Fig. 3. Example of source code conversion to graphs followed by vector generation.

our subsequent graph classification task [47,53], because entire graphs can be viewed as subgraphs of themselves, allowing subgraphs within the same class to exhibit greater similarity than those in different classes. Section 4.4 describes this in more detail.

To formalize the pretraining step, given a graph G, we extract contextual subgraph triplets (S_v, S_a, S_b) for node v in G. Here, S_v and S_a are positive candidates, while S_v and S_b are negative candidates. Node a is chosen from the neighboring nodes of v, whereas node b is selected as a node not linked to v. We then form subgraphs S_v, S_a, and S_b by including all nodes and edges within a fixed number of hops from the original node. This results in the contextual subgraphs capturing rich contextual information alongside information about itself [16]. This process is repeated for all graphs in the label-free preprocessed dataset $\mathcal{T}_{processed}$ and accumulated into a training dataset $\mathcal{T}_{pre}$. Lines 2–4 in Algorithm 2 represent these steps. Then we use this data to optimize our loss function (line 6) such that it minimizes the distance between positive subgraphs while maximizing the distance between the negative ones. The loss function is as follows:

$$\mathcal{L}_{pre}(\Theta) = -\sum\nolimits_{(v,a,b)\epsilon\mathcal{T}_{pre}} \ln \frac{\exp(sim(s_v^{\Theta}, s_a^{\Theta})/\tau)}{\sum_{u\epsilon(a,b)} \exp(sim(s_v^{\Theta}, s_u^{\Theta})/\tau)} \quad (1)$$

where τ is the temperature hyperparameter used to regulate the output. The subgraph embeddings s_x^{Θ} are given by aggregator function, f_{ag}, which combines the node embeddings of the subgraphs using sum-pooling. Parameter Θ is the weight set of the GNN model that gives the node embeddings used in f_{ag}. The

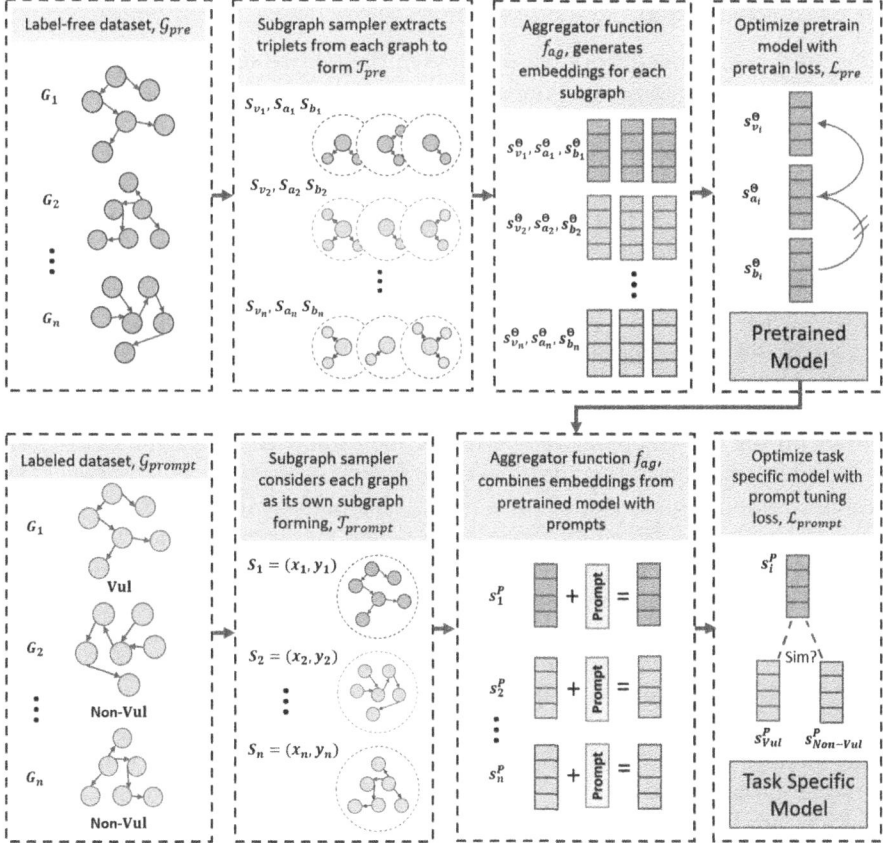

Fig. 4. Overview of VulPrompt. Top row indicates the pretraining steps whereas the bottom row is for the prompt tuning process.

goal is to get the weights that minimize the loss. These weights can then be used as the initial weights of the model for the downstream task.

The dataset for this stage does not need to be specific to vulnerability detection; any label-free graph dataset suitable for link prediction suffices. This means, it can be independent of the dataset used for the few-shot prompt tuning task, liberating us from dependency on large labeled domain-specific datasets for pretraining. With just a small labeled dataset for prompt tuning containing the same number of data points in each class as the shots required, our approach can effectively detect vulnerabilities. As a result, there's no necessity for excessively large vulnerability datasets like BigVul [7] and DiverseVul [4], as we only utilize a specific number of instances for training, irrespective of the total number of instances in the dataset. Well-known datasets like Reveal [3] and FFM-Peg+Qemu [54] are sufficient for training the prompt tuning stage.

4.3   Prompt Design

Prior work uses manually crafted prompts in the textual domain to guide down-stream tasks, leveraging a common task template to extract prior knowledge from pretrained models [2]. However, this approach is infeasible for graphs due to their abstract nature. This led to the emergence of learnable prompts [22,33]. GraphPrompt [34] integrates a learnable prompt p into the aggregation function f_{ag}, responsible for generating subgraph embeddings during the downstream phase. This method involves an element-wise multiplication between the prompt and the node embeddings of the nodes in the subgraph. In our case, since we use sum-pooling to aggregate all node embeddings within the subgraph to generate its embedding, utilizing the prompt implies seeking a feature-weighted summation of node representations. We call this version of the aggregator function $f_{weighted\_ag}$.

But during our comparative analysis of the prompting technique introduced by GraphPrompt and other baselines in a few-shot scenario, we observed that GraphPrompt often struggled to surpass the performance of those models. Inspired by *universal prompt tuning* [8], we therefore devised a domain-specific variant of the approach that employs multiple prompts instead of a singular one, aiming to enhance guidance for the downstream task. The objective is to introduce adaptable elements in the input feature space, enabling the pretrained model's weights to better align with the requirements of the downstream task.

Given frozen pretrained weights Θ and a labeled downstream graph dataset $\mathcal{G}_{prompt}$ consisting of graphs G_i with node features $V_i = v_1, v_2, ..., v_m$ (where m is the number of nodes in G_i), our objective is to optimize a set of n task-specific learnable prompt basis vectors $p_1, p_2, ..., p_n$. During evaluation, these prompts are added to the node features of the test instances, yielding a set of new features $\hat{V}_i = v_1 + p_1, \ v_2 + p_2, \ ..., \ v_n + p_n$. These transformed features serve as inputs to our model. $\mathcal{G}_{prompt}$ differs from $\mathcal{G}_{pre}$ in that the former contains labels and is preprocessed using Algorithm 1 into $\mathcal{T}_{processed}$ before it is used for training. The primary aim is to maximize the likelihood of accurately predicting the labels y for the newly generated $\hat{V}_i$, while keeping the weights Θ unchanged.

However, employing n prompts introduces challenges related to the varying node counts in graphs, necessitating dynamic adjustments for individual graphs. Moreover, memory constraints can arise due to the model's memory usage scaling with graph sizes. One straightforward solution to this problem is to leverage attention mechanisms, which excel in handling variable-length sequences. They can adaptively attend to different parts of the sequence based on their importance, without being constrained by fixed-length windows or recurrence. Building on this idea, we propose an alternate method utilizing k independent basis vectors p^b along with k learnable linear projections a. These elements are attentively aggregated to compute the prompts p_i for nodes i. The aggregation of the prompts follows the formula:

$$p_i = \sum_j^k a_{i,j} p_j^b \qquad a_{i,j} = \frac{\exp(a_j^T v_i)}{\sum_l^k \exp(a_l^T v_i)} \qquad (2)$$

Algorithm 3: Prompt tuning pipeline of VulPrompt

Data: processed graph dataset, $\mathcal{T}_{processsed}$
      pretrained weights, Θ
      number of shots, H
Result: set P of k prompt basis vectors $p_1^b, p_2^b, ..., p_k^b$

1 **Initialize:** $p_1^b, p_2^b, ..., p_k^b$
2 $\mathcal{T}_{prompt} \leftarrow \emptyset$;
3 **while** $len(\mathcal{T}_{prompt}) < 2H$ **do**
4      $G_{vul}, G_{nonvul} \leftarrow$ randomly select one pdg from each class in $\mathcal{T}_{processsed}$
5      $S_1, S_2 \leftarrow$ consider G_{vul}, G_{nonvul} as subgraphs of itself
6      $p_1, p_2, ..., p_n \leftarrow$ for each node in S_1, S_2 compute prompts using Equation 2
7      $v_1 + p_1, v_2 + p_2, ..., v_n + p_n \leftarrow$ add prompts to node features in S_1, S_2
8   append the graphs to $\mathcal{T}_{prompt}$
9 **end**
10 $\mathcal{T}_{test} \leftarrow$ remaining $\mathcal{T}_{processsed}$ set aside for testing phase
11 $p_1^b, p_2^b, ..., p_k^b \leftarrow$ optimize $\mathcal{L}_{prompt}$ from Equation 3 using $\mathcal{T}_{prompt}$ and aggregator
    function, $f_{weighted\_ag}$
12 **return** $p_1^b, p_2^b, ..., p_k^b$

The parameter count, k, can be adjusted based on the specific requirements of the downstream task, which we found to yield the most optimal outcome when set to three. This approach grants VulPrompt the advantage of deriving the prompts p_i from a reduced set P of k basis vectors, $p_1^b, p_2^b, ..., p_k^b$, effectively minimizing the number of trainable parameters in the model. Additionally, it eliminates the risk of prompts overfitting to the limited set of training instances used for the few-shot learning task, as the number of trainable parameters is small. This makes the model well-suited to be used when labeled data is scarce.

4.4   Prompt Tuning

Having established our prompt design, our next step involves formulating a loss function for the downstream vulnerability detection task. Algorithm 3 summarizes this process. The task involves optimizing the learnable prompts, similar to the pretrained model's template. Inspired by the subgraph similarity concept in the pretrained task, our focus for graph classification leads us to consider the entire graph as a subgraph of itself (line 5) when computing subgraph embeddings. This approach enables the retention of similarity and knowledge transfer from the pretrained model to the downstream task. Furthermore, we introduce the prompts to the nodes of the subgraphs following the guidelines outlined in Sect. 4.3 to generate our transformed input (lines 6–7). This transformed input is fed into the pretrained model to optimize the prompt loss $\mathcal{L}_{prompt}$ (line 11).

Formally, given a labeled training dataset $\mathcal{T}_{prompt} = \{(x_1, y_1), (x_2, y_2), ...\}$ where x_i is a graph instance and y_i is its corresponding label belonging to the set of classes Y, the prompt tuning loss can be defined as:

$$\mathcal{L}_{prompt}(P) = -\sum_{(x_i,y_i)\epsilon \mathcal{T}_{prompt}} \ln \frac{\exp(sim(s^P_{x_i}, \hat{s}^P_{y_i})/\tau)}{\sum_{c\epsilon Y} \exp(sim(s^P_{x_i}, \hat{s}^P_c)/\tau)} \qquad (3)$$

where τ is the temperature hyperparameter. The graph embeddings s^P_j are given by f_{weight_ag} which is the same sum-pooling aggregator function, f_{ag}, used in Sect. 4.2 but with the new transformed node embeddings. These embeddings are given by combining the original node embeddings with the set of learnable basis vectors P using Eq. 2. The goal it to optimize P such that we get the lowest loss, $\mathcal{L}_{prompt}$. Since this is a few-shot learning task, $\mathcal{T}_{prompt}$ contains as many instances in each class as the number of shots (line 4). This means that for a 10-shot learning task, both vulnerable and non-vulnerable classes will have 10 instances each in both the training and validation dataset. The rest of the data points are used as testing data $\mathcal{T}_{test}$ (line 10) during evaluation phase. We consider the embeddings of the class prototypical subgraphs $\hat{s}^P_c$ for all classes c within the dataset, which in the context of our task comprises vulnerable and non-vulnerable categories. They are computed using the formula:

$$\hat{s}^P_c = \frac{1}{k} \sum_{\substack{(x_i,y_i)\in Y, \\ y_i \in c}} s^P_{x_i} \qquad (4)$$

where k is the number of instances in each class. During this phase, we also optimize linear projections $a_1, a_2, ..., a_k$ along with the basis vectors for the prompts. All weights associated with the pretrained model remain unchanged throughout this process.

5   Evaluation

In this section, we conducted a quantitative comparison of VulPrompt's performance against other vulnerability detection baselines. Firstly, we chose to compare with Devign [54], as it is a pioneering work in this domain. Additionally, we selected VulCNN [49] for comparison, as it has outperformed other well-known baselines such as SySeVr [28], VulDeePecker [29], and VulDeeLocator [26]. Furthermore, we considered two state-of-the-art Large Language Model (LLM) based approaches, namely CodeBERT-base [9] and CodeT5-base [48], which were pretrained on large code bases to handle tasks such as code defect detection and code clone detection. We followed the approach outlined in DiverseVul [4], implementing a linear layer on top of the embeddings from the last decoder layer of the LLMs to derive probabilities for the two classes of our classification task.

Additionally, recognizing the influence of GraphPrompt on our method's development, we incorporate its results to emphasize the performance enhancements attained through the innovations detailed in Sect. 4. The key research questions addressed through this comparative analysis are:

- **RQ1:** How well does *Pretrain-Prompt* strategies perform compared to *Pretrain-Finetune* strategies for few-shot learning tasks in this domain?
- **RQ2:** How does using multiple prompts affect prompt-based model performance compared to using a single prompt as in GraphPrompt? If there is improvement with multiple prompts, what is the optimal number of basis vectors to maintain manageable trainable parameters?
- **RQ3:** How does VulPrompt's performance compare against other baseline methods? How does their performance vary across various numbers of shots?

5.1   Datasets

We use three distinct datasets: (1) Software Assurance Reference Dataset (SARD) [37], (2) FFMPeg+Qemu [54], and (3) Reveal [3]. Each dataset is comprised of snippets of source code extracted from various program functions. The compute resources, experimental setup and the metrics used are detailed in Appendix A.

SARD, overseen by the National Institute of Standards and Technology (NIST), is a comprehensive dataset comprising various synthetic, academic, and production vulnerabilities alongside non-vulnerable instances, totaling 12,303 vulnerable C/C++ functions and 21,057 non-vulnerable ones. However, due to the synthetic nature of the dataset, we conducted further evaluations with real-world datasets like Reveal and FFMPeg+Qemu. These were chosen due to their popularity, consistent maintenance, and substantial history of reported vulnerabilities. Reveal includes vulnerabilities from Linux Debian Kernel and Chromium, providing 1664 labeled vulnerable instances out of 18,169 programs. This accounts for less than 10% of data, which is a realistic portrayal of the difficulty of labeling vulnerable instances in real-world scenarios. FFMPeg+Qemu contains code snippets from FFMPeg and Qemu, manually labeled by domain experts, yielding 10,066 vulnerable instances and 12,295 non-vulnerable ones.

As discussed in Sect. 4.2, we selected these datasets instead of larger ones like Big-Vul or Diverse-Vul because we only require a fixed number of labeled data points for our few-shot learning task, regardless of the overall dataset size. Also, using a small dataset allowed us to comfortably run experiments within the limitations of our computational resources. All reported results in subsequent sections are the average of five runs with different seed values to enhance reliability, with standard deviation indicating score variation between these runs.

5.2   Main Results

RQ1: The highlighted cells in Table 1 positively answer our research question that investigates whether prompt tuning consistently outperformes finetuning for training downstream tasks in a few-shot setting. Each experiment utilized 10 shots, indicating that both the training and validation datasets consist of 10 instances per class, while the remaining data points are allocated for testing. Consequently, the baseline methods achieved slightly lower scores than those

Table 1. Comparison of VulPrompt with other baselines using datasets in a 10-shot learning scenario. Results are reported in percent with standard deviations. VulPrompt results are generated using 3 basis vectors.

Dataset	Method	Parameters	Precision	Recall	F1
SARD	CodeT5	222 M	28.43 ± 0.41	18.43 ± 0.34	21.27 ± 0.34
	CodeBERT	124 M	36.80 ± 0.10	36.80 ± 0.12	36.80 ± 0.07
	VulCNN	676,802	39.63 ± 2.35	47.42 ± 1.28	40.57 ± 2.39
	Devign	515,662	39.61 ± 2.89	47.32 ± 2.31	40.14 ± 1.67
	GraphPrompt	96	39.32 ± 2.00	49.25 ± 3.86	44.19 ± 1.07
	VulPrompt	771	$\mathbf{42.75 \pm 1.61}$	$\mathbf{55.42 \pm 2.15}$	$\mathbf{45.62 \pm 0.59}$
FFMPeg+Qemu	CodeT5	222 M	42.22 ± 0.34	26.50 ± 0.60	31.19 ± 0.52
	CodeBERT	124 M	51.73 ± 1.19	31.21 ± 3.51	37.52 ± 3.10
	VulCNN	676,802	53.50 ± 0.44	50.69 ± 2.54	50.80 ± 2.13
	Devign	515,662	53.59 ± 0.82	39.88 ± 3.59	45.03 ± 3.84
	GraphPrompt	96	53.55 ± 0.34	47.50 ± 3.92	49.34 ± 2.57
	VulPrompt	771	$\mathbf{53.90 \pm 0.14}$	$\mathbf{61.90 \pm 3.35}$	$\mathbf{57.56 \pm 2.12}$
Reveal	CodeT5	222 M	32.72 ± 1.38	20.87 ± 1.35	24.14 ± 1.01
	CodeBERT	124 M	42.16 ± 0.42	32.41 ± 3.25	35.79 ± 2.42
	VulCNN	676,802	40.09 ± 0.65	48.90 ± 1.63	48.49 ± 0.91
	Devign	515,662	43.89 ± 3.30	46.47 ± 4.38	45.07 ± 3.32
	GraphPrompt	96	49.59 ± 0.64	43.44 ± 1.37	44.44 ± 2.83
	VulPrompt	771	$\mathbf{50.12 \pm 0.49}$	$\mathbf{60.97 \pm 3.32}$	$\mathbf{53.20 \pm 2.80}$

reported in their original work, as we had to employ a limited number of data points in the training dataset to simulate the few-shot learning setting.

The results reveal that LLM-based approaches perform the poorest across all datasets, which is expected given the inadequacy of data points to effectively train a large number of parameters. In contrast, GNN-based methods exhibit better performance, with VulCNN slightly outperforming Devign on the real-world datasets Reveal and FFMPeg+Qemu. While GraphPrompt shows promise with the synthetic SARD dataset, it falls short in recall and F1 scores for the real-world datasets. This limitation arises from the inadequacy of a single prompt to handle lengthy programs, as observed in the real-world dataset. This issue is effectively addressed by VulPrompt, which surpasses all other methods by achieving an approximate 10% boost in recall and 5% in F1 scores across all datasets. This highlights our method's proficiency in detecting true positives, a crucial aspect in the cybersecurity domain. Moreover, there is a notable overall improvement in VulPrompt's performance across real-world datasets compared to synthetic data. This enhancement can be attributed to the abundance of features present in longer codes, which are necessary to effectively train multiple prompts.

Table 1 also shows that when considering pretrained models, finetune-based methods like CodeT5, CodeBERT, VulCNN, and Devign require significantly

Table 2. Analysis of VulPrompt on the FFMPeg+Qemu dataset, with different basis vector counts for prompts, reported in percents with standard deviations

Method	Vectors	Params	Precision	Recall	F1
GraphPrompt	-	96	53.55 ± 0.34	47.50 ± 3.92	49.34 ± 2.57
VulPrompt	1	257	53.73 ± 0.22	47.12 ± 3.02	50.09 ± 2.63
	3	771	53.90 ± 0.14	61.90 ± 3.35	57.56 ± 2.12
	5	1285	$\mathbf{54.06 \pm 0.39}$	$\mathbf{64.16 \pm 2.89}$	$\mathbf{58.33 \pm 1.29}$

more parameters with counts of 222 million, 123 million, 676,802, and 515,662 respectively. Prompt tuning on the other hand utilizes only 96 and 771 parameters respectively. This substantial disparity in parameter count arises because finetuning demands the adjustment of all its weights, unlike prompt tuning where only the prompts undergo tuning. This allows us to safely consider the pretrained model as separate from the prompt tuning task resulting in the exclusion of the pretrained model parameters from the total parameter count. As a result, the pretrained model need not be specifically tailored to vulnerability detection; rather, it can be any model trained on a link prediction task.

RQ2: Table 2 investigates whether multiple prompts work better for detecting vulnerabilities than using just a single prompt as in GraphPrompt. The results show that using a single basis vector p^b for VulPrompt yields almost identical results to GraphPrompt. However, as the number of basis vectors increases, VulPrompt shows further improvements over its already good results from RQ1. This implies that while utilizing a single basis vector for VulPrompt mirrors the performance of GraphPrompt, leveraging multiple basis vectors empowers the model to extract more useful features, thereby achieving enhanced performance.

However, this improvement comes with the trade-off of computing a few extra features due to the attention mechanism. This means that a larger parameter count necessitates a more substantial dataset for effective training, thereby escalating training time proportionately. That is why we opted for three basis vectors to compute prompts for our model to balance performance against these factors. All experiments were conducted using the FFMPeg+Qemu dataset for two reasons: it yielded the best results in RQ1 for VulPrompt, and it contains vulnerabilities that closely resemble those encountered in real-world scenarios.

RQ3: To analyze the impact of varying the number of shots on model performance, we evaluated recall and F1 scores across four different shot counts: 5, 10, 20, and 50, using the FFMPeg+Qemu dataset. Figure 5 demonstrates a general performance increase across most methods as the number of shots increases, aligning with the expectation that more data points facilitate better model learning and representation. However, this trend does not hold true for CodeT5, which has the highest number of parameters among all models. The

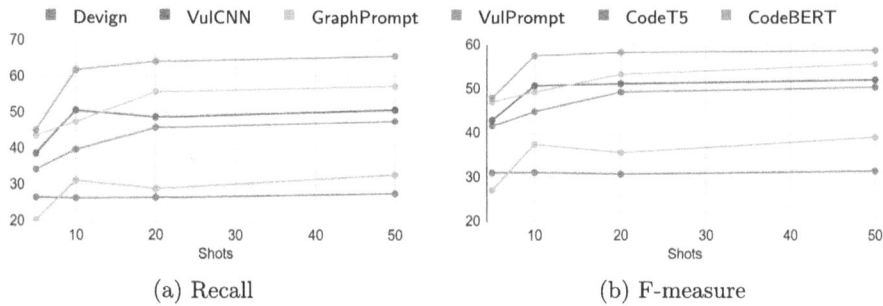

Fig. 5. Comparative analysis of model performance for different numbers of shots. Results reported for VulPrompt are generated using 3 basis vectors in each case.

consistent performance across different shot counts, as illustrated by the orange line, is expected, since a large model requires a considerably larger dataset than just 50 or lower data points in each training class for effective training.

VulPrompt consistently displays the most optimal results across all shot counts, as depicted by the green line, highlighting the model's suitability for downstream tasks even when the dataset is limited to a few instances-a common scenario in vulnerability detection datasets. One notable anomaly in the results is the poor performance of Graphprompt compared to VulCNN for 10 shots. This discrepancy could be attributed to the selection of data points for each class, which may have been too long to be efficiently learned using just a single prompt. This hypothesis is supported by the fact that VulPrompt achieves better results for the same number of shots, leveraging multiple prompts to more effectively learn features instead of just relying on a single prompt.

6   Limitations

One limitation of our current approach is the lack of specific vulnerability categorization, as we conduct binary classification by grouping all vulnerabilities into a single class. This may confuse the model, leading to suboptimal results. Future work should consider multi-label datasets to address this issue. Our work is limited to function-level vulnerability detection, which we aim to extend to detect vulnerabilities at the line-level for greater usability and granularity in machine-assisted code auditing. Improvements can also be made by adopting better strategies for generating node embeddings beyond the use of sent2vec embeddings, such as incorporating more comprehensive neighborhood information to enrich node representations. Exploring heterogeneous graph learning methods by categorizing nodes based on their content could further enhance performance.

7   Conclusion

We present a novel prompt-based model for detecting vulnerabilities in source code graphs, employing few-shot learning-a crucial advancement in addressing real-world challenges of limited labeled datasets. Our method outperforms GraphPrompt and other models, as shown through diverse experiments. Utilizing an attention mechanism, our model generates multiple prompts from a limited set of basis vectors, tuning them with the downstream dataset. Unlike traditional finetuning methods, we freeze weights from pretraining, adapting input data to the pretrained model rather than fitting the model to the task. This reduces trainable parameters and diminishes the need for labeled data, making VulPrompt highly effective for vulnerability detection in the face of data scarcity.

Acknowledgments. This research was supported in part by ARO award W911NF-21-1-0032, DARPA awards N6600121C4024 and 140D04-23-C-0070, NIST Award 60NANB23D007, NSF awards DMS-1737978, DGE-2039542, OAC-1828467, OAC-1931541, and DGE-1906630, ONR awards N00014-17-1-2995 and N00014-20-1-2738, and the National Center for Transportation Cybersecurity and Resiliency (TraCR).

A   Compute Resources and Experimental Setup

We conducted experiments on an in-house cluster, using a single Nvidia Quadro RTX 8000 GPU with 48 GB of memory. Each experiment ran for 100 epochs and was repeated five times with varied random seed values.

Depending on the number of shots n, we randomly selected n instances from each class for training and validation, reserving the remainder for testing. Hyperparameters varied: VulPrompt and GraphPrompt used a learning rate of 0.01 and a batch size of 512, while VulCNN and Devign used a learning rate of 0.001 and a batch size of 32 due to their large parameter counts. CodeBERT and CodeT5 used a learning rate of 0.0005 and a batch size of 8.

We evaluated our results using three key metrics: Precision, Recall, and F-measure, chosen for their ability to comprehensively quantify model performance, particularly in cybersecurity. For example, Recall measures the ratio of correctly classified true positives, critical for identifying undetected vulnerabilities.

References

1. Bahng, H., Jahanian, A., Sankaranarayanan, S., Isola, P.: Exploring visual prompts for adapting large-scale models. arXiv Preprint arXiv:2203.17274 (2022)
2. Brown, T., et al.: Language models are few-shot learners. In: Proceedings of 34th Conference on Neural Information Processing Systems (NeurIPS), pp. 1877–1901 (2020)
3. Chakraborty, S., Krishna, R., Ding, Y., Ray, B.: Deep learning based vulnerability detection: are we there yet. IEEE Trans. Software Eng. (TSE) **48**, 3280–3296 (2021)
4. Chen, Y., Ding, Z., Alowain, L., Chen, X., Wagner, D.: Diversevul: a new vulnerable source code dataset for deep learning based vulnerability detection. In: Proceedings of 26th International Symposium on Research in Attacks, Intrusions and Defenses (RAID), pp. 654–668 (2023)
5. Cheng, X., Wang, H., Hua, J., Xu, G., Sui, Y.: DeepWukong: statically detecting software vulnerabilities using deep graph neural network. IEEE Trans. Softw. Eng. Methodol. (TOSEM) **30**(3), 1–33 (2021)

6. Duan, X., et al.: VulSniper: focus your attention to shoot fine-grained vulnerabilities. In: Proceedings of 28th International Joint Conference on Artificial Intelligence (IJCAI), pp. 4665–4671 (2019)

7. Fan, J., Li, Y., Wang, S., Nguyen, T.N.: A C/C++ code vulnerability dataset with code changes and CVE summaries. In: Proceedings of 17th International Conference on Mining Software Repositories, pp. 508–512 (2020)

8. Fang, T., Zhang, Y., Yang, Y., Wang, C., Chen, L.: Universal prompt tuning for graph neural networks. In: Proceedings of 37th Conference on Neural Information Processing Systems (NeurIPS) (2023)

9. Feng, Z., et al.: CodeBERT: a pre-trained model for programming and natural languages. In: Findings Association for Computational Linguistics (EMNLP), pp. 1536–1547 (2020)

10. Fu, M., Tantithamthavorn, C.: LineVul: a transformer-based line-level vulnerability prediction. In: Proceedings of 19th International Conference on Mining Software Repositories (MSR), pp. 608–620 (2022)

11. Hanif, H., Maffeis, S.: VulBERTa: simplified source code pre-training for vulnerability detection. In: Proceedings of International Joint Conference on Neural Networks (IJCNN) (2022)

12. He, Y., Wang, W., Sun, H., Zhang, Y.: Vul-mirror: a few-shot learning method for discovering vulnerable code clone. EAI Endorsed Trans. Secur. Saf. **7**(23) (2020)

13. Hin, D., Kan, A., Chen, H., Babar, M.A.: LineVD: statement-level vulnerability detection using graph neural networks. In: Proceedings of 19th International Conference on Mining Software Repositories (MSR), pp. 596–607 (2022)

14. Hu, W., et al.: Strategies for pre-training graph neural networks. In: Proceedings of 7th International Conference on Learning Representations (ICLR) (2019)

15. Hu, Z., Dong, Y., Wang, K., Chang, K.W., Sun, Y.: GPT-GNN: generative pre-training of graph neural networks. In: Proceedings of 26th ACM SIGKDD International Conference on Knowledge Discovery & Data Mining (KDD), pp. 1857–1867 (2020)

16. Huang, K., Zitnik, M.: Graph meta learning via local subgraphs. Adv. Neural. Inf. Process. Syst. **33**, 5862–5874 (2020)

17. Jang, J., Agrawal, A., Brumley, D.: ReDeBug: finding unpatched code clones in entire OS distributions. In: Proceedings of 33rd IEEE Symposium on Security & Privacy (S&P), pp. 48–62 (2012)

18. Jia, M., et al.: Visual prompt tuning. In: Farinella, G.M., Hassner, T. (eds.) ECCV 2022. LNCS, vol. 13693, pp. 709–727. Springer, Cham (2022). https://doi.org/10.1007/978-3-031-19827-4_41

19. Boland, Jr., F.E., Black, P.: The Juliet 1.1 C/C++ and Java test suite. Computer **45**(10) (2012)

20. Kim, S., Woo, S., Lee, H., Oh, H.: VUDDY: a scalable approach for vulnerable code clone discovery. In: Proceedings of 38th IEEE Symposium Security & Privacy (S&P), pp. 595–614 (2017)

21. Kipf, T.N., Welling, M.: Variational graph auto-encoders. arXiv Preprint arXiv:1611.07308 (2016)

22. Lester, B., Al-Rfou, R., Constant, N.: The power of scale for parameter-efficient prompt tuning. In: Proceedings of Conference on Empirical Methods in Natural Language Processing, pp. 3045–3059 (2021)

23. Li, J., Ernst, M.D.: CBCD: cloned buggy code detector. In: Proceedings of 34th International Conference on Software Engineering (ICSE), pp. 310–320 (2012)

24. Li, X.L., Liang, P.: Prefix-tuning: optimizing continuous prompts for generation. In: Proceedings of 59th Annual Meeting Association for Computational Linguistics and 11th International Joint Conference on Natural Language Processing (ACL-IJCNLP), vol. 1, pp. 4582–4597 (2021)

25. Li, Y., Wang, S., Nguyen, T.N.: Vulnerability detection with fine-grained interpretations. In: Proceedings of 29th ACM Joint Meeting European Software Engineering Conference and Symposium on Foundations Software Engineering (ESEC/FSE), pp. 292–303 (2021)

26. Li, Z., Zou, D., Xu, S., Chen, Z., Zhu, Y., Jin, H.: VulDeeLocator: a deep learning-based fine-grained vulnerability detector. IEEE Trans. Dependable Secure Comput. (TDSC) **19**(4), 2821–2837 (2021)

27. Li, Z., Zou, D., Xu, S., Jin, H., Qi, H., Hu, J.: VulPecker: an automated vulnerability detection system based on code similarity analysis. In: Proceedings of 32nd Annual Conference on Computer Security Applications (ACSAC), pp. 201–213 (2016)

28. Li, Z., Zou, D., Xu, S., Jin, H., Zhu, Y., Chen, Z.: SySeVR: a framework for using deep learning to detect software vulnerabilities. IEEE Trans. Dependable Secure Comput. (TDSC) **19**(4), 2244–2258 (2021)

29. Li, Z., et al.: VulDeePecker: a deep learning-based system for vulnerability detection. In: Proceedings of 25th Annual Network & Distributed System Security Symposium (NDSS) (2018)

30. Lin, G., Zhang, J., Luo, W., Pan, L., Xiang, Y.: POSTER: vulnerability discovery with function representation learning from unlabeled projects. In: Proceedings of 24th ACM Conference on Computer and Communications Security (CCS), pp. 2539–2541 (2017)

31. Liu, P., Yuan, W., Fu, J., Jiang, Z., Hayashi, H., Neubig, G.: Pre-train, prompt, and predict: a systematic survey of prompting methods in natural language processing. ACM Comput. Surv. **55**(9), 1–35 (2023)

32. Liu, X., et al.: P-tuning: prompt tuning can be comparable to fine-tuning across scales and tasks. In: Proceedings of 60th Annual Meeting Association for Computational Linguistics (Volume 2), pp. 61–68 (2022)

33. Liu, X., et al.: GPT understands, too. AI Open (2023)

34. Liu, Z., Yu, X., Fang, Y., Zhang, X.: GraphPrompt: unifying pre-training and downstream tasks for graph neural networks. In: Proceedings of 32nd International World Wide Web Conference (WWW), pp. 417–428 (2023)

35. Lu, Y., Jiang, X., Fang, Y., Shi, C.: Learning to pre-train graph neural networks. In: Proceedings of 35th AAAI Conference on Artificial Intelligence (AAAI), pp. 4276–4284 (2021)

36. Mirsky, Y., et al.: VulChecker: graph-based vulnerability localization in source code. In: Proceedings of 32nd USENIX Security Symposium, pp. 6557–6574 (2023)

37. NIST: Software assurance reference dataset (2023). https://samate.nist.gov/SARD

38. Pagliardini, M., Gupta, P., Jaggi, M.: Unsupervised learning of sentence embeddings using compositional n-gram features. In: Proceedings of Conference on North American Chapter Association for Computational Linguistics: Human Language Technologies (NAACL), vol. 1, pp. 528–540 (2018)

39. Pham, N.H., Nguyen, T.T., Nguyen, H.A., Nguyen, T.N.: Detection of recurring software vulnerabilities. In: Proceedings of 25th IEEE/ACM International Conference on Automated Software Engineering (ASE), pp. 447–456 (2010)

40. Russell, R., et al.: Automated vulnerability detection in source code using deep representation learning. In: Proceedings of 17th IEEE International Conference on Machine Learning and Applications (ICMLA), pp. 757–762 (2018)

41. Sajnani, H., Saini, V., Svajlenko, J., Roy, C.K., Lopes, C.V.: SourcererCC: scaling code clone detection to big-code. In: Proceedings of 38th International Conference on Software Engineering (ICSE), pp. 1157–1168 (2016)

42. Satorras, V.G., Estrach, J.B.: Few-shot learning with graph neural networks. In: Proceedings of 6th International Conference on Learning Representations (ICLR) (2018)

43. Shankar, U., Talwar, K., Foster, J.S., Wagner, D.: Detecting format string vulnerabilities with type qualifiers. In: Proceedings of 10th USENIX Security Symposium (2001)

44. Shar, L.K., Briand, L.C., Tan, H.B.K.: Web application vulnerability prediction using hybrid program analysis and machine learning. IEEE Trans. Dependable Secure Comput. (TDSC) **12**(6), 688–707 (2014)

45. Sun, M., Zhou, K., He, X., Wang, Y., Wang, X.: GPPT: graph pre-training and prompt tuning to generalize graph neural networks. In: Proceedings of 28th ACM SIGKDD Conference on Knowledge Discovery and Data Mining (KDD), pp. 1717–1727 (2022)

46. Synopsys: Open source security and risk analysis report (OSSRA) (2023). https://www.synopsys.com/content/dam/synopsys/sig-assets/reports/rep-ossra-2023.pdf

47. Togninalli, M., Ghisu, E., Llinares-López, F., Rieck, B., Borgwardt, K.: Wasserstein Weisfeiler-Lehman graph kernels. In: Proceedings of 33rd Conference on Neural Information Processing Systems (NeurIPS), pp. 6436–6446 (2019)

48. Wang, Y., Wang, W., Joty, S., Hoi, S.C.: CodeT5: identifier-aware unified pre-trained encoder-decoder models for code understanding and generation. In: Proceedings of Conference on Emperical Methods in Natural Language Processing (EMNLP), pp. 8696–8708 (2021)

49. Wu, Y., Zou, D., Dou, S., Yang, W., Xu, D., Jin, H.: VulCNN: an image-inspired scalable vulnerability detection system. In: Proceedings of 44th International Conference on Software Engineering, pp. 2365–2376 (2022)

50. Yamaguchi, F., Golde, N., Arp, D., Rieck, K.: Modeling and discovering vulnerabilities with code property graphs. In: Proceedings of 35th IEEE Symposium on Security & Privacy (S&P), pp. 590–604 (2014)

51. Yamaguchi, F., Maier, A., Gascon, H., Rieck, K.: Automatic inference of search patterns for taint-style vulnerabilities. In: Proceedings of 36th IEEE Symposium on Security & Privacy (S&P), pp. 797–812 (2015)

52. Yao, H., et al.: Graph few-shot learning via knowledge transfer. In: Proceedings of 34th AAAI Conference on Artificial Intelligence (AAAI), pp. 6656–6663 (2020)

53. Zhang, M., Cui, Z., Neumann, M., Chen, Y.: An end-to-end deep learning architecture for graph classification. In: Proceedings of 32nd AAAI Conference on Artificial Intelligence (AAAI) (2018)

54. Zhou, Y., Liu, S., Siow, J., Du, X., Liu, Y.: Devign: effective vulnerability identification by learning comprehensive program semantics via graph neural networks. In: Proceedings of 33rd Conference on Neural Information Processing Systems (NeurIPS) (2019)

55. Zou, D., Wang, S., Xu, S., Li, Z., Jin, H.: μVulDeePecker: a deep learning-based system for multiclass vulnerability detection. IEEE Trans. Dependable Secure Comput. (TDSC) **18**(5), 2224–2236 (2019)

All Your LLMs Belong to Us: Experiments with a New Extortion Phishing Dataset

Fatima Zahra Qachfar[✉] and Rakesh M. Verma

University of Houston, Houston, TX 77004, USA
fqachfar@uh.edu, rmverma2@central.edu

Abstract. In the last decade, there has been a dramatic rise in phishing emails including business email compromise, and extortion attacks. *Ransomware* and *blackmail* are examples of extortion threats, where attackers force victims to follow orders, send money, or share sensitive information. To our knowledge, no phishing dataset that includes text-based extortion attacks has been made available to the security research community. To address this problem, we present "TExtPhish" consisting of a sentence-level subset and a full-body email-level subset that can be used for multiple classification and regression tasks. We also provide another challenging subset with homograph text perturbations to address a specific Unicode NLP attack targeting primarily LLMs, causing them to hallucinate and significantly degrade their performance. We show this by conducting multiple experiments including extortion classification, sentiment analysis, and language identification. Our findings indicate that DistilBERT is most susceptible to homograph attacks at sentence level, resulting in a 94.9% decrease in F1-score while DeBERTa's performance decreased 94.1% at email level.

Keywords: LLM · NLP · Trustworthy AI · Homograph · Attack · Perturbation · Extortion · Phishing · Blackmail · Ransomware · Sextortion

1 Introduction

The FBI's Internet Crime Complaint Center (IC3) [25] reported that the top three infection vectors of ransomware attacks are (i) phishing email campaigns, (ii) software vulnerabilities, and (iii) remote desktop control. According to IC3 2023 annual report [26], the losses from extortion attacks have increased significantly, with adjusted losses of more than $74.8 million from $48,223$ victims in 2023 including $2,825$ complaints identified as ransomware with adjusted losses of more than $59.6 million across 14 different sectors. To our knowledge, there are no quality public datasets that examine the link between phishing emails and ransomware attacks. In this work, we will try to fill this gap by creating a well-designed dataset with multiple subsets.

A Russian ransomware named "Netwalker" was first discovered in September 2019. This attack evolved through time, introducing Covid-themed phishing emails to trick employees into injecting malware into their company networks. This Russian attack is

Supplementary Information The online version contains supplementary material available at (https://doi.org/10.1007/978-3-031-65172-4_15).

A. L. Ferrara and R. Krishnan (Eds.): DBSec 2024, LNCS 14901, pp. 241–258, 2024.
https://doi.org/10.1007/978-3-031-65172-4_15

a Ransomware as a Service (RaaS) model ready for criminals to easily deploy. RaaS offers highly sophisticated cyberattacks, not requiring any background hacking knowledge. Other examples of RaaS include "REvil" [5] and "Ryuk" [6]. In July 2020, The FBI issued a FLASH (FBI Liaison Alert System) warning "MI-000130-MW" [17] about these RaaS attacks due to their extreme severity. The FLASH reports convey industry-specific details about highly advanced and emerging cyber threats to warn potential victims and reduce risks.

Most of these ransomware attacks start from phishing emails. For instance, REvil [5] and Ryuk [6] ransomware attacks have been delivered through legitimate-looking emails such as document signing services "DocuSign" containing links to malicious documents with embedded macros. These emails were either accompanied by phishing links that redirect the victim to malicious content or contained malware as email attachments.

The knowledge base of adversary tactics and techniques MITRE ATT&CK [7] provides an exhaustive list of all malware attacks that were initiated using spear-phishing emails with an attachment, with a link, or as a service. Some malware actors like WannaCry or Cuba ransomware [27] do not rely on phishing emails as a medium to upload malware, but instead they use operating system exploits, backdoor access, or disk encryption. In this work, we focus on extortion scams that employ emails to initiate attacks, and we provide a textual email dataset that highlights extortion behavior in blackmail, ransomware, and sextortion scams[1]. We show that extortion emails in the majority contain anticipation, fear, anger sentiments. We also take a deeper look into homograph visual attacks on phishing emails.

In summary, we make the following contributions:

– We provide an extortion dataset that contains three types of extortion attacks (*blackmail*, *ransomware*, and *sextortion*) at both email and sentence levels (Sect. 3).
– We ensure a sequential data curation process from collection to sanitization to categorization with data quality control (Sect. 3).
– We show that LLMs struggle when confronted with homograph attacks in various NLP tasks (Sect. 4, and Sect. 5).

First, we review prior research work related to extortion and homograph attacks in Sect. 2. Then, we introduce our extortion dataset "TExtPhish". We provide a detailed process of collecting, sanitizing, and categorizing the datasets in Sect. 3. We conduct different experiments using this dataset in Sect. 4 including sentiment analysis, and extortion classification. We further investigate the impact of homograph attacks on LLMs beyond TExtPhish extending to large language identification datasets in Sect. 5 and we conclude the paper with future work in Sect. 6.

2   Related Works

We discuss previous work under two categories of attacks: extortion and homograph.

---

[1] ***Disclaimer :*** This paper includes profane and vulgar language. It is only for research purposes.

2.1 Extortion Attacks

IC3 [26] defined extortion as an unlawful extraction of money or property through intimidation or undue exercise of authority. It includes different types of attacks, causing physical harm, criminal prosecution, or public exposure. Extortion can take many forms (*blackmail, ransomware, sextortion*) depending on the attacker's purpose using different abusive means (*vocals* [19], *emails*, *SMS messages* [39]). Extortion emails aim to extort money, personal property, or other confidential information from victims via an explicit email that may contain a malicious attachment or link.

One type of extortion email is *blackmail*, considered as a threatening abuse of authority bound by time which is a form of scarcity. Blackmailing is not always money-driven, but can also include other forms of extortion like revenge or power abuse [12]. While *ransomware* is an extortion attack designed to extort money from victims using malware [49]. Researchers in [57] studied the decisions made by victims that were threatened to pay the ransomware. They showed that many victims consider paying the ransom to minimize disruption and the risk of further financial loss. Another form of extortion that involves sexual and sensitive content is *sextortion*. The targets are threatened to have their lives ruined unless they pay untraceable money using cryptocurrencies like Bitcoin [42].

These extortion attacks can be carried out by implementing automated botnets that can download malicious payloads and send extortion attacks in mass quantities. Initially discovered in 2018, "MyloBot" was considered one of the most evasive botnets at the time, with sophisticated techniques such as code injection and anti-sandboxing. Another version surfaced in 2022 that downloads malicious payloads from the command-and-control (C&C) server and attaches them to extortion emails. The C&C servers serve as command centers for malware that stores stolen data or downloads commands to execute targeted attacks.

The "MyloBot" botnet is designed to abuse the victim by sending extortion emails based on their online usage. Another botnet called "Phorpiex" was reported to have sent over 30,000 *sextortion* emails per hour. While another attack delivered by "Phorpiex" called "BitRansomware" distributed a swarm of ZIP archive files containing ransomware in malicious executables. In "BitRansomware", during the encryption process, all compromised files are appended with the".readme" extension. The ransom demand informs its victims that all their databases, documents, photos, and other important files have been encrypted.

In sum, many extortion attacks are initiated using phishing emails, which makes a dataset like "TExtPhish" important for security researchers to help them analyze and study the patterns of such attacks. Before a victim opens an email attachment or unintentionally installs malware, this dataset can have a significant role in detecting ransomware attacks as soon as an email is received, and before it can do damage.

2.2 Homograph and Related Attacks

Homograph attacks are visual attacks based on look-alike textual characters called "confusables"[2] used to fool victims into thinking that they are accessing legitimate content.

---

[2] https://util.unicode.org/UnicodeJsps/confusables.jsp.

For example, the mathematical alpha character "α" looks confusingly similar to "a" in some italic fonts. These confusable characters might seem identical, but they have different Unicodes that identify them.

Several studies have proposed detection methods for homograph attacks on domain names, but there is no available dataset that contains homograph attacks on long text.

Although most literature examined homograph attacks in light of International Domain Names IDNs, other researchers investigated the potential effect of homographs not only in IDNs but also in text and images [32,46,52,53]. Authors in [32] proposed two novel hidden backdoor attacks on language models, the "homograph attack" and the "dynamic sentence attack." The homograph attacks are highlighted in the case of tokenization with large language models (LLMs) where homographs are replaced with unrecognizable tokens [UNK]. Typically, [UNK] tokens are accepted as legitimate inputs and disregarded during classification which can lead to false negatives. Researchers in [41] presented an injection attack called *Zero-Width* (ZeW) based on injecting nonprintable Unicode characters into sentences evading ML classification.

For example, CatchPhish [52] identified brand names using homographs on phishing websites. While authors in [1] detected plagiarism in texts obfuscated with homographs. Another study [45] investigated homograph attacks on LLMs, specifically targeting sentiment analyzers for the Maghreb region. They show that these attacks result in a 65.3% decrease in sentiment classification from an F1-score of 0.95 to 0.33 when data is written in an "Arabizi" style (Arabic in ASCII script form only). Researchers in [46] detect Unicode visual spoofing based on the context of suspicious characters with a probabilistic Bayesian model. The approach highlights in red color the homograph characters in a given string to prevent end users from being deceived.

To measure character similarity, researchers in [20] use a Unicode homograph list based on the amount of visual and semantic similarity between each pair of characters. They apply regular expressions to detect phishing patterns by using a simple pixel-overlapping method, while another group [48] proposed the NCD (Normalized Compression Distance) to measure the degree of similarity. The lower the NCD, the more similar the glyphs ($0 \leq NCD \leq 1$). In [48], researchers calculated NCD for a set of around 6,200 Unicode characters selected from more than 40 languages.

Another visual deceptive method of bypassing email security is to insert invisible unicode characters within the text, such as soft hyphens or word joiners [38] [41]. When a soft hyphen is inserted into an email, it normally renders as invisible. Nevertheless, security email software that scans for malicious content will detect it as a Unicode character. In the "TExtPhish" dataset, we focus on *visual homograph*, which uses visually look-alike characters that have a different Unicode. Several tools were also developed to help generate homographs within the text such as [15,54]. More details on the implementation are in Sect. 3.3.

2.3 Data Privacy

Recent studies showed that memorization of the training data presents significant privacy risks when it contains sensitive personal information. Researchers in [10] attacked large pretrained language models to show how models' memorization can affect data privacy. They extracted with the best attack configuration about 67% of candidate samples that were memorized verbatim in training data, including employee identities and organization entities [18].

In order to guarantee data privacy, we have considered techniques like *k-anonymity* [51]. However, only tabular data benefits from similar methods in contrary to textual data. A solution for textual data anonymization is the de-identification tool "PrivateAI" [43] which removes personally identifiable information (PII).

PrivateAI [43] also provides a protection layer to mitigate ChatGPT privacy concerns to create privateGPT based on privacy norms. Open AI's ChatGPT has recently raised significant privacy concerns regarding breaches of European legislation on personal data processing and protection. Due to this, ChatGPT was banned in Italy on April 1st 2023, by the Italian Data Protection Authority (DPA) [8,31]. Personal information including credit card numbers was exposed in its first data leak.

3   Dataset Description

In this section, we present "TExtPhish" starting with the *email collection* process followed by *data sanitization* and *anonymization* to ensure data privacy, and concluding with *data categorization*.

3.1   Data Collection

The collection process to obtain extortion attacks and benign samples is as follows.

Extortion Samples. In May 2022, we scraped seven Reddit thread websites from the phishing campaign "The Blackmail Email Scam" [36]. The last Reddit thread is active today, and victims are still posting extortion attacks. We used the Reddit API python wrapper for developers to scrape all the posts and comments that contain extortion emails. Emails extracted from the Reddit posts date from July 2018 to March 2022.

We also select a total of 150 extortion emails that were used to initiate a ransomware attack from evasive botnets and RaaS including Hancitor [3], Phorpiex (BitRansomware), Emotet [4], TeslaCrypt, Qakbot [37], CryptoLocker, NetWalker [17], REvil [5], and Ryuk [6]. For the purpose of collecting ransomware emails spanning the period of 2013 to 2022, we gathered data from multiple botnets.

We observe that the collected extortion samples can be categorized into three types (i) *Blackmail* which can be defined as a recurrent threat where attackers forbid victims from engaging in any lawful occupation, (ii) *Ransomware* which can be defined as emails containing malicious file attachments or links leading to malware. A potential breach or loss of data can result from this malware encrypting files and documents, and (iii) *sextortion*[3] which can be defined as emotional, mental and sexual harassment that threatens victims to expose sexual images to coerce them to provide additional pictures, follow unlawful orders or pay money.

---

[3] Upon closer examination of the collected sextortion emails, we discovered that some samples were sent through the Phorpiex botnet.

According to [12], the most bitcoin-enabled cybercrime is sextortion followed by blackmail scams then ransomware in a total of 21,650 reported bitcoin addresses linked to criminal activities from 2017 to 2022. These addresses were extracted from thousands of *BitCoin Abus* reports, a service platform for victims to disclose Bitcoin addresses of criminals blackmailing them asking for untraceable payments. Table 1 shows some examples of these attack types.

Table 1. Examples of Extortion Attack Types

Attack Type	Examples from Sentence-Level Subset
Blackmail	*- I will delete the corresponding recording and I will not blackmail you ever again.*
Ransomware	*- Send the above amount on my BTC wallet (bitcoin).*
Sextortion	*- In case you ignore me, within 96 h, ur sex tape will be posted on the net.*

Benign Samples. We select benign email samples from publicly available datasets, such as Enron [30] and SpamAssassin [2]. To make the dataset challenging, we keep only the most semantically similar benign emails to the extortion attacks. For semantic textual similarity, we first applied sentence transformers (SBERT) [47] to get contextual sentence embeddings of sanitized benign and extortion samples. Then, we apply the Facebook AI Similarity Search (FAISS) [28] measure to search for similar benign instances to extortion attacks. To achieve this, we index the extortion data and apply a similarity search based on dot product vector comparison to find the top two similar benign samples to each extortion sample. The indexing is saved for later use if we add more benign samples to apply semantic text search.

FAISS is a fast and efficient similarity measure unlike using cosine similarity, which scales very badly in larger datasets. In addition, we choose to apply the semantic similarity on benign samples to preserve the collected extortion attacks which have more priority and fewer examples.

Table 2 below presents the dataset statistics for extortion at sentence-level and email-level.

Table 2. Statistics of Email-level and Sentence-level Subsets

Extortion Samples

Source	#Emails	#Sentences
Reddit Posts [36]	1,113	17,393
Botnet Emails	150	1,510
Total	1,263	18,903

Benign Samples

Source	#Emails		#Sentences	
	BS [a]	**AS** [b]	**BS** [a]	**AS** [b]
Enron [30]	4,207	1,360	35,059	26,835
SpamAssasin [2]	2,750	1,010	15,537	12,348
Total	6,957	2,370	50,596	39,183

[a] *BS : Before Similarity.*   [b] *AS : After Similarity.*

3.2   Data Sanitization and Anonymization

Sanitization Process.   The first step we took after collecting the data is removing all unnecessary extraneous user comments that are not considered part of an attack email. FAISS semantic similarity was applied to cleaned and sanitized attack emails.

Once we have a collection of extortion emails only without additional Reddit user comments, we process the content of these emails and remove any *Personally Identifiable Information* (PII) like names, emails, phone numbers, and *Payment Card Industry* (PCI) information like credit cards and bank account credentials. To ensure data anonymization, we identified sensitive information such as emails, names, SSNs, and passwords by applying both regular expressions and named entity recognition combined with "PrivateAI" tool [43] PII and PCI de-identification processes.

We use PrivateAI to replace sensitive information with synthetic information by using their *synthetic PII generator* that never sees our original data, eliminating sensitive data leakage. The resulting text would reduce re-identification risk making the adversary in need to determine which PII is real. In extortion, attackers demand money from victims through untraceable cryptocurrencies like Bitcoin (btc), DASH (dsh) and Ethereum (eth). We have found multiple wallet hashes that we redact using regular expressions.

Observations.   During the data sanitization phase, we found that many of the newest attacks have homographs (attacks dating from 2018 to 2022). Some of the victims shared that their passwords came from LinkedIn and Dropbox's data breaches. LinkedIn's data breach leaked over 700 million user credentials [22] while Dropbox affected nearly 68 million users [21] leaking very sensitive information. This huge amount of data containing personal information enabled attack vectors like social engineering attacks on targeted users.

Furthermore, researchers in [55] assessed the risk of stolen credentials in phishing and malware attacks. According to their study, millions of victims worldwide are exposed to targeted keylogging and phishing attacks because of data breaches. Hence, our dataset must be thoroughly sanitized before any release so it cannot be used for any social engineering attacks. Several of the extracted emails date back to March 2022, meaning some users' credentials might still be active on other websites if the victims failed to redact them before posting on the Reddit thread.

3.3   Data Categorization

We provide three different subsets in the "TExtPhish" dataset: (i) *email-level subset*, (ii) *sentence-level subset* and (iii) *homograph perturbed subset*. Figure 1 illustrates the process of creating email-level and sentence-level subsets.

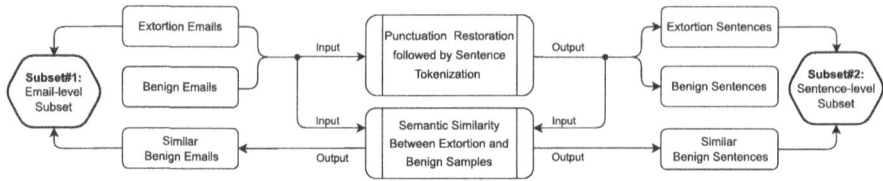

Fig. 1. Subset Creation Process After Email Collection

Email-Level Subset. To construct this dataset, we simply clean and sanitize the originally collected data. We keep the full email body without metadata. The cleaning of these samples includes the removal of any email metadata, encoded attachments, or embedded style sheets. These emails are followed by the FAISS semantic similarity process described above to select similar samples. As illustrated in Fig. 1 (starting from the left side), extortion and benign emails are given as input to semantic similarity to find and concatenate the most similar benign emails to extortion attacks.

Sentence-Level Subset. To construct this dataset, we used sentence tokenization from the NLTK toolkit [35] ("sent_tokenize") on the email samples. Non-punctuated samples are sometimes not well handled by this tokenizer. Therefore, we use the "*deep-multilingualpunctuation*" model [23] to restore the punctuation and extract properly the sentences. This model was trained on the Europarl dataset from the first shared task on Sentence End and Punctuation Prediction in NLG Text (*SEPP-NLG 2021*), which contains transcripts in multiple languages. It can restore text punctuation in up to four languages including English. After punctuation restoration and sentence tokenization, the FAISS semantic similarity process is also applied to sentences to keep only similar benign sentences.

Homograph Perturbed Subset. To construct this dataset, we apply homograph perturbation at character-level using non-Ascii characters and Unicode tables on both email-level and sentence-level subsets to generate homograph text attack examples in different settings as shown in Fig. 2. To achieve this subset, we use the PySameSame tool [54] on the email-level subset with a 30% perturbation on each email sample.

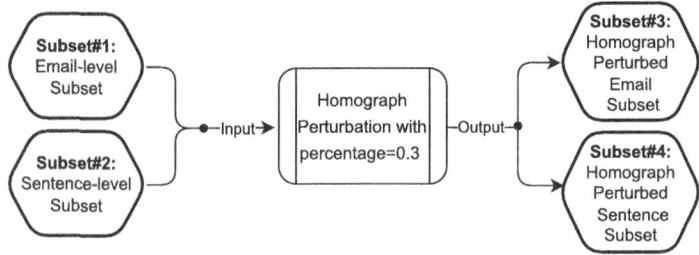

Fig. 2. Homograph Perturbed Subset Process

Upon cleaning the dataset and removing some unnecessary comments associated with the email scam messages, we discovered that many of the recent attacks use homographs and invisible characters for text padding. This is consistent with what the Microsoft 365 Defender Threat Intelligence team reported in its blog post [38] publish in August 2021. In addition, we found that these type of samples embedded with homographs throw off language detectors and machine translation models that do not predict accurately the language. This raises a great challenge for machine translation tasks since most language identification datasets do not consider homographs.

4   Experiments

We ran experiments on a 512 GB RAM machine with an Intel(R) Xeon(R) E5-2667 v4 (3.20 GHz) processor with 16 cores and 32 threads running Linux Red Hat Enterprise Server 7.6 (with Nvidia Tesla M10 GPU for semantic search with *FAISS-GPU*). To expose the weaknesses of LLMs against homograph attacks, we conduct two NLP tasks.

4.1   Extortion Multi-classification

All the extortion subsets were put through rigorous experimentation using a total of eight models, consisting of four baseline algorithms and four deep learning language transformers. To perform these experiments, a splitting percentage of 80% was applied to the data for training and 20% for testing after shuffling with a random seed of 42.

For the baseline models, we implement a pipeline object that extracts the TF-IDF features and vectorizes textual content using unigram and bigram count vectorizer. We have defined a grid of hyperparameters to search over using GridSearchCV with two cross-validation splits. Hyperparameter tuning include parameters like regularization strength (C), number of maximum iteration (*max_iter*), and the number of estimators (*n_estimators*) reported in the supplementary material. The best estimator is used to

Table 3. F_1 Score of Phishing Extortion Multi-Classification at **Email-Level** On *TExtPhish* Testing Sets. Results in **bold** represent the best performing model.

	Models	Email-Level Subset Without Perturbation	Trained and Tested on Homograph Perturbation	Only Tested on Homograph Perturbation
Baselines	SVM [14]	0.853	0.776 (9.0% ▽)	**0.732** (14.2% ▽)
	RF [9]	0.814	0.551 (32.3% ▽)	0.410 (49.6% ▽)
	LR [56]	0.839	0.746 (11.1% ▽)	0.668 (20.4% ▽)
	XGB [11]	0.826	0.705 (14.6% ▽)	0.563 (31.8% ▽)
Transformers	BERT [16]	0.915	0.810 (11.5% ▽)	0.218 (76.2% ▽)
	DeBERTa [24]	0.984	0.807 (18.0% ▽)	0.058 (94.1% ▽)
	RoBERTA [34]	**0.985**	**0.834** (15.3% ▽)	0.147 (85.1% ▽)
	DistilBERT [50]	0.952	0.807 (15.2% ▽)	0.208 (78.2% ▽)

Table 4. F_1 Score of Phishing Extortion Multi-Classification at **Sentence-Level** On *TExtPhish* Testing Sets. Results in **bold** represent the best performing model.

	Models	Sentence-Level Subset Without Perturbation	Trained and Tested on Homograph Perturbation	Only Tested on Homograph Perturbation
Baselines	SVM [14]	0.808	0.639 (20.9% ▽)	**0.527** (34.8% ▽)
	RF [9]	0.754	0.531 (29.6% ▽)	0.474 (37.1% ▽)
	LR [56]	0.799	0.624 (21.9% ▽)	0.518 (35.2% ▽)
	XGB [11]	0.756	0.560 (25.9% ▽)	0.495 (34.5% ▽)
Transformers	BERT [16]	0.814	0.654 (19.7% ▽)	0.087 (89.3% ▽)
	DeBERTa [24]	**0.875**	0.728 (16.8% ▽)	0.051 (94.2% ▽)
	RoBERTA [34]	0.867	**0.736** (15.1% ▽)	0.104 (88.0% ▽)
	DistilBERT [50]	0.855	0.663 (22.5% ▽)	0.044 (94.9% ▽)

predict the testing set. For the large language models, we use four high performing transformers for sequence classification in their base, uncased setting.

Table 3 presents F1 macro scores for multi-classification of phishing extortion at the email level. Our experiments reveal that ML models' performance deteriorates drastically, when exposed to homograph perturbations solely during testing without training.

When trained and tested on homographs, RF algorithm suffers the most with a decrease percentage of 32.3% at email-level and 29.6% at sentence-level. When not trained on homograph perturbations, DeBERTa and DistilBERT transformers suffer the most with a decrease percentage of 94.1% at the email-level and 94.9% at the sentence-level respectively. This staggering decline in LLMs' performance is attributed to their lack of exposure to homographs during training. This results in an increase in the total count of unknown word instances, commonly referred to as Out of Vocabulary (OOV) tokens which are represented as [UNK] tokens by the transformers. This substitution of OOV tokens with [UNK] leads to the loss of crucial information, impacting the accuracy of classifiers. Transformers heavily rely on contextual information to make predictions. When multiple [UNK] tokens emerge within a sequence, the model may struggle to grasp the overall context of the perturbed text, resulting in diminished performance.

For the best performing models, when we train with perturbation RoBERTa [34] achieves an $F_1 score$ of 0.834 and 0.736 at email and sentence-level. We observe that SVM baseline shows a significant resilience to the homograph attack with and without training on character perturbations. This is due to the margin of SVM which enhances the ability to generalize to unseen data points, making it more robust to noise and OOV words compared to transformers which struggle with larger amounts of unknown tokens.

Our conclusion is that all transformers experience substantial difficulties when they have not been previously trained on perturbed data. Tables 3 and 4 provide evidence that training LLMs with perturbed samples serves as a defense strategy against these attacks.

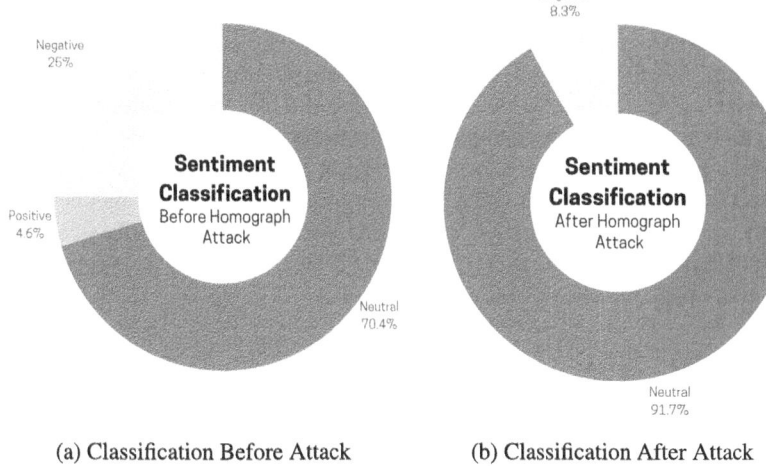

(a) Classification Before Attack   (b) Classification After Attack

Fig. 3. Homograph Attack Impact on Sentiment Classification at Sentence-Level

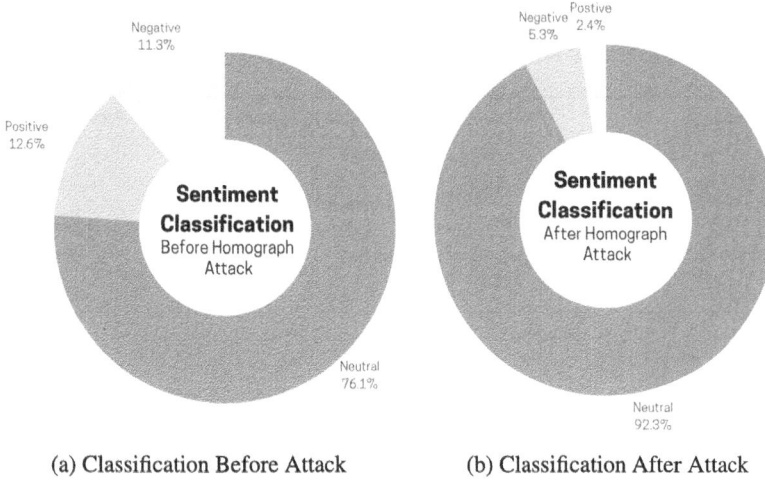

(a) Classification Before Attack   (b) Classification After Attack

Fig. 4. Homograph Attack Impact on Sentiment Classification at Email-Level

4.2 Sentiment Analysis

We investigate the impact of homograph perturbations using the sentiment classifier BERTweet [40,44] on TExtPhish. BERTweet is fine-tuned on Twitter data for sentiment classification, capable of analyzing and categorizing sentiments expressed in text into three labels (Negative, Postive, or Neutral).

In Fig. 3b, we observe that BERTweet failed to identify any positive instances at sentence-level. It only captures 8.3% of negative ones instead of 25% in Fig. 3a. As a result, the majority of negative examples were misclassified as neutral because the model was unable to capture offensive and negative terms obfuscated using homograph characters. At the email-level, as shown in Fig. 4b, 92.3% of TExtPhish emails are labeled neutral when only 76.1% were detected as neutral prior to the attack in Fig. 4a.

This becomes concerning in the context of harmful and toxic content classification, as homograph attacks can successfully evade classifiers, yet they remain visually harmful to human users even though machines may not detect the harm. Such attacks can enable the circumvention of automated bans in groups and on social media platforms, potentially facilitating the spread of propaganda, hate speech, and offensive language. We conduct these experiments to further highlight the importance of examining homograph attacks on LLMs across various NLP tasks.

5   Discussion

5.1   LLM Weakness Beyond TExtPhish

We further investigate the impact of homograph attacks on different LLMs FastText [29], and XLM-V [33] trained and tested on multilingual identification datasets.

Table 5. Experiment of Language Identification on Homoglyph Perturbation Subset

Examples	LID Model	Language Detected	Prediction Score
that ıs ą fɑır deąl, aⁿd the prıce ıѕ relatıvely loщ, consıderıng tʜąt i Hąve ƀeeⁿ cʜecкıng oʋt yoʋr profıle ąnd traffıc for sᴏme tıⱥe ƀy ɳᴏw.	FastText	Dinka (Sudan)	0.16319
	XLM-V-base	Maltese	0.9922
I ąccęѕѕęd to ąłł your ąccountѕ, ѕociął nętworkѕ, ęmąıł, browѕing hiѕtory	FastText	Polish	0.9976
	XLM-V-base	Maltese	0.99547

Facebook Language Identification. To highlight the importance of building a homograph perturbation subset in our extortion dataset. We test the two most popular language identification models in the Hugging-Face Hub that were trained on large multilingual datasets. This training data does not consider the impact of homoglyph perturbation on language prediction.

In Table 5, Facebook language identification model [29] inaccurately predicts the homoglyph perturbed *English* sentence as *Dinka* sentence as the first classification language label with the highest score of 0.16. This weak score shows that this perturbation drastically decreases the prediction confidence of the language identification model while increasing the error rate of predicted labels.

The "XLM-V-base" model [33] confidently predicts both *English* sentences as *Maltese* with a shocking prediction score of 0.99. The results demonstrate that large language models are susceptible to homograph attacks.

Google Fleurs Language Identification. We address these challenges by applying our homograph perturbation implementation on the validation set that was earlier employed to evaluate the performance of the presumed highly confident "XLM-V-base" model [33]. The model delivered an accuracy rate of 0.9930 and *1* on the complete Google Fleurs validation set (102 languages) [13] and the English validation set, respectively. It is important to note that the complete evaluation set includes languages such as Arabic, Japanese, and Korean, which were not accounted for in our character level perturbation. In this multilingual scenario, the model's performance only experiences a 20% decline, in contrast to the English validation subset where "XLM-V-base" classification plummets significantly by 89%, resulting in an accuracy of 0.1091.

In a nutshell, assessing language models' resilience to textual perturbations is of utmost importance. Our experiments have unequivocally demonstrated the criticality of analyzing perturbed subsets and their impact on the models. It is imperative that we thoroughly evaluate the models' ability to withstand such perturbations. As illustrated in Fig. 5, the model delivered an accuracy rate of 0.9930 and 1.

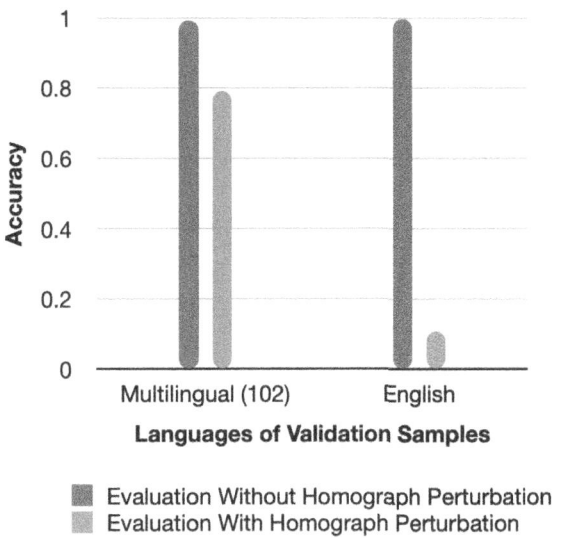

Fig. 5. Impact of Homograph Attack on Language Identification with 30% Perturbation

5.2 Perturbation Percentage

During the homograph perturbation process described in Fig. 2, we identify the minimal perturbation percentage that results in a substantial degradation of the model's performance, illustrating the sensitivity of large models to even minor alterations. As this perturbation percentage increases, the task becomes more challenging for LLMs, leading to a decline in their predictive accuracy and overall model behavior as illustrated in Fig. 6 on Google Fleurs multilingual dataset.

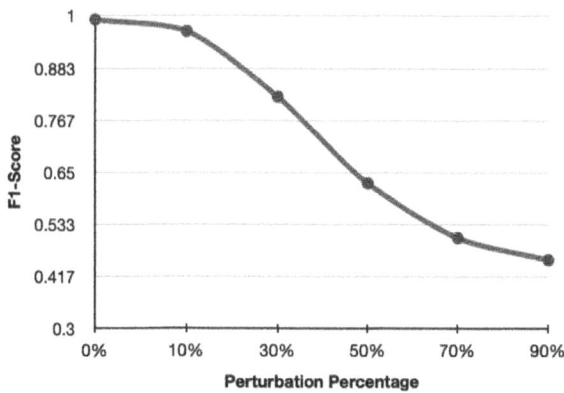

Fig. 6. Varying Perturbation Percentage of Homographs on Language Identification Task

6   Conclusions and Future Work

This paper provides the first email extortion dataset named "TExtPhish" that includes a total of 1,263 unperturbed extortion emails including 251 *blackmail*, 280 *ransomware*, and 732 *sextortion* cyberattacks in the English Language. "TExtPhish" can be used to train a model for phishing classification, and offensive speech recognition, which are multi-class text classification tasks.

This dataset also includes a homograph perturbed subset to draw more attention to this backdoor attack on language models to study and address such attacks on a broader research scale. Char-based phishing classification tasks could also benefit greatly from this subset. Language models that accept raw Unicode input can be attacked with homograph substitution as an adversarial method.

In the future, the dataset could be extended by including ransomware demand messages sent to victims after establishing the encryption phase. To address the scalability of "TExtPhish", we can augment it using prompt engineering and ChatGPT type models and periodically scrape Reddit and other relevant sites to maintain the dataset.

Limitations and Challenges

We list below the "TExtPhish" dataset limitations:

- Phishing attacks are constantly evolving, with new targeted tactics and extortion techniques being developed by attackers. This means that phishing datasets can quickly become outdated especially with the era of large language generative models.
- "TExtPhish" exhibits a disproportionate distribution of benign and phishing instances, with the latter being overwhelmingly composed of sextortion attacks.

We also list some challenges of *homograph perturbations and LLMs*:

– Low-resource languages might struggle most of such attacks.
– LLMs, such as prompt-based models that rely on user input for retraining, could potentially introduce a substantial amount of noise to their data when confronted with homograph attacks.

Ethics Statement. The authors of this dataset acknowledge the potential harm caused by misuse and manipulation of the information within it, and they caution against utilizing it in ways that promote deceit or harm.

With the fast-growing field of natural language generation and adversarial attacks, we are aware that this data might be used unlawfully. Therefore, we license the data to avoid any unintended use of the dataset. The "TExtPhish" collection may only be used for linguistic research and technology development including but not limited to text classification, natural language processing, machine learning, phishing detection, data privacy, and like fields.

A portion of the dataset was downloaded using Reddit's API Wrapper through the PRAW python package. Re-use of this data is subject to Reddit API terms:
– Users shall not encourage or promote illegal activity throughout the use of this dataset.
– Users shall not use this dataset with the intent of introducing any viruses, worms, defects, Trojan horses, malware, or any other items of a destructive nature.
– Users shall not sell, lease, or sublicense this data whether for direct commercial or monetary gain.

To ensure the anonymity of "TExtPhish", we followed a rigorous data sanitization and anonymization process described in Sect. 3.2.

Acknowledgements. Research partly supported by NSF grants 2210198 and 2244279, and ARO grants W911NF-20-1-0254 and W911NF-23-1-0191. Verma is the founder of Everest Cyber Security and Analytics, Inc.

References

1. Alvi, F., Stevenson, M., Clough, P.: Plagiarism detection in texts obfuscated with homoglyphs. In: Jose, J.M., et al. (eds.) ECIR 2017. LNCS, vol. 10193, pp. 669–675. Springer, Cham (2017). https://doi.org/10.1007/978-3-319-56608-5_64
2. Apache: Spam Assassin. Apache Software Foundation (2018). https://spamassassin.apache.org/old/publiccorpus/
3. ATT&CK, M.: Hancitor Malware Description: Techniques used. MITRE Corporation (2020). https://attack.mitre.org/software/S0499/
4. ATT&CK, M.: Hancitor Malware Description: Techniques used. MITRE Corporation (2020). https://attack.mitre.org/software/S0367/
5. ATT&CK, M.: REvil Ransomware Description: Techniques used. MITRE Corporation (2020). https://attack.mitre.org/software/S0496//
6. ATT&CK, M.: Ryuk Ransomware Description: Techniques used. MITRE Corporation (2020). https://attack.mitre.org/software/S0446/
7. ATT&CK, M.: Techniques and Tactics of Phishing: Spearphishing Attachment. MITRE Corporation (2020). https://attack.mitre.org/techniques/T1566/001/
8. BBC: ChatGPT banned in Italy over privacy concerns. BBC news (2023). https://www.bbc.com/news/technology-65139406

9. Breiman, L.: Random forests. Mach. Learn. **45**, 5–32 (2001)
10. Carlini, N., et al.: Extracting training data from large language models. In: 30th USENIX Security Symposium, pp. 2633–2650 (2021)
11. Chen, T., et al.: Xgboost: extreme gradient boosting. R Package Version 0.4-2 **1**(4), 1–4 (2015)
12. Cong, L.W., Harvey, C.R., Rabetti, D., Wu, Z.Y.: An anatomy of crypto-enabled cybercrimes. In: Working Paper 30834, National Bureau of Economic Research (2023). https://doi.org/10.3386/w30834. http://www.nber.org/papers/w30834
13. Conneau, A., et al.: Fleurs: Few-shot learning evaluation of universal representations of speech. arXiv preprint arXiv:2205.12446 (2022)
14. Cortes, C., Vapnik, V.: Support-vector networks. Mach. Learn. **20**, 273–297 (1995)
15. Crenshaw, A.: Homoglyph Attack Generator. IronGreek (2012). http://www.irongeek.com/homoglyph-attack-generator.php
16. Devlin, J., Chang, M.W., Lee, K., Toutanova, K.: BERT: pre-training of deep bidirectional transformers for language understanding (2019)
17. FBI: Netwalker Ransomware, FBI FLASH: MI-000130-MW. Federal Bureau of Investigation, Cyber Division (2020). https://www.aha.org/system/files/media/file/2020/08/FLASH-MI-000130-MW.pdf
18. Feldman, V.: Does learning require memorization? a short tale about a long tail. In: Proceedings of the 52nd Annual ACM SIGACT Symposium on Theory of Computing. STOC 2020, pp. 954–959. Association for Computing Machinery, New York (2020). https://doi.org/10.1145/3357713.3384290
19. Forbes: Fraudsters cloned company director's voice in $35 million bank heist, police find (2021). https://www.forbes.com/sites/thomasbrewster/2021/10/14/huge-bank-fraud-uses-deep-fake-voice-tech-to-steal-millions/?sh=5f81d9537559
20. Fu, A.Y., Deng, X., Wenyin, L.: Regap: a tool for unicode-based web identity fraud detection. J. Digital Forensic Pract. **1**(2), 83–97 (2006)
21. Gibbs, S.: Dropbox hack leads to leaking of 68m user passwords on the internet. The Guardian (2016). https://www.theguardian.com/technology/2016/aug/31/dropbox-hack-passwords-68m-data-breach
22. Gibson, B., Townes, S., Lewis, D., Bhunia, S.: Vulnerability in massive API scraping: 2021 linkedin data breach. In: 2021 International Conference on Computational Science and Computational Intelligence (CSCI), pp. 777–782. IEEE (2021)
23. Guhr, O., Schumann, A.K., Bahrmann, F., Böhme, H.J.: Fullstop: multilingual deep models for punctuation prediction. In: Proceedings of the Swiss Text Analytics Conference 2021. CEUR Workshop Proceedings, Winterthur, Switzerland (2021). http://ceur-ws.org/Vol-2957/sepp_paper4.pdf
24. He, P., Liu, X., Gao, J., Chen, W.: DeBERTa: decoding-enhanced BERT with disentangled attention (2021)
25. IC3: Annual Internet Crime Report Released. Federal Bureau of Investigation (2021). https://www.ic3.gov/Media/PDF/AnnualReport/2021_IC3Report.pdf
26. IC3: Annual Internet Crime Report Released. Federal Bureau of Investigation (2023). https://www.ic3.gov/Media/PDF/AnnualReport/2023_IC3Report.pdf
27. Federal Bureau of Investigation, C.D.: Cuba Ransomware, FBI FLASH: CU-000156-MW. Federal Bureau of Investigation (2021). https://www.ic3.gov/Media/News/2021/211203-2.pdf
28. Johnson, J., Douze, M., Jégou, H.: Billion-scale similarity search with GPUs. IEEE Trans. Big Data **7**(3), 535–547 (2019)
29. Joulin, A., Grave, E., Bojanowski, P., Mikolov, T.: Bag of tricks for efficient text classification. arXiv preprint arXiv:1607.01759 (2016)

30. Klimt, B., Yang, Y.: The enron corpus: a new dataset for email classification research. In: Boulicaut, J.-F., Esposito, F., Giannotti, F., Pedreschi, D. (eds.) ECML 2004. LNCS (LNAI), vol. 3201, pp. 217–226. Springer, Heidelberg (2004). https://doi.org/10.1007/978-3-540-30115-8_22

31. Kreitmeir, D.H., Raschky, P.A.: The unintended consequences of censoring digital technology–evidence from Italy's chatgpt ban. arXiv preprint arXiv:2304.09339 (2023)

32. Li, S., et al.: Hidden backdoors in human-centric language models. In: Proceedings of the 2021 ACM SIGSAC Conference on Computer and Communications Security, pp. 3123–3140 (2021)

33. Liang, D., et al.: XLM-V: Overcoming the Vocabulary Bottleneck in Multilingual Masked Language Models. arXiv e-prints arXiv:2301.10472 (2023). https://doi.org/10.48550/arXiv.2301.10472

34. Liu, Y., et al.: RoBERTa: a robustly optimized BERT pretraining approach (2019)

35. Loper, E., Bird, S.: Nltk: the natural language toolkit. arXiv preprint cs/0205028 (2002)

36. Members, U.R.: The Blackmail Email Scam. Reddit Inc. (2018). https://www.reddit.com/r/Scams/comments/n00kg3/the_blackmail_email_scam_part_7/

37. Microsoft: A closer look at qakbot's latest building blocks (2021). https://www.microsoft.com/en-us/security/blog/2021/12/09/a-closer-look-at-qakbots-latest-building-blocks-and-how-to-knock-them-down/

38. Microsoft: Trend-spotting email techniques: How modern phishing emails hide in plain sight (2021). https://www.microsoft.com/en-us/security/blog/2021/08/18/trend-spotting-email-techniques-how-modern-phishing-emails-hide-in-plain-sight/

39. Mulliner, C., Golde, N., Seifert, J.P.: Sms of death: from analyzing to attacking mobile phones on a large scale. In: USENIX Security Symposium, San Francisco, CA, vol. 168 (2011)

40. Nguyen, D.Q., Vu, T., Nguyen, A.T.: BERTweet: a pre-trained language model for English Tweets. In: Proceedings of the 2020 Conference on Empirical Methods in Natural Language Processing: System Demonstrations, pp. 9–14 (2020)

41. Pajola, L., Conti, M.: Fall of giants: how popular text-based mlaas fall against a simple evasion attack. In: 2021 IEEE European Symposium on Security and Privacy (EuroS&P), pp. 198–211 (2021). https://doi.org/10.1109/EuroSP51992.2021.00023

42. Paquet-Clouston, M., Romiti, M., Haslhofer, B., Charvat, T.: Spams meet cryptocurrencies: sextortion in the bitcoin ecosystem. In: Proceedings of the 1st ACM Conference on Advances in Financial Technologies. AFT 2019, pp. 76–88. Association for Computing Machinery, New Yor (2019). https://doi.org/10.1145/3318041.3355466. https://doi-org.ezproxy.lib.uh.edu/10.1145/3318041.3355466

43. PrivateAI-3.2.1: Identify, Redact, and Replace Personally Identifiable Information in Unstructured Text. PrivateAI (2023). https://www.private-ai.com/text/

44. Pérez, J.M., Giudici, J.C., Luque, F.: pysentimiento: a python toolkit for sentiment analysis and SocialNLP tasks (2021)

45. Qachfar, F.Z., Verma, R.M.: Homograph attacks on maghreb sentiment analyzers. arXiv preprint arXiv:2402.03171 (2024)

46. Qiu, B., Fang, N., Wenyin, L.: Detect visual spoofing in unicode-based text. In: 2010 20th International Conference on Pattern Recognition, pp. 1949–1952. IEEE (2010)

47. Reimers, N., Gurevych, I.: Sentence-bert: sentence embeddings using siamese bert-networks. In: Proceedings of the 2019 Conference on Empirical Methods in Natural Language Processing. Association for Computational Linguistics (2019)

48. Roshanbin, N., Miller, J.: Finding homoglyphs - a step towards detecting unicode-based visual spoofing attacks. In: Bouguettaya, A., Hauswirth, M., Liu, L. (eds.) WISE 2011. LNCS, vol. 6997, pp. 1–14. Springer, Heidelberg (2011). https://doi.org/10.1007/978-3-642-24434-6_1

49. Salvi, M.H.U., Kerkar, M.R.V.: Ransomware: a cyber extortion. Asian J. Conver. Technol. (AJCT) **2** (2016). ISSN 2350–1146
50. Sanh, V., Debut, L., Chaumond, J., Wolf, T.: DistilBERT, a distilled version of BERT: smaller, faster, cheaper and lighter (2020)
51. Sweeney, L.: k-anonymity: a model for protecting privacy. Int. J. Uncertain. Fuzz. Knowl.-Based Syst. **10**(05), 557–570 (2002)
52. Teixeira, L.C., De Barros, J.C.G., Fernandes, B.J.T., Da Silva, C.M.R.: Catchphish: model for detecting homographic attacks on phishing pages. In: 2022 International Joint Conference on Neural Networks (IJCNN), pp. 01–08 (2022). https://doi.org/10.1109/IJCNN55064.2022.9892525
53. Thao, T.P.: Improving Homograph Attack Classification (2020). https://doi.org/10.48550/ARXIV.2009.08006. https://arxiv.org/abs/2009.08006
54. TheTarquin: PySameSame (2018). https://github.com/DissectMalware/PySameSame
55. Thomas, K., et al.: Data breaches, phishing, or malware? understanding the risks of stolen credentials. In: Proceedings of the 2017 ACM SIGSAC Conference on Computer and Communications Security, CCS 2017, pp. 1421–1434. Association for Computing Machinery, New York (2017). https://doi.org/10.1145/3133956.3134067
56. Wright, R.E.: Logistic regression. Reading and understanding multivariate statistics (1995)
57. Connolly, A.Y., Borrion, H.: Reducing ransomware crime: analysis of victims' payment decisions. Comput. Secur. **119**, 102760 (2022). https://doi.org/10.1016/j.cose.2022.102760. https://www.sciencedirect.com/science/article/pii/S0167404822001559

Adaptive Image Adversarial Example Detection Based on Class Activation Mapping

Xiujuan Wang⑩, Qipeng Li⁽✉⁾, and Shuaibing Lu⑩

Beijing University of Technology, Beijing 100124, China
`921356743@qq.com`

Abstract. With the development of deep learning technology, convolutional neural network (CNN) has been widely used in many fields such as face recognition, automatic driving, biomedicine, etc., replacing human beings to complete complex and redundant work, which brings great convenience to people's lives. However, the discovery and development of adversarial examples have created a greater threat to image recognition. In this paper, we propose an adaptive image adversarial example detection method based on class activation mapping, which utilizes the hot zone discovery results of the Grad-CAM algorithm to perform adaptive noise reduction on images and analyzes the differences in the classification results of images before and after the noise reduction in the same benchmark network, including the KL dispersion, the label change, the label confidence, etc., to achieve the detection of adversarial examples on the ImageNet-1000 dataset. The experimental results show that the algorithm proposed in this paper achieves better detection results, and the F1 reaches 0.82 in detecting the generated FGSM adversarial examples with $\epsilon = 0.3$, which is better than the baseline model.

Keywords: deep learning · computer vision · interpretable artificial intelligence · image adversarial example detection

1 Introduction

With the development of deep learning technology, convolutional neural networks (CNN) are widely used in the fields of image recognition, face recognition, automatic driving, and other technologies that have brought great convenience to people's lives. Adversarial examples can attack all types of CNN models. Recognition of traffic signs in automatic driving [1,3], identity authentication for face recognition [15], and medical imaging systems [4,12] all have high requirements on image processing results and accuracy, and once these systems are attacked and give incorrectly predicted classification results, it will pose a serious threat to our property and even life safety. Therefore, it is important to ensure the security and reliability of image recognition systems.

This paper proposes an adversarial example detection method based on image noise reduction disparity. The method performs adaptive noise reduction on the

© IFIP International Federation for Information Processing 2024
Published by Springer Nature Switzerland AG 2024
A. L. Ferrara and R. Krishnan (Eds.): DBSec 2024, LNCS 14901, pp. 259–266, 2024.
https://doi.org/10.1007/978-3-031-65172-4_16

input image based on the heat map obtained by the class activation mapping algorithm and utilizes the difference between normal and adversarial examples before and after noise reduction to achieve adversarial example detection.

The main contributions of the research in this paper include:

1. proposing a method for generating heat maps using adaptive noise reduction based on the class activation mapping algorithm;
2. proposing to realize adversarial example detection based on the KL divergence, label change rate, and label confidence of classification results before and after noise reduction.

2   Related Works

2.1   Adversarial Example Generation

In recent years, under the joint efforts of many researchers and scholars, many emerging adversarial example generation techniques have emerged, among which the most widely used methods include FGSM, JSMA, C&W, and DeepFool.

The FGSM (Fast Gradient Sign Method) method is a method used to generate adversarial examples, which was proposed by Ian Goodfellow et al. in 2015 [6]. The main idea is to add an opposite perturbation in the direction of the backpropagation gradient on the input image, which increases the loss function of the model on the image, thus leading to incorrect predictions in the model output.

The JSMA (Jacobian-based Saliency Map Attack) method is a gradient-based adversarial example generation method [11], which does not require a large number of iterations and can generate effective adversarial examples in a shorter period of time compared to other adversarial example generation methods.

The C&W attack algorithm (Carlini and Wagner attack algorithm) is a more powerful adversarial example generation technique, first proposed by Carlini N and Wagner D in their 2017 paper [2]. The C&W attack algorithm is an optimization-based attack approach that aims to achieve high attack accuracy and low adversarial perturbation as a balance between the two objectives.

The DeepFool method proposed by Moosavi-Dezfooli S. M. et al. [10] Generate adversarial examples by finding the smallest perturbation that can misclassify the classifier, which is based on iterative linearization of the classifier to generate the smallest perturbation that is sufficient to change the classification label.

2.2   Adversarial Example Detection

The current mainstream image adversarial example detection methods fall into two main categories: statistic-based detection methods and detection methods for constructing auxiliary models.

Differences in statistical properties detected by statistics-based detection methods are often reflected as differences in the data distribution or data flow shape of the training data output. Zheng et al. [16] investigated the hidden state distribution of a deep learning model with clean training data, used a Gaussian Mixture Model (GMM) to approximate the intrinsic hidden state distribution of each class, and finally calculated the threshold to construct the detector. Feinman et al. [5] found that the subspace density of the adversarial examples is usually lower than that of the clean samples, especially when the input samples are far away from the stream shape. Therefore, they proposed to perform Kernel Density Estimation (KDE) on the training data for each category and then train the detector using the density and uncertainty features of the clean, noisy, and adversarial examples. Ma et al. [9] used Local Intrinsic Dimensionality (LID) as an alternative to KDE to compute the distribution of distances from the input samples to their neighborhoods.

Detection methods that build auxiliary models are devoted to constructing classifiers by using auxiliary models to abstract the different features of the adversarial examples from the clean samples.

Feinman et al. [5] utilized uncertainty as a feature and proposed Bayesian Uncertainty (BU) as a metric to detect adversarial examples in the vicinity of a class flow shape using Monte Carlo Dropout (M-C Dropout) to estimate the uncertainty. Subsequently, Smith et al. [13] used a mutual information approach to accomplish this task. Zuo et al. [17] first erased and reconstructed some of the pixel points of the input samples, then classified the samples before and after the processing separately using CNN classifiers and performed the adversarial example detection based on the before-and-after confidence level. Sotgiu et al. [14] propose to use the output of the last n layers of the CNN classifier to construct n SVM classifiers and combine these classifiers to accomplish the classification of adversarial examples and clean samples.

3 Proposed Method

3.1 Principles of the Algorithm

We propose an adaptive image adversarial example detection method based on class activation mapping, which first generates its class activation mapping map for the input samples and performs adaptive noise reduction according to it: a high level of noise reduction is applied to the regions with a large contribution to the classification, and a low level of noise reduction is applied to the regions with a small contribution to the classification.

The method proposed in this section is shown schematically in Fig. 1:

The part of adaptive noise reduction in the figure is realized by using JPEG compression with different qualities. The quality factor Q of JPEG compression takes a value in the range of [1,100]. The larger Q, the lower the noise reduction level, and vice versa, the smaller Q, the higher the noise reduction level. Then, the trained baseline CNN classification network is selected to classify the samples before and after noise reduction, respectively, and the difference F_s between the two classification results is obtained and used to train the detector.

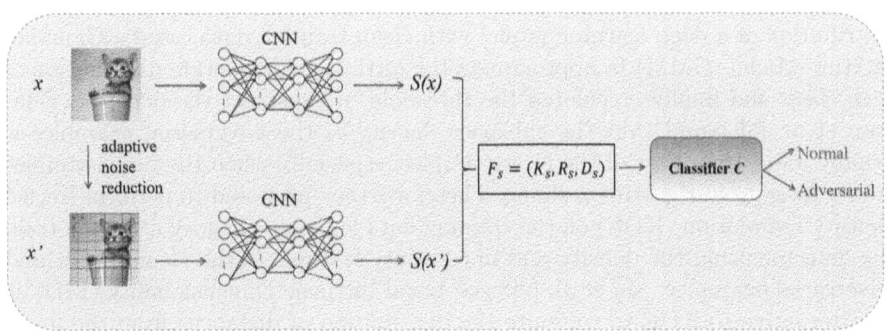

Fig. 1. Method framework

3.2 Detailed Design

The compression quality Q is used for different regions of the image to achieve adaptive noise reduction. The class activation mapping map C of the input image x is divided into $n \times n$ grid regions, and for each grid $k \in K = \{1, 2, 3, ..., n \times n\}$ region, the compression quality Q_k that should be taken for that grid is calculated as follows:

$$Q_k = \left(1 - \frac{p_k - p_{min}}{p_{max} - p_k}\right) \times \alpha + 20 \tag{1}$$

where $p_k \in P = \{p_1, p_2, p_3, ..., p_{n \times n}\}$ is the sum of the activation values of each pixel in the kth grid region, $p_{min} = min\{p_1, p_2, p_3, ..., p_{n \times n}\}$, $p_{max} = max\{p_1, p_2, p_3, ..., p_{n \times n}\}$, α is hyperparameter by changing the size of which the degree of adaptive noise reduction of the image can be adjusted.

3.3 Differential Feature Selection

In this paper, we experimentally verified that there are large differences in the KL divergence, whether the classification labels are changed, and the label confidence change between the CNN classification result distribution of the adversarial samples and the normal samples after noise reduction. The formula for the calculation of the KL divergence, K, is given as:

$$K(x) = \Sigma S(x) log \frac{S(x)}{S(x')} \tag{2}$$

where $S(x)$ is the classification result distribution of sample x in the baseline CNN network before noise reduction, and $S(x')$ is the classification result distribution of sample x' in the baseline CNN network after noise reduction.

The label change for input sample x, $shift$, is given by:

$$shift(x) = \begin{cases} 0, argmax\,(S(x)) = argmax\,(S(x')) \\ 1, argmax\,(S(x)) \neq argmax\,(S(x')) \end{cases} \tag{3}$$

where $argmax\,(S(x))$ is the subscript of the maximum value in $S(x)$. That is, $shift = 0$ when the categorical label is unchanged and $shift = 1$ when the categorical label is changed.

The variation of the labeling confidence of sample x, $delta$, is calculated as:

$$delta(x) =\mid S_c(x) - S_c(x') \mid \tag{4}$$

where $c \in C$ is the categorization label of sample x before noise reduction, $S_c(x)$ is the confidence that x is classified into class c, the cth term of $S(x)$; and $S_c(x')$ is the confidence that x' is classified into class c, the cth term of $S(x')$.

4   Experimental Design

4.1   Experimental Data

FGSM, an existing fast gradient symbolic method that is more mature and has a better attack effect, is used in this paper to attack the downloaded data from the public dataset in order to generate adversarial examples. 10,000 images are randomly selected from the ILSVRC2012_val dataset as experimental data. Firstly, the adversarial perturbation is added to half of them as adversarial examples, and the other half are the original normal samples. The combination of 6250 normal samples and adversarial examples is used as a training set, 1250 as a validation set, and the rest as a test set.

4.2   Setup of the Experiment and Evaluation of the Results

Equation 1 is used to calculate the adaptive noise reduction level of each image, and $\alpha = 80$, $\alpha = 60$, $\alpha = 40$, and $\alpha = 20$ are taken for comparison experiments.

In this paper, three main sets of experiments are designed to test and analyze the parameter settings of the algorithm and the performance of the algorithm respectively:

Experiment 1. Effect of adaptive noise reduction hyperparameter α on detector performance

In order to verify the effect of α used in adaptive noise reduction on the detector performance, ResNet50 is selected as the benchmark CNN network to classify the samples before and after noise reduction, and the FGSM adversarial perturbation of the adversarial examples is taken when $\epsilon = 2/255$. The classification effect of the adversarial example detector with different α selected for noise reduction of the samples is shown in Table 1:

From the data in the table, it can be seen that the classifier at $\alpha = 40$ is the most effective. This is because when the value of α is too large, the adaptive noise reduction is not enough to eliminate the effect of adversarial perturbation, and when the value of α is too small, the level of image noise reduction is too high, resulting in a decrease in the recognition rate of normal samples as well.

Table 1. Impact of different α on classification effect

	Acc	P	R	F1
$\alpha = 80$	0.723	0.708	**0.750**	0.728
$\alpha = 60$	0.703	0.702	0.727	0.714
$\alpha = 40$	**0.750**	**0.776**	0.713	**0.746**
$\alpha = 20$	0.721	0.733	0.624	0.674

Experiment 2. Classification Effect of Classifier on FGSM Adversarial Examples with Different ϵ

In order to verify the classification effect of the classifier on FGSM adversarial examples with different ϵ, ResNet50 is selected as the baseline CNN network to classify the samples before and after noise reduction, and $\alpha = 40$ is used for adaptive noise reduction of the samples, and the classification effect of the adversarial example detector on the adversarial examples with different ϵ is shown in Table 2:

Table 2. The classification effect of the detector on adversarial examples of different ϵ

	Acc	P	R	F1
$\epsilon = 1/255$	0.660	0.681	0.595	0.635
$\epsilon = 2/255$	0.750	0.776	0.713	0.746
$\epsilon = 0.15$	0.754	0.760	0.753	0.756
$\epsilon = 0.3$	0.820	0.814	0.826	0.820

The adversarial example detector we have proposed has a better detection effect on the FGSM adversarial examples with lower perturbations $\epsilon = 1/255$ and $\epsilon = 2/255$ as well as higher perturbations $\epsilon = 0.15$ and $\epsilon = 0.3$, and the more adversarial perturbations are added, the better the detection effect.

When attacking an image using adversarial perturbations, the more perturbations added to the generated adversarial examples, the better the attack on the image, but also the easier it is to be detected by the detector.

Experiment 3. Comparison with baseline algorithms

In order to evaluate the performance of the adaptive image adversarial example detection method based on class-activation mapping, this experiment compares the selected algorithms with some adversarial example detection methods that have been proposed as:

(1) The VisionGuard (VG) method proposed by Kantaros et al. [7]
(2) The SSF method proposed by Liang et al. [8] selects different noise reduction methods based on the image entropy of the input samples.

Figure 2 shows the Acc, P, R, and F1 obtained by the proposed algorithm and the above two algorithms when classifying the FGSM adversarial examples with $\epsilon = 2/255$. From the experimental results, it can be seen that the proposed algorithm obtains a higher Acc of 0.75, a higher P of 0.776, a higher R of 0.713, and a higher F1 of 0.746 than the above two algorithms.

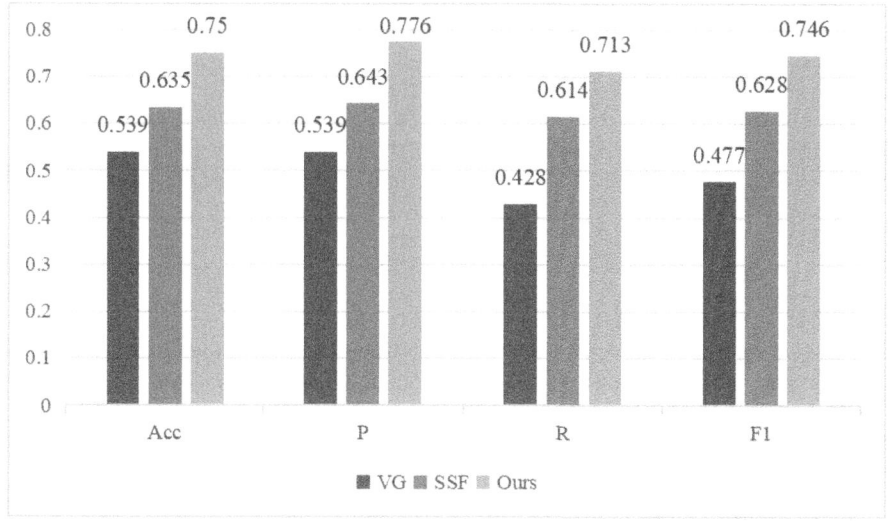

Fig. 2. Comparison Of Various Algorithms

5   Conclusion

In this paper, an adaptive image adversarial example detection method based on class activation mapping is proposed. The method performs adaptive noise reduction on the original image based on the heat map generated by the Grad-CAM algorithm, adopts a lower level of noise reduction for the regions with smaller contributions, and utilizes the characterization of the changes in KL divergence, label change, and label confidence in the classification results of the baseline CNN before and after noise reduction to achieve adversarial example detection, which improves the detection performance. The experimental results show that the algorithm proposed in this paper can achieve better detection results than the baseline model. However, this paper only adopts the three parameters of the distribution of classification results before and after noise reduction as the basis of classification through experimental statistical analysis and does not prove it theoretically, which can be analyzed theoretically to find better features to describe the differences in the distribution of the classification results in order to improve the classification effect of the classifier in the subsequent research.

References

1. Bojarski, M., et al.: End to end learning for self-driving cars. arXiv preprint arXiv:1604.07316 (2016)
2. Carlini, N., Wagner, D.: Towards evaluating the robustness of neural networks. In: 2017 IEEE Symposium on Security and Privacy (SP), pp. 39–57. IEEE (2017)
3. Darapaneni, N., et al.: Autonomous car driving using deep learning. In: 2021 2nd International Conference on Secure Cyber Computing and Communications (ICSCCC), pp. 29–33. IEEE (2021)
4. Esteva, A., et al.: Dermatologist-level classification of skin cancer with deep neural networks. Nature **542**(7639), 115–118 (2017)
5. Feinman, R., Curtin, R.R., Shintre, S., Gardner, A.B.: Detecting adversarial samples from artifacts. arXiv preprint arXiv:1703.00410 (2017)
6. Goodfellow, I.J., Shlens, J., Szegedy, C.: Explaining and harnessing adversarial examples. arXiv preprint arXiv:1412.6572 (2014)
7. Kantaros, Y., Carpenter, T., Sridhar, K., Yang, Y., Lee, I., Weimer, J.: Real-time detectors for digital and physical adversarial inputs to perception systems. In: Proceedings of the ACM/IEEE 12th International Conference on Cyber-Physical Systems, pp. 67–76 (2021)
8. Liang, B., Li, H., Su, M., Li, X., Shi, W., Wang, X.: Detecting adversarial image examples in deep neural networks with adaptive noise reduction. IEEE Trans. Dependable Secure Comput. **18**(1), 72–85 (2018)
9. Ma, X., et al.: Characterizing adversarial subspaces using local intrinsic dimensionality. arXiv preprint arXiv:1801.02613 (2018)
10. Moosavi-Dezfooli, S.M., Fawzi, A., Frossard, P.: Deepfool: a simple and accurate method to fool deep neural networks. In: Proceedings of the IEEE Conference on Computer Vision and Pattern Recognition, pp. 2574–2582 (2016)
11. Papernot, N., McDaniel, P., Jha, S., Fredrikson, M., Celik, Z.B., Swami, A.: The limitations of deep learning in adversarial settings. In: 2016 IEEE European Symposium on Security and Privacy (EuroS&P), pp. 372–387. IEEE (2016)
12. Shen, D., Wu, G., Suk, H.I.: Deep learning in medical image analysis. Annu. Rev. Biomed. Eng. **19**, 221–248 (2017)
13. Smith, L., Gal, Y.: Understanding measures of uncertainty for adversarial example detection. arXiv preprint arXiv:1803.08533 (2018)
14. Sotgiu, A., et al.: Deep neural rejection against adversarial examples. EURASIP J. Inf. Secur. **2020**, 1–10 (2020)
15. Sun, Y., Liang, D., Wang, X., Tang, X.: Deepid3: face recognition with very deep neural networks. arXiv preprint arXiv:1502.00873 (2015)
16. Zheng, Z., Hong, P.: Robust detection of adversarial attacks by modeling the intrinsic properties of deep neural networks. In: Advances in Neural Information Processing Systems, vol. 31 (2018)
17. Zuo, F., Zeng, Q.: Exploiting the sensitivity of L2 adversarial examples to erase-and-restore. In: Proceedings of the 2021 ACM Asia Conference on Computer and Communications Security, pp. 40–51 (2021)

Security User Studies

From Play to Profession: A Serious Game to Raise Awareness on Digital Forensics

Sabrina Friedl$^{(\boxtimes)}$ ⓘ, Tobias Reittinger ⓘ, and Günther Pernul ⓘ

University of Regensburg, Regensburg, Germany
{sabrina.friedl,tobias.reittinger,guenther.pernul}@ur.de

Abstract. With the increasing digitization and interconnectivity of organizations, the frequency of cyberattacks is rising. These attacks have serious consequences for data security and critical infrastructure. However, the persistent lack of cybersecurity specialists represents a major challenge for organizations. One approach to address this issue is to use serious games for career orientation in schools to steer people interested in security or digital forensics (DF) at an early age. Thus, we introduce a serious game called *Digital Detectives*, designed to familiarize students and young IT professionals with DF practices. Players adopt the role of forensic investigators tasked with probing a cyberattack on a company and acquire hands-on expertise by gathering evidence, identifying perpetrators, and recovering encrypted data. We evaluated the serious game with 102 young students in a pre- and post-test. We find that our serious game has a great game experience and significantly increases learning outcomes and awareness of DF. Our results indicate that the games' learning outcomes and game experience are strongly correlated to increased awareness of DF. In sum, the serious game is an effective means to aid students in career orientation and raise awareness of DF.

Keywords: Digital Forensics · Serious Game · Career Orientation · Cybersecurity Professional Shortage

1 Introduction

The increasing connectivity of organizations is shaping today's digital era. Emerging trends like the Internet of Things (IoT), sensors, communication devices, and other networked technologies permeate almost every industry, whether in healthcare, transportation, or homes. The immense amount of data generated enables companies and institutions to gain profound insights and provide more efficient services [14]. However, this connectivity also brings significant security risks. The increasing number of IoT devices creates an enormous attack surface for cybercriminals. Hackers and attackers can exploit vulnerabilities to penetrate networked systems, steal sensitive data, sabotage services, or even compromise critical infrastructure [35]. Amid these threats, cybersecurity is gaining a central role. The importance becomes even more apparent when considering the current state of cybersecurity. According to (ISC)^2 Cybersecurity

© IFIP International Federation for Information Processing 2024
Published by Springer Nature Switzerland AG 2024
A. L. Ferrara and R. Krishnan (Eds.): DBSec 2024, LNCS 14901, pp. 269–289, 2024.
https://doi.org/10.1007/978-3-031-65172-4_17

Workforce Study [22], it is estimated that the global cybersecurity workforce lacks 3.4 million workers [22]. This workforce gap poses a serious challenge and requires comprehensive action to offset this substantial skill shortage. One of the most important steps is to raise awareness of the various career fields in this area and to get young people excited about the many opportunities and tasks in cybersecurity at an early age. Digital forensics (DF) is one such career field that is taking on an increasingly important role in the security landscape. DF scientists work to identify, collect, and analyze digital evidence to investigate cyberattacks and identify the perpetrators. They play a critical role in solving crimes and are essential to the security of businesses and society [5]. However, teaching this specialized knowledge is often challenging, as theoretical knowledge and practical application must be sensibly combined [20]. A promising solution is to combine teaching with gamification approaches. Thus, we answer the following research questions.

RQ1: *Can a serious game raise awareness of the DF profession?*

RQ2: *Do game experience and learning outcomes of a serious game influence the awareness of the DF profession?*

Contributions. We develop a real serious game, which is a game that strikes a balance between entertainment and learning, aiming not just to entertain but also to convey knowledge [1,43]. It has classic game elements like achievements, collectibles, and mini-games in a point-and-click game character. The game called *Digital Detectives*, is built on a theoretical model, a custom-created design, and an exciting storyline. The creation of this serious game bridges the gap between theory and practice. It presents essential topics in DF and allows users to gain significant yet realistic insights in a fun way.

- We developed the serious game *Digital Detectives* based on the Experiential-Gaming-Model [26] and the Event-Based-Learning Model [8] combining game design and learning mechanisms for DF.
- We implemented the game with an open-source cloud-based platform for web game creation called JAWA [23]. The serious game is publicly accessible.[1]
- We evaluated our game with 102 young students using a pre- and post-test. The results show that the game significantly increases the learning outcome. Participants perceive a highly positive game experience, and 34% of participants continued to play *Digital Detectives* after the study concluded.
- The evaluation shows that the serious game significantly increases students' awareness of the DF profession. Furthermore, the results indicate a strong correlation between the learning outcome and the game experience with increased awareness.

The paper is structured as follows. In Sect. 2, essential concepts are defined for this work, and related work is discussed. After the methodology is outlined in Sect. 3, Sect. 4 describes the step-by-step development of *Digital Detectives* is provided. In Sect. 5, the evaluation is presented. Further, Sect. 6 discusses the developed game and future work. Finally, Sect. 7 concludes our work.

---

[1] Link to Digital Detectives Game: https://jawa.games/play/6192.

2 Background and Related Work

This section defines essential concepts regarding developing and implementing the presented serious game and related work.

2.1 Serious Game and Gamification Elements

Games have always been integral to human culture, offering structured play with clear outcomes. Digital games on devices like computers and smartphones have become prevalent with the advent of technology. Serious games, a subset of digital games, are designed for fun and additional purposes such as education or health [1, 12, 43]. Their definition focuses on the player's intent, and they can evolve from mere entertainment to educational tools, exemplified by a farming simulator that teaches agricultural practices [17]. The use of serious games in learning is growing, as they blend digital interaction with real-world activities and are increasingly employed for teaching and training [2]. They engage users through game-like elements and challenges, enhancing motivation and the effectiveness of learning experiences [6, 49]. Serious games facilitate practical learning and problem-solving by providing a safe, interactive environment, leading to sustained educational success [6]. Point-and-click games are known for their interactive storytelling, collectible items, and puzzle-solving elements that captivate players with unexpected plot twists to maintain engagement [11, 33, 44].

2.2 Digital Forensics

Classical forensics involves searching for clues at crime scenes and analyzing evidence in the lab. DF focuses on digital evidence, particularly in digital crime scenes, as a subset of classical forensics [13]. DF is commonly structured into four phases: identification, collection, analysis, and presentation [32]. During this investigation process, each step has its challenges. As discussed earlier, problems regarding the cybersecurity workforce also affect DF and awareness of DF investigations and challenges. Therefore, education and training are essential to generate interest and provide appropriate skills, knowledge, and consciousness.

2.3 Related Work

Developing serious games in DF is crucial for effective learning and training in this field. Studies like those by Reyes et al. [41] demonstrate that a point-and-click game can enhance the cognitive flexibility of preschool children, with these benefits extending to other cognitive tasks. Additionally, Katsaounidou et al. [24] highlight the effectiveness of serious games in enhancing critical thinking skills and creating an engaging learning environment. In the field of DF, serious games play a key role in improving the comprehension of forensic concepts and practices, as well as in training investigators and analysts. Here, we delve into existing learning approaches (game/game-based) for DF, summarized in Table 1.

Table 1. Game-based learning and teaching approaches with DF context.

Publication		Description	
Source	Year	Learning Approach	Target Group
[38]	2012	Game-based Course Modules	College Students
[39]	2015	Game-based Course Modules	College Students
[37]	2017	Game-based Course Modules	College Students
[28]	2020	Game-based Course Modules	DF Students
[36]	2014	Game-based E-Learning Tool	Students
[31]	2016	Capture the Flag	Students and Teachers
[15]	2020	Capture the Flag	Graduates and Professionals
[4]	2016	Serious Game	DF Students
[3]	2018	Mobile Adventure Game	DF Students
[19]	2020	Virtual Reality Game	High School Graduates
[40]	2018	Cyber Range	Military Learners
[16]	2022	Cyber Range	Postgraduate Learners
[18]	2023	Cyber Range	Postgraduate Learners

Only one paper on a game-based learning approach for DF is over 10 years old [38], with the latest paper being from last year [18]. 10 of the 13 results are from 2016 or later, meaning they were published within the last seven years. The increase in recent publications highlights the modern nature of DF education.

Learning Approaches. In DF education, serious games and game-based learning have emerged as innovative teaching methods. While many approaches incorporate game-like elements, only a select few, such as those by Blazic et al. [4], Blauw and Leung [3], and Hassenfeldt et al. [19], are designated as actual games. These range from virtual reality and mobile adventure games to comprehensive simulations like the "FOREVICA" project. Other methods, like Leung's [28], employ gamification to enrich computer forensics modules. Workshops integrating serious game concepts into Capture The Flag (CTF) challenges are also gaining traction, as seen in the work of Englbrecht and Pernul [15] and McDaniel et al. [31]. Friedl et al. [16] uniquely addresses IoT forensics through a serious game concept, while Raybourn et al. [40] and Pan et al. [37–39] apply game frameworks to develop course modules. Nordhaug et al. [36] highlight the benefits of game-based e-learning tools in DF education, demonstrating the diverse applications of gaming principles in this field.

Target Groups. Students in college [37–39], as well as those in middle school [31] and high school [19], are primarily identified as the target audience for these educational methods. While Friedl et al. [16] and Englbrecht and Pernul [15] also acknowledge postgraduate students, Raybourne et al. [40] notably define the target audience based on a military background. An interesting classification of target groups involves the consideration of necessary prior knowledge. Friedl et al. [16] suggest a cybersecurity background, while courses designed for DF students [3,6,28] imply a certain level of prerequisite knowledge. On the other hand, Pan et al. [38,39] note that forensic courses are typically tailored for advanced learners and propose an innovative approach with their game-based forensics course. Similarly, Englbrecht and Pernul [15] specifically emphasize the need to train not only IT experts but also managers and young employees in forensic and awareness subjects. McDaniel et al. [31] highlight the low level of prior knowledge among CTF participants in cybersecurity. Raybourn et al. [40] attempt to cater to all knowledge levels by including beginners, advanced learners, and professionals in the target audience of their cybersecurity program. While around 45% of the participants of Hassenfeld et al. [19] had educational backgrounds in criminal justice, individuals from diverse fields such as music or nutrition also participated. This suggests that the virtual reality game may not require prior knowledge of forensics or cybersecurity. Finally, Nordhaug et al. [36] do not specify the target audience or required prior knowledge.

Although there are some game-based learning and teaching methods for DF, there is a distinct lack of actual games. In contrast to existing work, our approach is an actual serious game. Additionally, we detail the development of the serious game very thoroughly and methodically. We extract and utilize the underlying game and learning models from the literature.

3   Methodology

We describe our methodology and outline the game development, evaluation, ethical considerations, and limitations in the following.

Development. We systematically develop a serious game, using the *Experiential Gaming Model* by Kiili [26] to design the game environment and the *Event-Based-Learning Model* by Carrier and Spafford [8] to incorporate learning objectives. Building on this, we establish the conceptual foundation of the game. Next, we implement our serious game in JAWA [23], an open-source platform for creating point-and-click adventures. Our serious game, *Digital Detectives*, is then published online to facilitate future education and research.

Evaluation. We conducted a pilot study with 11 students to ensure the serious game is displayed correctly on all devices and the questionnaire is comprehensible. Subsequently, we recruited 102 students through purposive sampling by contacting local schools in the region of Regensburg (Germany). As summarized in Table 2, participants are between 18 and 22 years old and primarily in high school without prior experience in cybersecurity or DF. We argue that these young students are an ideal target group for identifying a serious game's effect on DF career orientation. The participants voluntarily engaged in the study without incentives. We followed the Kirkpatrick model [27,42] for the evaluation. Participants completed the pre-test, played the serious game, and answered the post-test. The pre-and post-tests are established methodologies to evaluate knowledge gain. The questionnaire of the pre- and post-test is listed in Appendix II and was divided into three sections: *Learning outcome* (12 questions, single choice), *awareness of digital forensics* (8 statements, 5-point Likert scale), and *game experience* (10 statements, 5-point-Likert scale). The first two sections were used in both the pre- and post-test, and the third section was only used in the post-test. We did not provide the participants with the solution in the pre-test, and participants did not communicate with each other between the tests to not bias the post-test. The average duration of gameplay was 45 min (incl. the questionnaire). Finally, we analyzed the collected data using SPSS 29.0.1 [21] to facilitate multiple statistical tests (paired t-test, Wilcoxon signed-rank test, ANCOVA regressions).

Table 2. Participants' demographics.

	Variable	Participants
Gender	Female	45.1 (46)
	Male	54.9 (56)
	Other	0 (0)
Age	18 years	27.5 (28)
	19 years	41.2 (42)
	20 years	22.5 (23)
	21 years	5.9 (6)
	22 years	2.9 (3)
Education	High school	80.4 (82)
	College	19.6 (20)

Note. We describe percentages (with frequencies).

Ethics. Ethical considerations are fundamental for our research. Before conducting the study, the German Association for Experimental Economic Research validated our methodology, and we obtained an Institutional Review Board (IRB) certificate.[2] Participants could abort the study at any time and were assigned a unique random ID to connect the pre- and post-test.

Limitations. We acknowledge our limitations. First, the pre- and post-test surveys are self-reported and might not be fully accurate. Second, due to purposive sampling, the recruitment might not be fully random. Third, the geographic scope of participants is limited to Germany, which might restrict the generalizability of our findings. Future research could verify our results for other regions.

4   Development of Serious Game

After highlighting the significance of educational games, this section introduces the models employed in crafting the *Digital Detectives* game.[3] The *Experiential Gaming Model* [26] and the *Event-Based-Learning Model* [8] are notably

---

[2] The IRB certificate can be viewed online: https://gfew.de/ethik/BEFAhReP.
[3] Link to *Digital Detectives* game: https://jawa.games/play/6192.

prevalent in the domain of DF (Sect. 2.3). Consequently, we merge these models to create a point-and-click game for DF, offering players a fun and interactive foundation on fundamental DF methods and processes. While the *Experiential Gaming Model* [26] (Sect. 4.1) serves as a blueprint for game design, features, and engagement, the *Event-Based-Learning Model* [8] (Sect. 4.2) establishes the framework and order of learning objectives.

4.1   Experiential Gaming Model

The *Experiential Gaming Model* [26], as shown in Fig. 1, comprises an idea and experience loop and a challenge. The model's functional principle can be likened to the human vascular system. The challenges aligned with educational objectives serve as the core of the model. Similar to the heart's role in the vascular system, the primary function of these challenges is to sustain the player's motivation and engagement by offering appropriate challenges [26]. A well-defined and structured explanation of learning objects is crucial in the *Experiential Gaming Model* [26] to immerse players in the game story and enhance their learning experience. This immersive state, known as flow, boosts learning and exploration. To achieve flow, goals must be established, immediate feedback must be provided, and player control must be granted. It is also essential to stimulate player enjoyment, maintain appropriate speed and usability, and tailor challenges to the player's skill level for continuous skill development. Setting goals that increase motivation and engagement is key. Players must engage in the idea generation loop to overcome challenges, starting with the creative preemptive phase and progressing to structured idea development, considering game rules and resources available in the game world [26].

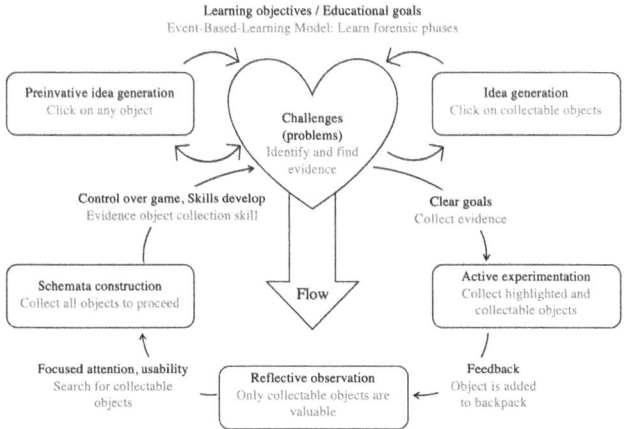

Fig. 1. Representation of the *Experiential Gaming Model* based on Kiili [26].

4.2   Event-Based Learning Model

In DF investigations, the process consists of several phases. Initially, the focus is on preserving the crime scene to prevent contamination and safeguard evidence. The next phase involves searching, documenting digital evidence, and looking for relevant information. Data is extracted, compared, and analyzed systematically. The third phase includes reconstructing the digital incident and scrutinizing each piece of evidence to establish the sequence of events. Hypotheses are formulated and tested, leading to the presentation of findings to the organization in the closure phase. The *Event-Based Learning Model* [8] can define learning objectives in the *Experiential Gaming Model* [26], with each phase containing additional objectives cycled through the model. DF phases and tasks become challenging learning objectives with clear goals. Ongoing feedback helps track progress and enhances immersion, potentially leading to a flow state.

4.3   Game Design

The *Experiential Gaming Model* [26] serves as a framework for both the design and evaluation of educational games. We emphasize the significance of gameplay, yet equal attention should be given to crafting an engaging story, appropriate graphics and sounds, game balance, and optimizing cognitive load. The *Event-Based Learning Model* [8] assists in defining learning goals and deriving challenges.

Gameplay. It describes the features of a game and how it is played [26]. We incorporated various knowledge transfer and interaction elements to generate an engaging gameplay and learning environment.

Interaction with Characters. At the start, the player discovers on the initial screen that they are a DF investigator aiding a company in identifying perpetrators and the nature of the game they are playing. Subsequently, they are thrust into the heart of the game's action. The storyline was crafted to immerse the DF expert in an interactive gaming world. The player gathers details about the cyber attack and the DF expert's activities by engaging with game characters and objects. Through interactions with various characters, the player acquires crucial information, clues, and guidance to navigate and resolve the scenario.

Interactive Bookshelf. Passive elements can also be discovered. For instance, by clicking on the bookshelf, a book containing information about DF opens up. The books provide details and definitions of different DF and IoT forensics subjects, highlighting key terms and the investigation process.

Incentives. Trophies were integrated into the game to enhance the game experience and provide motivation. These trophies are granted upon reaching significant milestones or completing challenging tasks successfully. They serve as a form of acknowledgment for the player's accomplishments and are prominently showcased throughout the game. For instance, a trophy can be earned for deciphering encrypted data, discovering crucial evidence, or finishing a mini-game.

Each trophy symbolizes a notable achievement in the gameplay and is linked to a specific challenge or accomplishment. The game visually presents the trophies, which can be accessed by the player in a dedicated trophy overview, which exhibits both earned trophies and those yet to be unlocked. Attaining key milestones and receiving trophies offers a sense of gratification and acknowledgment, heightening the player's motivation and sustaining their involvement. Thus, trophies serve as an added incentive and reward mechanism, enriching the gaming experience with diversity and excitement.

Backpack. This is a key gameplay feature where the player collects various items (e.g., USB sticks). The backpack is always accessible during the game, enabling the player to view the collected items and utilize them to take action. For instance, a key card is required to unlock the server room, given to the player by a character in the game. This action is simply performed by dragging and dropping the key card onto the security lock of the server room. Then, the player enters the server room.

Mini Games. We developed and integrated various mini-games into the game to add variety and challenges. These include a code deciphering game, sorting collected items, answering quiz questions, and communicating with the hacker. We ensured these mini-games were seamlessly integrated into the game world and fit naturally into the gameplay.

Time Constraint. The game aims for a maximum duration of one hour, emphasizing digital crime scene investigation phases, particularly evidence collection and crime reconstruction, to consider cognitive processing. To incorporate time as a gameplay factor, a countdown clock from one hour to zero was integrated, creating simulated time pressure for the player.

Feedback and Progression. While the learning objects are divided into challenges and goals, various elements are incorporated into the game to enhance the player's chances of entering the flow state. An illustration of this is the ongoing feedback provided to the player regarding their decisions. One aspect of this feedback is the overlay of game progress, which advances when the player makes decisions to overcome challenges. Further, provide game characters and objects with hints, descriptions, and clues for the player to ease game progression.

Settings Bar. This game feature is located in the top right corner of the game, enabling the player to adjust game settings. It includes an information tab and a save tab to save the game online or download the current score. Moreover, players can capture screenshots, adjust music and character voice volume, enter fullscreen mode, pause, exit, and restart the game.

Engaging Story. The developed serious point-and-click game deals with the solution of a ransomware malware attack from the perspective of a DF expert. The IoT company *SafeHome24*, which produces digital door locks for private customers, was attacked in the game. After stealing customer data and encrypting the company database with sensitive customer data, the hacker now demands a ransom from the customers and the company. In addition to names and addresses, the hacker has also stolen the customers' passwords, which are used to secure the door locks. He uses these as leverage against customers and *SafeHome24*. The main scenes that structure and guide the story (DF Office, Entrance Lobby, Meeting Room, Server Room, IT Manager Office, DF Laboratory) are depicted in Fig. 2. The story flow (Fig. 3 in the Appendix) illustrates the player's ideal click path.

Scene 1 - DF Office. In the game, the player assumes the role of the DF expert in a "first-person" view. The initial scene unfolds in the player's office, where they receive a call from the concerned CEO regarding an attack on the company *SafeHome24*. The CEO requests immediate assistance. During this scene, the player must orient themselves and identify useful items in the office to solve the case. They can gather initial IT and IoT forensics information from the bookshelf, have coffee, collect a USB flash drive, and download the delivery status from the email inbox. Once all crucial items are gathered, the player leaves the office to head to the *SafeHome24* company.

Scene 2 - Lobby. The second scene starts in the entrance lobby, with three doors. The game limits players from entering the meeting room door only, and they are guided towards the meeting room through a call at the game's start.

Scene 3 - Meeting Room. In the *SafeHome24* meeting room, the player encounters the manager and the head of customer service. They both share their perspectives on the event in a dialogue, providing crucial details about the unfolding situation. The primary objectives are to repair the damage from the attack and identify the perpetrators. Each character is depicted with distinct mannerisms to enhance the scene's realism and depth. The manager is composed yet

Fig. 2. Main scenes of the *Digital Detectives* game (tools: JAWA [23], Canva [7]).

informative, while the customer service head displays a determined demeanor when handling angry customer complaints. The player is prompted to connect emotionally with the characters, heightening immersion in the game's narrative. Additionally, the manager entrusts the player with a key card for server room access, instructing them to engage with the staff member there and communicate with the hacker through the server PC.

Scene 4 - Server Room. The player can access the server room again through the entrance lobby by combining the keycard with the reader next to the door. Before encountering the hacker, the player must demonstrate knowledge of the DF investigation process by interacting with the service agent. This interaction provides crucial information on the steps involved in the DF investigation. The sequence of crime scene preservation, evidence searching, and event reconstruction phases are tested through a single-choice quiz. This helps the player understand the correct order of phases in a DF process. Subsequently, a chat dialogue between the player and the hacker begins. Upon entering the chat window, a message from the attacker appears, and the player can respond by choosing from various options. The hacker reacts accordingly to the player's chosen responses. The dialogue concludes with the hacker demanding $1 million to release the customer data, regardless of the player's response.

Scene 5 - IT Manager Office. In the following scene, the player finds themselves in the IT manager's office. He explains the situation from his point of view and outlines his next steps. By interacting with the IT manager's PC screen, the player initiates the analysis of the cyber attack. The player is tasked with documenting and gathering evidence in this scene. The player uncovers a social engineering attack upon discovering a phishing email on the IT manager's computer posing as a message from the CEO urging an immediate update. The IT manager complied with the request to install the new security update and unwittingly downloaded ransomware onto the company's internal system by clicking the link. The player must identify the link between the malware and the encrypted customer data, gather the phishing email as evidence, and locate and contain the malware in the download directory. Furthermore, the player must identify that the customer data was transmitted to the attacker through an email generated automatically by the malware. Consequently, the customer data was encrypted within the company's database. In this scene, the player gathers various pieces of evidence and encounters the first mini-game, where they must decrypt the encrypted corporate database containing sensitive customer information. The player can decrypt the data using a DF tool by inputting a random five-letter combination from A to E.

Scene 6 - DF Lab. After collecting all data, the player exits Safe-Home24 through the lobby and returns to their office to analyze the data in their DF lab. To access the lab, the player needs four objects in their inventory: the malware file, phishing email, encrypted customer data, and stolen customer data. The player is now in the final phase of the event-based learning model. To pinpoint the attacker's location, the player must scrutinize network activities using Wireshark analysis.

The player can identify the attacker's IP address by filtering the network activity with the malware file. Inputting the correct IP address into the IP address finder reveals the hacker's location and identity as the rival company *EntranceSecure GmbH*. The player must then document their findings and investigations in a report. The next challenge is structuring the evidence in a mini-game to represent the crime's sequence. Starting the mini-game involves discarding irrelevant items like coffee or USB sticks. The case is solved once the report is completed and the player submits the documentation to the manager. Shocked by the findings, the manager insists on a chat confrontation with the attacker. The game exchanges messages with the attacker, where the player decides on the retaliation strategy. One option is to contact the authorities directly or threaten the attacker with the immediate release of all data publicly.

Appropriate Graphics and Sounds. The background images and characters should immerse players in the game world, providing a realistic yet playful interaction environment while maintaining the game's character. Secondary images and acoustic details add authenticity and enable viewers to understand emotions and actions more deeply. Background music and character voice audio recordings were carefully integrated to enhance the atmosphere and provide an immersive experience. Music selection matched the mood and gameplay, while audio recordings authentically portrayed dialogues. Music supported the action and scenes without distracting players, ensuring seamless integration for smooth gameplay. Sound effects reinforced player interactions with the game world, such as phone ringing or chat typing, providing feedback for correct or incorrect clicks. Volume and balance of music and sound effects were adjusted for optimal listening experience, avoiding dominance over dialogue or important audio elements. This integration enhanced the game's emotional impact, creating a fully immersive gaming experience through a combination of visual and auditory elements, enabling a holistic gaming experience.

Game Balance. To ensure the player found the optimal click path (Fig. 3) through the game and progressed the plot in a balanced way, appropriate conditions were defined. The inclusion of conditions helped make the gameplay structured and logical, creating tension and challenge. Important events, information, and key moments within the game plot were identified to set the conditions. It analyzed what actions or knowledge the player must have to unlock scenes, solve puzzles, mini-games, or obtain dialogue options. Conditions were integrated into the game, defined by objects or previously set variables. Once met, new actions were unlocked, such as setting variables to save progress or opening new locations. Care was taken to provide hints without making the game too easy or hard, balancing challenge and playability. Feedback mechanisms are included to guide the player through visual cues, character dialogue, or new action options. These conditions help guide the player while allowing freedom to make decisions and explore the game world.

Cognitive Load Optimization. Early in the game, progress is facilitated by simple decisions, but as the forensic investigation progresses, it becomes more challenging. This shift supports the evolution from initial idea generation to more complex ideation. Initially, players are given cues on the significance of certain facts to advance in the game. The objective is to guide players from an unstructured approach to a systematic process that helps them devise solutions to the presented challenges. Moreover, the game incorporates various paths and decision-making opportunities to grant players autonomy. This non-linear approach allows for diverse outcomes and endings, enhancing the interactive experience. Players can chart their course in the game, actively shaping the narrative and feeling immersed in the virtual world, thus fostering motivation to progress and engage further with the storyline.

The integration of all gameplay elements creates an immersive and comprehensive gaming experience. Players are guided through intuitive gameplay that prompts them to utilize forensic techniques to gather, analyze, and interpret digital evidence. In the final version of the developed DF game, hints were added to maintain game flow. These hints notify players that they may have overlooked relevant content and aspects of the current game scene, motivating them to conduct further investigations to uncover crucial evidence and information. Overall, the game incorporates diverse elements to immerse players in a flow state. Realistic yet playful scenes, along with sound and voices mirroring the daily life of a DF investigator, enhance the game's immersion and authenticity. Notably, including interactive objects and establishing rules and conditions are essential for providing players a challenging experience.

4.4   Implementation of Game

Technical Development. The game was developed using JAWA [23], starting with a brainstorming session to devise an engaging story focusing on a ransomware attack at an IoT company. The narrative was designed to captivate students, featuring interactive character development and plot twists. Scene sketches were created collaboratively, outlining the action settings and interaction points. These sketches informed the design of the game's backgrounds and object placements, highlighting the distinct features of each IoT company area and ensuring scenes met visual standards effectively.

Background Scenes. The initial sketches led to a web search for appropriate background images, which often varied in style, posing a challenge for visual consistency in the game. This led the team to Canva [7], a graphic design tool that simplified creating uniform and engaging background scenes. With Canva's diverse templates and elements, the scenes were crafted with attention to detail and atmosphere, ensuring a realistic and interactive game world. The final backgrounds (PNG files) were then integrated into JAWA [23].

Game Character Creation. As the game progressed, various characters were developed, such as CEO Samantha, the Head of IT, Customer Service, and IT employee Torsten. All visual representations of the characters were created using Canva [7]. Secondary images of the characters were included to enhance realism, allowing for changes in facial expressions or positions during dialogues or interactions with the player.

Character Voices, Object Sounds and Background Music. Furthermore, all character dialogues have been enhanced with soundtracks to elevate the gaming experience. Giving a voice to the characters enhances the player's immersion in the game's world and enables a more profound portrayal of the characters. JAWA [23] allows recording your audio or uploading an mp3 file. The latter choice was selected. The audio tracks were created using the Murf.ai [34] application, an AI speech generator that transforms text into voice recordings through text-to-speech technology. This allows users to select from a variety of voices. To enhance the atmosphere further, suitable freely available background music from JAWA [23] was integrated.

Interactive Elements. To enhance the point-and-click game, interactive elements like computers, documents, and books were integrated into scenes, aiding in the ransomware attack resolution. These elements were strategically placed to align with the narrative and gameplay, ensuring a seamless player experience. A list of objects was curated to support the plot and encourage exploration. Even non-essential items, such as a coffee machine, were added for a touch of realism and to challenge players. Players sift through collected items in the game's finale, keeping only those pertinent to the case. The game featured two object types per JAWA [23]: scene objects with designated interaction zones or image insertions and inventory objects, which required images for both scene display and inventory inclusion.

Interactions Between Elements. In addition to object placement, links and interactions between objects and characters were added to drive gameplay and storyline. Players could interact with objects to receive hints and progress further. This included starting dialogues with characters, collecting information and objects, and solving tasks to track down and defeat the ransomware attack. Players could also combine objects to gain new information or complete tasks.

Mini-Games. Two distinct methods were employed for mini-game implementation. Initially, custom background scenes were crafted with objects and interactions symbolizing the game. Conversely, specific templates with pre-set functionalities were accessible on the development platform. These templates could be imported and customized to fit the game's context and design. Developing and adapting mini-games greatly enhanced game depth and long-term engagement. By introducing additional challenges to players, their interest was piqued, and their skills were honed. The mini-games helped effectively convey crucial information.

5   Evaluation

We present the outcomes of evaluating a serious game with 102 participants, focusing on learning outcomes, gaming experience, and awareness of the DF profession. We also analyze how learning and experience influence awareness.

5.1   Learning Outcome

We evaluated the learning outcome of participants from the pre-test in comparison to the post-test with a paired t-test [30]. We used Cohen's d [9] as the effect size (ES). Thus, we conducted a comparative analysis of the average rates of correctly answered questions across each group and overall, with the findings detailed in Table 3. In detail, participants had a significantly strong learning outcome in *digital forensics* ($t = 9.48$, $p < .001$, $ES = .94$) and *digital forensic tools* ($t = 9.86$, $p < .001$, $ES = .98$). Furthermore, there was a significantly medium effect of learning outcome for the categories *digital forensic phases* ($t = 7.42$, $p < .001$, $ES = .74$) and *digital forensics evidence* ($t = 7.42$, $p < .001$, $ES = .74$). The results indicate that the serious game significantly increases learning outcomes in all areas, as overall knowledge increased from 69% in the pre-test to 83% in the post-test ($t = 17.71$, $p < .001$, $ES = 1.75$). As the effect size is above 0.8, the effect is strong [9]. In sum, our results suggest that *Digital Detectives* is an effective tool to increase knowledge about DF.

5.2   Awareness of Digital Forensics Profession

We calculated a Wilcoxon signed-rank test [47] to evaluate the effects of the serious game on the awareness of the DF profession. Because the awareness is measured through a Likert scale, the Wilcoxon signed-rank test is suitable [10]. We identified a statistically significant increase in the awareness of the DF profession after playing the serious game ($z = 7, 32$, $p < .001$, $ES = .72$). As the effect size (ES) is above 0.5, the correlation is statistically strong [9]. Thus, the results suggest that the serious game can significantly increase the awareness of participants in the DF profession. Upon visually inspecting the histogram, the distribution of differences appeared symmetrical. This fulfills the prerequisites for interpreting the test results.

Table 3. Results of the learning outcome with 102 participants.

Category	Pre		Post		t	df	Sig. (2-tailed)	ES
	M	SD	M	SD				
Digital forensics	.76	.28	.92	.14	9,48	101	<.001	.94
Digital forensic phases	.71	.30	.82	.26	7,42	101	<.001	.74
Digital forensic evidence	.73	.35	.82	.26	7,42	101	<.001	.74
Digital forensic tools	.55	.30	.75	.27	9,86	101	<.001	.98
Overall	.69	.27	.83	.22	17,71	101	<.001	1.75

Note. We outline Cohen's d [9] as *Effect Size (ES)*.

Table 4. Results of the game experience with 102 participants.

Evaluation Criteria	Strongly Agree	Agree	Neutral	Disagree	Strongly Disagree
Gameplay & Mechanics	24%	46%	16%	14%	0%
Storyline	29%	45%	13%	9%	4%
Visuals	36%	35%	18%	6%	5%
Topic	28%	32%	17%	12%	11%
Overall	29%	40%	17%	11%	5%

5.3   Game Experience

We measured the game experience for each question with a Likert scale and outlined the results in Table 4. 70% of the participants (fully) agree that gameplay and mechanics were enjoyable and easy to understand. Furthermore, 74% of participants (fully) agree that the storyline was appealing, and 71% of participants (fully) find the visuals appealing. Finally, 60% of participants (fully) agree that the presented topic is interesting. Put simply, the results of the learning outcome and game experience suggest the game has a good balance between educational input and fun gameplay. Furthermore, we find in the play statistics that an impressive 34% of participants returned to the game after the study was completed, and 21% of participants shared the game with their friends.

5.4   Effect of Learning Outcome and Game Experience on Awareness

We conducted two ANCOVA regressions [25], a blend of ANOVA, to analyze the effect of learning outcome and game experience on awareness of the DF profession. In both regressions, we adjusted for the prior awareness of participants in the pre-test. For the learning outcome, the ANCOVA revealed a significant strong correlation between learning outcome and awareness ($F(5, 95) = 29.55$, $p < .001$, partial $\eta^2 = .609$), as well as for game experience and awareness ($F(4, 96) = 51.98$, $p < .001$, partial $\eta^2 = .684$). Hence, the results suggest that an enhanced learning outcome and game experience positively correlate with increased awareness of the DF profession.

6   Discussion and Future Work

The findings from our study indicate that the *Digital Detectives* game significantly enhances the competence and awareness of students. Recognizing its potential as an educational tool, we have integrated the game into school curriculums for career guidance, aiming to spark interest in the DF profession from an early age. In line with our commitment to education and accessibility, we have made the game open-source, allowing for low-cost adoption and adaptation. Addressing the critical shortage of cybersecurity professionals, our game serves as an essential building block in the foundational training of future experts in this field. It is designed not only to pique the interest of young learners but also to reinforce their knowledge through increased awareness [29, 48], which is a key factor in cultivating a robust cybersecurity workforce. Looking ahead, our future work involves developing alternative scenarios to test the potential for even greater educational outcomes. Additionally, we plan to expand the game's scope, taking a proactive step toward the practical training of aspiring forensic scientists. This strategic extension aims to further bridge the gap in cybersecurity expertise, contributing to a safer digital future. Moreover, future research could compare the serious game with traditional learning materials and replicate the study in different geographical settings to validate and extend our conclusions.

7   Conclusion

Despite the importance of cybersecurity, the workforce gap remains a challenge. To address this gap, promote career orientation, and raise awareness of the DF profession, we employ a serious game called *Digital Detectives*. The game facilitates the effective transmission of specialized knowledge in a playful way. Thus, this serious game serves as an educational tool representing a link between theory and practice, shedding light on the field of DF. Ultimately, the evaluation indicates that the serious game significantly increases knowledge and awareness of the DF profession. Additionally, the results suggest a strong correlation between learning outcomes and game experiences with increased awareness. Thus, our thorough game development has emerged as a key factor. Put simply, the serious game can aid young students from *play to profession* of DF.

Use of Generative AI. We utilized DeepL, Grammarly, and ChatGPT (Version 4) to enhance the language and readability of this paper and take responsibility for the content.[4]

Acknowledgment. This research is part of the CONTAIN project, funded by the Federal Ministry of Education and Research as part of the German government's "Research for Civil Security" framework program (13N16586).

Appendix

I. Optimal Click Path of Digital Detectives

To ensure that the player finds the optimal click path (see Fig. 3) through the game and progresses through the story, appropriate conditions were defined. The inclusion of conditions helped to make the gameplay structured and logical. By requiring the player to fulfill certain conditions or perform certain actions in a certain order to progress, a certain level of tension and challenge was created.

---

[4] We transparently report AI Usage Cards [46]: https://ai.uversy.com/serious-game.

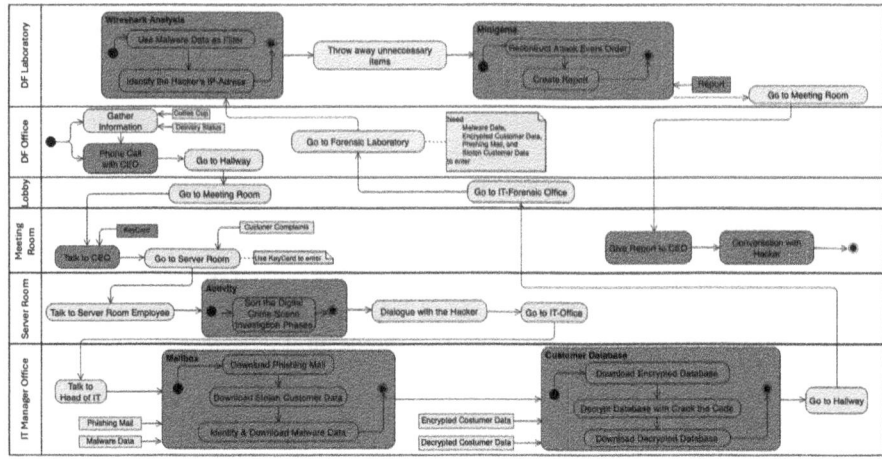

Fig. 3. UML diagram of the ideal click path created with Visual Paradigm [45]

II. Questionnaire of the Evaluation

1. **Learning Outcome** *(Single Choice with five options per question)*
   1.1 Digital Forensic
      - What is digital forensics?
      - How is the IoT defined?
      - Why is Digital Forensics (DF) relevant in criminal investigations?
   1.2 Digital Forensic Phases
      - What are the four phases of a forensic investigation?
      - Why is it crucial to follow these phases?
      - Can the forensic phases be directly transferred to IoT forensics?
   1.3 Digital Forensic Evidence
      - On what layers can evidence be found in IoT forensics?
      - How is ransomware defined?
      - What factors influence the reliability of evidence?
   1.4 Digital Forensic Tools
      - Why is it important to use forensic tools to support investigations?
      - What is the tool Wireshark used for in DF?
      - Which tools can be used for investigations?
2. **Awareness of Digital Forensics Profession** *(5-point Likert scale)*
   - I have heard about the profession of digital forensics.
   - I know what a digital forensic investigator does.
   - I have knowledge about the duties of a digital forensics professional.
   - I am familiar with the role of a digital forensic analyst.
   - I know of the digital forensics profession and its importance.
   - I understand the functions and significance of a career in digital forensics.
   - I've been introduced to the career of digital forensics.
   - I am aware of what it entails to work in digital forensics.
3. **Game Experience** *(5-point Likert scale)*
   3.1 Gameplay & Mechanics
      - The game as a whole was enjoyable.
      - I was focused on the game and did not lose my concentration.
      - The gameplay and mechanics felt enjoyable.
      - The game mechanics were easy to understand.
   3.2 Storyline
      - The overall storyline was appealing.
      - The activities and narrative in the game kept me motivated to keep playing.
      - The feedback for each performed action (answering quiz, collecting items, solving tasks) was appropriate and satisfying.
   3.3 Visuals
      - The visuals were appealing.
   3.4 Topic
      - The presented topics in the game were interesting.
      - I think the game draws a realistic picture of a forensic investigation.

References

1. Abt, C.: Serious games. University press of America (1987)
2. Ahmad, S., Umirzakova, S., Jamil, F., Whangbo, T.K.: Internet-of-things-enabled serious games: a comprehensive survey. Futur. Gener. Comput. Syst. **136**, 67–83 (2022)
3. Blauw, F.F., Leung, W.S.: ForenCity: a playground for self-motivated learning in computer forensics. In: Drevin, L., Theocharidou, M. (eds.) WISE 2018. IAICT, vol. 531, pp. 15–27. Springer, Cham (2018). https://doi.org/10.1007/978-3-319-99734-6_2
4. Blažič, A.J., Cigoj, P., Blažič, B.J.: Serious game design for digital forensics training. In: 2016 Third International Conference on Digital Information Processing, Data Mining, and Wireless Communications (DIPDMWC), pp. 211–215. IEEE (2016)
5. Blažič, B.J.: The cybersecurity labour shortage in europe: moving to a new concept for education and training. Technol. Soc. **67**, 101769 (2021)
6. Blažič, A.J., Cigoj, P., Blažič, B.J.: Serious game design for digital forensics training. In: 2016 Third International Conference on Digital Information Processing, Data Mining, and Wireless Communications (DIPDMWC), pp. 211–215 (2016)
7. Canva: Canva - the visual suite for all (2024). https://www.canva.com/de_de/. Accessed 05 Apr 2024
8. Carrier, B., Spafford, E.: An event-based digital forensic investigation framework. Digital Investigation (2004)
9. Cohen, J.: Statistical Power Analysis for the Behavioral Sciences, 2nd edn. L. Erlbaum Associates, Hillsdale (1988)
10. Derrick, B., White, P.: Comparing two samples from an individual likert question. Int. J. Math. Stat. **18**(3), 1–13 (2017)
11. Dickey, M.D.: Game design narrative for learning: appropriating adventure game design narrative devices and techniques for the design of interactive learning environments. Educ. Tech. Res. Dev. **54**, 245–263 (2006)
12. Dörner, R., Göbel, S., Effelsberg, W., Wiemeyer, J.: Introduction, pp. 1–34. Springer, Cham (2016)
13. Eckert, W.G.: Introduction to Forensic Sciences. CRC Press, Boca Raton (1996)
14. Empl, P., Hager, H., Pernul, G.: Digital twins for IoT security management. In: Atluri, V., Ferrara, A.L. (eds.) DBSec 2023. LNCS, vol. 13942, pp. 141–149. Springer, Cham (2023). https://doi.org/10.1007/978-3-031-37586-6_9
15. Englbrecht, L., Pernul, G.: A serious game-based peer-instruction digital forensics workshop. In: Drevin, L., Von Solms, S., Theocharidou, M. (eds.) WISE 2020. IAICT, vol. 579, pp. 127–141. Springer, Cham (2020). https://doi.org/10.1007/978-3-030-59291-2_9
16. Friedl, S., Glas, M., Englbrecht, L., Böhm, F., Pernul, G.: ForCyRange: an educational IoT cyber range for live digital forensics. In: Drevin, L., Miloslavskaya, N., Leung, W.S., von Solms, S. (eds.) WISE 2022, pp. 77–91. Springer, Cham (2022). https://doi.org/10.1007/978-3-031-08172-9_6
17. FS: Farming simulator (2024). https://www.farming-simulator.com/index.php?lang=de&country=uk. Accessed 08 Apr 2024
18. Glas, M., Vielberth, M., Reittinger, T., Böhm, F., Pernul, G.: Improving cybersecurity skill development through visual programming. Inf. Comput. Secur. **31**(3), 316–330 (2023)

19. Hassenfeldt, C., Jacques, J., Baggili, I.: Exploring the learning efficacy of digital forensics concepts and bagging & tagging of digital devices in immersive virtual reality. Forensic Sci. Int. Digit. Investig. **33**, 301011 (2020)
20. Humphries, G., Nordvik, R., Manifavas, H., Cobley, P., Sorell, M.: Law enforcement educational challenges for mobile forensics. Forensic Sci. Int. Digit. Investig. **38**, 301129 (2021)
21. IBM: IBM SPSS statistics - version 29.0.1 (2024). https://www.ibm.com/de-de/products/spss-statistics. Accessed 02 Mar 2024
22. ISC: ISC2 - cybersecurity workforce study (2024). https://www.isc2.org/research. Accessed 07 Apr 2024
23. JAWA: JAWA - create your own escape games (2024). https://jawa.games/. Accessed 05 Apr 2024
24. Katsaounidou, A., Vrysis, L., Kotsakis, R., Dimoulas, C., Veglis, A.: MAthE the game: a serious game for education and training in news verification. Educ. Sci. **9**(2), 155 (2019)
25. Keselman, H.J., et al.: Statistical practices of educational researchers: an analysis of their Anova, Manova, and Ancova analyses. Rev. Educ. Res. **68**(3), 350–386 (1998)
26. Kiili, K.: Digital game-based learning: towards an experiential gaming model. Internet High. Educ. **8**(1), 13–24 (2005)
27. Kirkpatrick, D.L.: Evaluating Training Programs: The Four Levels, 1st edn. Berrett-Koehler, San Francisco; Publishers Group West, Emeryville (1994)
28. Leung, W.S.: Encouraging equivocal forensic analysis through the use of red herrings. In: Drevin, L., Von Solms, S., Theocharidou, M. (eds.) WISE 2020. IAICT, vol. 579, pp. 184–197. Springer, Cham (2020). https://doi.org/10.1007/978-3-030-59291-2_13
29. Li, L., He, W., Xu, L., Ash, I., Anwar, M., Yuan, X.: Investigating the impact of cybersecurity policy awareness on employees' cybersecurity behavior. Int. J. Inf. Manage. **45**, 13–24 (2019)
30. Manfei, X., Fralick, D., Zheng, J.Z., Wang, B., Changyong, F., et al.: The differences and similarities between two-sample t-test and paired t-test. Shanghai Arch. Psychiatry **29**(3), 184 (2017)
31. McDaniel, L., Talvi, E., Hay, B.: Capture the flag as cyber security introduction. In: 2016 49th Hawaii International Conference on System Sciences (HICSS), pp. 5479–5486. IEEE (2016)
32. McKemmish, R.: What is forensic computing? Australian Institute of Criminology Canberra (1999)
33. Mostafa, M., Faragallah, O.S.: Development of serious games for teaching information security courses. IEEE Access **7**, 169293–169305 (2019)
34. MurfAI: Murf AI - AI voice generator: versatile text to speech software (2024). https://murf.ai/text-to-speech. Accessed 05 Apr 2024
35. Neupane, S., et al.: On the data privacy, security, and risk postures of IoT mobile companion apps. In: Sural, S., Lu, H. (eds.) DBSec 2022. LNCS, vol. 13383, pp. 162–182. Springer, Cham (2022). https://doi.org/10.1007/978-3-031-10684-2_10
36. Nordhaug, Ø., Imran, A.S., Alawawdeh, A.M., Kowalski, S.J.: The forensic challenger. In: 2014 International Conference on Web and Open Access to Learning (ICWOAL), pp. 1–6. IEEE (2014)
37. Pan, Y., Mishra, S., Schwartz, D.: Gamifying course modules for entry level students. In: Proceedings of the 2017 ACM SIGCSE Technical Symposium on Computer Science Education, pp. 435–440 (2017)

38. Pan, Y., Mishra, S., Yuan, B., Stackpole, B., Schwartz, D.: Game-based forensics course for first year students. In: Proceedings of the 13th Annual Conference on Information Technology Education, pp. 13–18 (2012)
39. Pan, Y., Schwartz, D., Mishra, S.: Gamified digital forensics course modules for undergraduates. In: 2015 IEEE Integrated STEM Education Conference, pp. 100–105. IEEE (2015)
40. Raybourn, E.M., Kunz, M., Fritz, D., Urias, V.: A zero-entry cyber range environment for future learning ecosystems. In: Cyber-Physical Systems Security, pp. 93–109 (2018)
41. Reyes, D.M., Santos, J.J., Tsai, A.L.: Designing an educational point-and-click adventure game about cognitive flexibility. In: Stephanidis, C., Antona, M., Ntoa, S. (eds.) HCII 2022. CCIS, vol. 1582, pp. 88–95. Springer, Cham (2022). https://doi.org/10.1007/978-3-031-06391-6_12
42. Smidt, A., Balandin, S., Sigafoos, J., Reed, V.: The kirkpatrick model: a useful tool for evaluating training outcomes. J. Intellect. Dev. Disabil. **34**, 266–274 (2009)
43. Susi, T., Johannesson, M., Backlund, P.: Serious games: an overview (2007)
44. Torrente, J., Marchiori, E.J., Moreno-Ger, P., Fernández-Manjón, B., Vallejo-Pinto, J.Á., Ortega-Moral, M.: Evaluation of three accessible interfaces for educational point-and-click computer games. J. Res. Pract. Inf. Technol. **45**(3/4), 267–284 (2013)
45. VP: Visual paradigm - agile project tools to improve productivity (2024). https://www.visual-paradigm.com/. Accessed 05 Apr 2024
46. Wahle, J.P., Ruas, T., Mohammad, S.M., Meuschke, N., Gipp, B.: AI usage cards: responsibly reporting AI-generated content. In: 2023 ACM/IEEE Joint Conference on Digital Libraries (JCDL), pp. 282–284. IEEE (2023)
47. Woolson, R.F.: Wilcoxon signed-rank test. Wiley Encyclopedia of Clinical Trials, pp. 1–3 (2007)
48. Yang, N., Singh, T., Johnston, A.: A replication study of user motivation in protecting information security using protection motivation theory and self determination theory. AIS Trans. Replication Res. **6**(1), 10 (2020)
49. Yerby, J., Hollifield, S., Kwak, M., Floyd, K.: Development of serious games for teaching digital forensics. Issues Inf. Syst. **15**(2) (2014)

User Perceptions of CAPTCHAs: University vs. Internet Users

Arun Reddy[1] and Yuan Cheng[2][(✉)] [ID]

[1] Intel Corporation, Santa Clara, CA 95054, USA
r3ddy.arun@gmail.com
[2] Grand Valley State University, Allendale, MI 49401, USA
chengy@gvsu.edu

Abstract. We surveyed over 250 participants from a university and Amazon Mechanical Turk to understand user perceptions of CAPTCHA security and usability. Users reported struggling with CAPTCHA challenges due to increasing difficulty, leading to frustration and a negative experience. Concerns about reliability and security were also noted. Our findings offer valuable insights for developing more secure and user-friendly CAPTCHA technologies.

Keywords: CAPTCHA · Authentication · Usability

1 Introduction

CAPTCHA is a challenge-response test to distinguish human users from bots, preventing abuse like false form submissions, fraudulent purchases, spam emails, and fake registrations. Since the late 1990s, CAPTCHAs have evolved into various forms, including text, image, audio, and video. However, as they have become more common, security concerns have arisen. An ideal CAPTCHA should balance security and usability. To defend against attacks, CAPTCHAs often introduce distortion and noise, making them harder for users to solve. Text-based CAPTCHAs often contain scattered lines, dots, and distorted characters that require extra user effort. These distortions can be especially unfriendly to users whose native language does not use the Latin alphabet [7]. Image-based CAPTCHAs present accessibility challenges for users with visual impairments or color blindness [7]. Audio and video-based CAPTCHAs face issues with large file sizes and limited user comprehension time [1,6].

Creating a secure yet user-friendly CAPTCHA system is challenging. We surveyed over 250 participants, including computer science students and Amazon Mechanical Turk users. We found that illegible texts/images and unclear instructions were the main reasons for failing CAPTCHAs. Human errors like lack of time and inattention also contributed to failures. Users often abandoned webpages if they could not solve CAPTCHAs on the first attempt, causing decreased traffic. The study showed that users view CAPTCHAs as a necessary burden. We also examined differences between the two user groups and explored factors contributing to these variances.

© IFIP International Federation for Information Processing 2024
Published by Springer Nature Switzerland AG 2024
A. L. Ferrara and R. Krishnan (Eds.): DBSec 2024, LNCS 14901, pp. 290–297, 2024.
https://doi.org/10.1007/978-3-031-65172-4_18

Our work explores user perspectives on CAPTCHA security and usability, with a focus on the latter. We integrate usability issues from Yan et al.'s work [7]. Unlike prior studies focusing on specific CAPTCHA types [3–7], our study covers all types. Similar to Fidas et al.'s survey [2], we examine security-usability trade-offs, attempts needed to solve CAPTCHAs, and language influences. However, we also investigate how user views have evolved over the past decade and compare opinions from two distinct participant groups. Conversely, Fidas et al.'s study did not specify their participant sources.

2   Methodology

We conducted an online survey to understand how users perceive and adopt CAPTCHA mechanisms. The study was reviewed and approved by our institution's Institutional Review Board.

2.1   Recruitment

We recruited participants from Amazon Mechanical Turk (MTurk) and university students majoring in Computer Science. We selected MTurk candidates with a Human Intelligence Task (HIT) rate of over 90% and more than 50 approved HITs, paying each $1 for the 5-minute survey. Within 30 min, we received 150 responses, primarily from the U.S. University users were upper-division computer science students at a public university in the U.S. with a technical background.

2.2   Study Design

We used Qualtrics for an anonymous survey, providing unique codes to prevent multiple submissions. The survey included seven demographic and eligibility questions, followed by 16 questions on CAPTCHA perceptions. Participants rated different CAPTCHA types on a five-point Likert scale and provided feedback regarding non-English CAPTCHAs. Usability questions covered experiences on small devices (phones or tablets), considering screen size, orientation, network bandwidth, and processor speed. We asked if they generally solved CAPTCHAs on the first attempt and, if not, the reasons behind their difficulties. We explored user frustration and whether they had abandoned a website due to CAPTCHAs. We assessed security with three questions on user views of CAPTCHA security and effectiveness. Participants needed prior CAPTCHA experience to proceed. We recorded completion times and discarded results completed in under 15 s, though none were disqualified. We received 154 MTurk responses, discarding four incomplete ones. All 109 responses from university users were valid.

2.3   Demographics

Table 1 summarizes respondents' demographics. Our survey included 259 valid respondents: 62% male and 32% female. Only 14% of university users were female, aligning with the U.S. national average (18%). Respondents' ages ranged from 19 to 70, averaging 32 with a standard deviation of 11.39. The largest age group (25–34) comprised 41% of respondents (107 out of 259), with 69 MTurk users and 38 university users. 60% of university users (65 out of 109) were aged 18–24. Over 70% of MTurk users held a bachelor's degree or higher, while 88% of university users were pursuing their first bachelor's degree. 86% (223) were native English speakers. University users had relatively more non-native English speakers (20%), reflecting the university's diversity.

Table 1. Demographics

	MTurk		Univ.		Total				MTurk		Univ.		Total	
	No.	%	No.	%	No.	%			No.	%	No.	%	No.	%
Gender								**Education**						
Male	82	55	90	83	172	66		Less than high school degree	2	1	0	0	2	1
Female	68	45	15	14	83	32		High school degree	9	6	5	5	14	5
Non-binary	0	0	2	2	2	1		Some college experience	17	11	55	50	72	28
Decline to say	0	0	2	2	2	1		Associate degree	15	10	36	33	51	20
Age								Bachelor's degree	85	57	13	12	98	38
18-24	6	4	65	60	71	27		Postgraduate degree	22	15	0	0	22	8
25-34	69	46	38	35	107	41		**Language**						
35-44	38	25	5	5	43	17		Native English speaker	136	91	87	80	223	86
45-54	19	13	1	1	20	8		Non-native speaker	14	9	22	20	36	14
55-64	12	8	0	0	12	5								
64 or older	6	4	0	0	6	2								

2.4 Hypotheses

This study tested the following null hypotheses: Between the two user groups,

- (H1) No significant difference in preferred CAPTCHA types.
- (H2) No significant difference in perception of solving CAPTCHAs on smaller devices.
- (H3) No significant difference in reasons for abandoning CAPTCHAs.
- (H4) No significant difference in perception of CAPTCHA difficulty.
- (H5) No significant difference in perception of CAPTCHA security.
- (H6) No significant difference in overall opinion of CAPTCHAs.

3   Findings

We analyzed the data from the Likert and closed-ended questions to determine differences between MTurk and university users' perspectives on CAPTCHAs.

Use and Preferences on CAPTCHA Types. We surveyed respondents on various CAPTCHA types: math, text identification, image recognition, image drag and drop, audio/video, and game/puzzle. Image recognition was the most prevalent (224 votes, 89%), followed by text identification (189 votes, 75%), with audio/video CAPTCHAs being the least used (57 votes, 23%). These results align with the widespread use of Google reCAPTCHA v2. Respondents ranked their preferences on a scale of 1 to 5. Figure 1 shows the preferences of MTurk users, and Fig. 2 presents those of university users. MTurk users preferred text identification (55 votes, 37%) and image recognition (45 votes, 31%), while university users favored image recognition (33 votes, 31%) and text identification (23 votes, 22%). Over half of university users never solved the other four CAPTCHA types, whereas MTurk users showed more diverse usage. An ANOVA test revealed significant differences in preferences between the groups ($p < 0.05$), possibly due to age, as most university users were under 25.

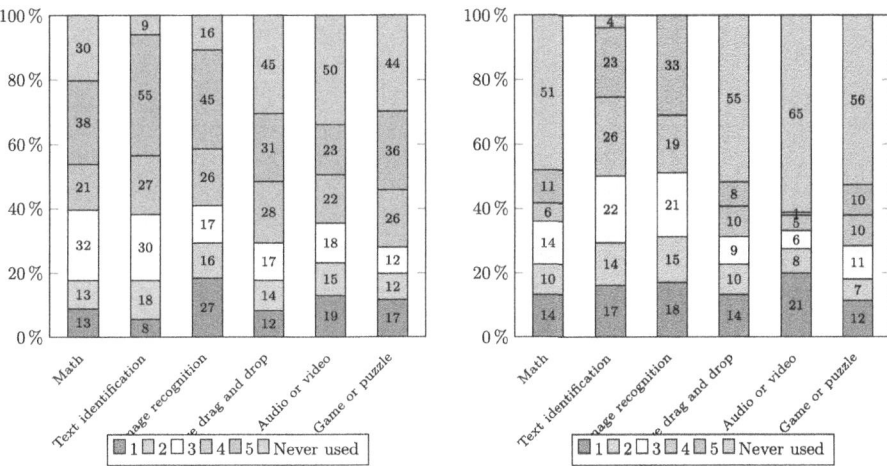

Fig. 1. Preferences of MTurk users   **Fig. 2.** Preferences of university users

CAPTCHAs in Non-English Languages. Only 3% of respondents encountered non-English CAPTCHAs, mostly in Portuguese (37.5%), followed by Hindi, Tamil, Turkish, French, Japanese, and Korean. Among these, 50% preferred text-based CAPTCHAs, 13% favored audio/video CAPTCHAs, and 25% liked both. However, more research is needed due to the small sample size.

Solving CAPTCHAs on a Smaller Device. 68% of respondents found solving CAPTCHAs easier on a computer than on a phone or tablet, while 32% saw no difference. A significant difference in opinions was noted between MTurk and university users (chi-squared $p = 0.0004$). Of the 172 respondents, 74% cited size and orientation as the main reasons (73% MTurk, 78% university), and 58% considered input method (Touch vs. Keyboard/Mouse) essential.

Failing a CAPTCHA Challenge on a Single Attempt. Solving CAPTCHAs can be frustrating. In our survey, 28% of respondents always solved CAPTCHAs on the first attempt, as shown in Table 2. However, 3% of respondents consistently struggled. We also gathered their failure experiences through an open-ended question.

Table 2. Solving CAPTCHAs in one attempt

		MTurk	Univ.	Total
Always	n	52	19	71
	%	35%	18%	28%
Frequently	n	90	85	175
	%	61%	80%	69%
Rarely	n	5	2	7
	%	3%	2%	3%
		147	106	253

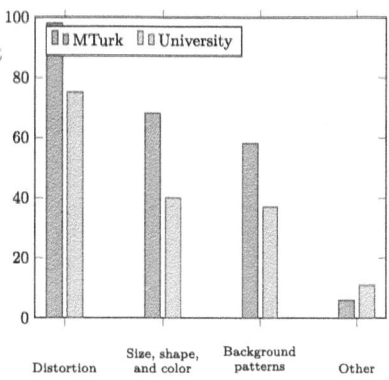

Fig. 3. Why CAPTCHAs are hard to use?

Respondents often cited image blurriness and unclear visuals as primary issues. For instance, P2 stated: *"I can't always see the images clearly."* P17 commented: *"Pictures too blurred or small to see properly."* Text clarity was also a problem, with many finding the font, font size, and formatting difficult to read. P25 said: *"Cannot recognize font."* P197 shared: *"CAPTCHA text or image can be distorted beyond recognition."* Some respondents had to refresh the CAPTCHA for a clearer version. Another common issue was the perceived similarity of letters and numbers, such as '9' vs. 'g,' "vv" vs. 'w,' 'I' vs. '1,' leading to frequent mistakes. P73 explained: *"Some letter number combos like nine g and one I l and s f and o zero and similar letter number combo are very difficult to solve."* These similarities often led to typing errors and failures on the first attempt. Human errors have caused many user failures, especially with image recognition and text identification CAPTCHAs. P259 shared: *"Mistaken letter or misheard a letter."* P32 noted: *"Sometimes miss a picture or put a letter in the wrong place in a text CAPTCHA."* Users often face difficulty deciding whether to select a particular image segment when only a small portion of the target object is visible. P163 highlighted: *"Normally incorrect image selected or character."* Other common issues included carelessness, typos, accidental clicks, and rushing. Case-sensitive text CAPTCHAs posed challenges, often leading to multiple attempts. P64 shared: *"Typing a lowercase while CAPTCHA needs an uppercase.*" This issue commonly leads to another problem: unclear instructions. P181 recounted: "*The image says to click all the stoplights, and I don't know if it wants me to click the little bit on a square that is part of it but not really the stoplights.*" Our findings suggest obscuring instructions may be a deliberate tactic to increase CAPTCHA difficulty for bots.

Why Did You Give Up? We investigated if respondents ever left a webpage due to CAPTCHA failures. 75% of university users persisted until success, while nearly 65% of MTurk users abandoned the page, with 19% leaving immediately after one failure compared to 7% of university users. We also asked if users left a webpage simply to avoid solving a CAPTCHA. 62% of university users remained on the webpage despite a CAPTCHA, while 15% sought bypass methods. In contrast, 45% of MTurk users left due to challenging or annoying CAPTCHAs, and 20% found alternative bypass methods. The two groups showed significantly different behaviors ($p < 0.0001$, chi-squared test). Active MTurk users tend to complete tasks or jobs as much as possible within a given time frame, showing less patience for CAPTCHAs. University students, being tech-savvy and goal-oriented, show more patience during the authentication process.

Perceived CAPTCHA Difficulty. We asked respondents to identify the feature that contributes most to CAPTCHA difficulty. Figure 3 presents the distribution of votes from our participants. Distortion emerged as the top factor, aligning with the insights gathered from user responses in Sect. 3. Notably, 98 MTurk users (65%) and 75 university users (69%) endorsed this option. Size, shape, and color were also identified as troublesome factors, with 68 votes from MTurk users (45%) and 40 votes from university users (37%). Background patterns ranked third, albeit with a narrow margin (58 votes from MTurk users and 37 votes from university users). We then inquired about the trend in CAPTCHA difficulty over time. 43% of MTurk users felt that CAPTCHAs are getting harder, while 49% of university users believed that CAPTCHA difficulty has remained consistent. Conversely, 27% of MTurk users and 19% of university users found CAPTCHAs easier. A chi-squared test revealed significant differences between the two groups ($p = 0.008$).

Accessibility Concerns. Accessibility is crucial for user experience, yet several respondents shared frustrations. P12 noted that "*audio files sometimes did not properly load when solving an audio CAPTCHA.*" P181 encountered issues on mobile devices, particularly in touchscreen mode, stating: "*I was on a mobile phone and tried to solve a CAPTCHA and when I couldn't click on the screen, I chose to make it an audio CAPTCHA and it failed to let me solve it.*" These complaints highlight common accessibility issues with CAPTCHAs on small screens.

How Secure Are Current CAPTCHAs? We assessed users' perceptions of CAPTCHA security. The results indicate that 49% of MTurk users praised

CAPTCHA effectiveness, while 45% agreed that CAPTCHAs do a decent job. Conversely, only 15% of university users expressed high confidence, with 70% thinking CAPTCHAs are okay. Regarding CAPTCHAs keeping up with advancing attacks, 69% of university users were uncertain, with only 25% confident. Similarly, 41% of MTurk users shared this uncertainty, while 44% were confident. These findings suggest university users are more aware of security threats since many of them have taken security courses and are tech-savvy. However, they also highlight the need for greater public awareness and education on cybersecurity.

Final Stance. The final question aimed to gather respondents' overall opinions on CAPTCHA security and usability. Table 3 shows that among university users, 44% found CAPTCHAs secure and 66% found them usable. In contrast, 64% of MTurk users considered CAPTCHAs secure, with 55% finding them usable. Only 6% of MTurk users and 8% of university users viewed CAPTCHAs as neither secure nor usable. A significant disparity in opinions between the two groups was evident (p = 0.021, chi-squared test).

Table 3. Final stance on CAPTCHAs

	MTurk	University
Secure and usable	35 (24%)	20 (19%)
Secure but usability is a concern	57 (40%)	27 (25%)
Usable but security is a concern	43 (30%)	50 (47%)
Neither secure nor usable	9 (6%)	9 (8%)

4 Discussions

All six null hypotheses were rejected, indicating significant variations in user perceptions of CAPTCHA issues between MTurk and university users.

We observed distinct preferences for CAPTCHA types and differing perspectives on CAPTCHA difficulty. While more MTurk users claimed to solve CAPTCHAs on the first attempt (35% vs. 18%), they also gave up more easily compared to university users, who showed greater persistence. Age played a role in these differences, with MTurk users generally older (96% aged 25 or older) than university users (60% under 25). Our results aligned with previous research [5], indicating younger individuals prefer different CAPTCHA schemes on smartphones. Moreover, our study supported the finding that older participants are more likely to solve CAPTCHAs on the first attempt [2].

Despite higher education levels (72% with a bachelor's or postgraduate degree), MTurk users showed less concern for CAPTCHA security (64% considered it secure) compared to university users, who were more aware of online security threats.

Both groups found solving CAPTCHAs on smaller devices equally challenging (100% for MTurk users, 99% for university users), with screen size, orientation, and input methods as top obstacles. Many CAPTCHA issues persisted or worsened on smaller devices. We support recommendations by Reynaga et al. [5] for CAPTCHAs on smaller devices to adopt cross-platform input mechanisms, maintain consistent orientation, and employ minimalist designs.

CAPTCHAs remain difficult to use and heavily rely on visual processing, rendering them inaccessible to millions with visual impairments. There is a need for simpler, less burdensome solutions. Google's reCAPTCHA v2 introduced *Invisible CAPTCHAs* in 2014, but privacy concerns remain. We advocate for an invisible, passive CAPTCHA scheme leveraging user device data and behavioral activities, with transparency about data use. Our future agenda includes developing and testing this new mechanism to address these concerns.

5   Conclusion

This study surveyed 259 respondents from a university and Amazon Mechanical Turk to examine user perceptions of CAPTCHA usability and security. Our findings revealed significant disparities between the two groups. We explored possible explanations for these differences, including age and technical background. Both groups struggled with current CAPTCHA schemes, leading to failure and frustration. Our insights offer recommendations for more user-friendly CAPTCHAs and can guide future research in developing more effective solutions.

References

1. Bursztein, E., Bethard, S., Fabry, C., Mitchell, J.C., Jurafsky, D.: How good are humans at solving CAPTCHAs? A large scale evaluation. In: 2010 IEEE Symposium on Security and Privacy, pp. 399–413 (2010)
2. Fidas, C.A., Voyiatzis, A.G., Avouris, N.M.: On the necessity of user-friendly CAPTCHA. In: the SIGCHI Conference on Human Factors in Computing Systems. CHI '11, pp. 2623–2626 (2011)
3. Gossweiler, R., Kamvar, M., Baluja, S.: What's up CAPTCHA? A CAPTCHA based on image orientation. In: the 18th International Conference on World Wide Web. WWW '09, pp. 841–850 (2009)
4. Krol, K., Parkin, S., Sasse, M.A.: Better the devil you know: a user study of two CAPTCHAs and a possible replacement technology. In: NDSS Workshop on Usable Security (USEC), vol. 10 (2016)
5. Reynaga, G., Chiasson, S., van Oorschot, P.C.: Exploring the usability of CAPTCHAS on smartphones: comparisons and recommendations. In: NDSS Workshop on Usable Security (USEC), vol. 112 (2015)
6. Sauer, G., Hochheiser, H., Feng, J., Lazar, J.: Towards a universally usable CAPTCHA. In: the 4th Symposium on Usable Privacy and Security (2008)
7. Yan, J., El Ahmad, A.S.: Usability of CAPTCHAs or usability issues in CAPTCHA design. In: the 4th Symposium on Usable Privacy and Security (2008)

Differential Privacy

Incentivized Federated Learning with Local Differential Privacy Using Permissioned Blockchains

Saptarshi De Chaudhury[1], Likhith Reddy[1], Matta Varun[1],
Tirthankar Sengupta[1], Sandip Chakraborty[1], Shamik Sural[1(✉)],
Jaideep Vaidya[2], and Vijayalakshmi Atluri[2]

[1] Indian Institute of Technology Kharagpur, Kharagpur, India
saptarshi_dechaudhury@kgpian.iitkgp.ac.in,
likhith26090@kgpian.iitkgp.ac.in , mailtisen03@kgpian.iitkgp.ac.in,
sandipc@cse.iitkgp.ac.in, shamik@cse.iitkgp.ac.in
[2] Rutgers University, Newark, USA
jsvaidya@business.rutgers.edu, atluri@rutgers.edu

Abstract. Federated Learning (FL) is a collaborative machine learning approach that enables data owning nodes to retain their data locally, preventing its transfer to a central server. It involves sharing only the local model parameters with the server to update a global model, which is then disseminated back to the local nodes. Despite its iterative convergence, FL has several limitations, such as the risk of single-point failure, inadequate incentives for participating nodes, and potential privacy breaches. While Local Differential Privacy (LDP) is often used to mitigate privacy concerns, the other challenges of FL have not yet been addressed comprehensively, even for Locally Differentially Private Federated Learning (LDP-FL). We propose an integrated approach that uses permissioned blockchains to guard against a single point of failure and a token-based incentivization (TBI) mechanism for encouraging participation in LDP-FL. In our scheme, participating nodes receive tokens upon sharing their model parameters, which can subsequently be used to access updated global models. The number of tokens awarded for parameter sharing is determined by ϵ - the privacy factor of LDP, ensuring that the nodes do not overly obfuscate the data they share. We demonstrate the feasibility of our approach by developing the *Blockchain-based TBI-LDP-FL framework* (hereinafter, referred to as BTLF) on HyperLedger Fabric. Extensive results of experimentation establish the efficacy of BTLF.

Keywords: Federated Learning · Blockchain · Local Differential Privacy · Incentivization · HyperLedger Fabric

1 Introduction

Machine learning (ML) models have achieved remarkable success in various domains due to the large-scale datasets available for robust training. Federated

© IFIP International Federation for Information Processing 2024
Published by Springer Nature Switzerland AG 2024
A. L. Ferrara and R. Krishnan (Eds.): DBSec 2024, LNCS 14901, pp. 301–319, 2024.
https://doi.org/10.1007/978-3-031-65172-4_19

Learning (FL) [25], which is a form of collaborative learning, has emerged as a flexible and dynamic technique facilitating algorithmic training through independent client sessions, each leveraging its own unique dataset. In this framework, a central server selects a model for training, subsequently communicating it to a set of nodes that own data, which in turn conduct local model training utilizing their respective datasets. Unlike traditional distributed learning, instead of the entire dataset, only the resulting model parameters are shared with the server in FL. The server utilizes the collected local parameters to update global model weights, which are then sent back to the local nodes [25].

FL has been studied extensively in recent years, and the field has developed rapidly [6,15]. However, the traditional FL frameworks still face several shortcomings that reduce the reliability of the whole system [4,31,32,36]. First, the FL framework typically relies on a central server to consolidate local training outcomes for updating the global model. However, the server's vulnerability to intentional manipulation, network failures, and external attacks poses a significant threat. A server compromise can lead to the failure of the entire FL system. Secondly, with a multitude of participants in FL, it becomes impractical to assume the absolute honesty of all clients in adhering to the FL protocols during local model training. Dishonest clients may introduce false data, significantly impacting the performance of the global model.

Further, despite the basic premise of FL to solely transmit local model parameters without sharing the complete data, exposure of private information due to inference attacks at the server still remains possible under certain conditions [37]. This susceptibility can trigger a range of privacy-related issues. Another critical concern is that standard FL models lack an incentivization structure, compelling participating clients to contribute their computational resources without receiving adequate compensation. Absence of incentives not only hampers honest participation but also limits the engagement of a sufficiently large number of clients, often found necessary for data-intensive tasks. The above challenges necessitate an integrated and robust framework that considers the existing limitations while ensuring the efficiency and correctness of the model training process. In order to address these, we present an innovative approach through the integration of Permissioned Blockchains [36], Local Differential Privacy (LDP) [9], and Token-Based Incentivization (TBI).

Inclusion of each of these facets in our framework is crucial for several reasons. Using a blockchain network instead of a central aggregator/server is robust against potential server crashes. With blockchain, model aggregation is distributed across multiple nodes, reducing the vulnerability to a single point of failure. It can also effectively mitigate model parameter tampering risks by malicious clients or the aggregated weights by the server as these are written to a ledger and are operated on by well formed smart contracts. Supporting LDP in the FL process minimizes the possibility of revealing private data to the server by the clients [8]. The local nodes can perturb their model parameter values pertaining to a privacy budget ϵ before writing to the blockchain ledger. On the other hand, the Token-Based Incentivization (TBI) mechanism awards tokens to the clients when the server/aggregator receives the model parameters from the corresponding nodes. These tokens can subsequently be used by the clients to access the updated global model parameters. The quantum of tokens awarded is a function of the privacy parameter ϵ of LDP used by a client, thereby ensuring that the nodes do not over-restrict the data they share. TBI promotes participation of clients in federated model building and at the same time discourages those who attempt to derive its benefits without meaningfully contributing.

In summary, this paper makes the following novel contributions.

i We use HyperLedger Fabric (HLF) blockchain as the underlying technology for uploading of local model parameters by clients, and updating and dissemination of global model parameters by the server in an FL setting.
ii We develop chaincodes for implementing Local Differential Privacy while sharing model parameters by the clients, thereby removing the need to trust the server chaincode performing aggregation in the FL process.
iii Besides carrying out local model training, the clients are also equipped with a token based incentivization mechanism that enables them to acquire new tokens while sharing their model parameters and use them at the time of reading the global parameters, both in accordance with a predefined policy.

To the best of our knowledge, there is no existing work in the literature that combines incentivization with locally differentially private federated learning in a permissioned blockchain setting.

2   Preliminaries

This section introduces some of the basic concepts underlying the work presented later in the paper.

2.1   Federated Learning

Federated learning orchestrates collaborative training of local deep neural network (DNN) models among distributed clients connected to a central server [25]. The process commences with the server randomly initializing the model parameters, denoted as μ_o, which are then distributed to the clients to initialize their respective model copies. Each client independently trains its model using its local dataset for multiple epochs, eventually producing updated model parameters, μ_u. The process of federated learning in each round includes averaging of all the μ_u's received from the clients to obtain μ_{fed} (the parameters of the global model). Such an iterative process, known as a federation round, continues through multiple rounds until μ_{fed} either converges or a predetermined number of rounds is completed [8]. It has been shown that μ_{fed} produces almost the same accuracy as that of a local model trained on the complete data [41].

In spite of the fact that the clients only share their model parameters with the server and not the actual data, privacy concerns in the context of Federated Learning have still been raised, especially due to its possible vulnerability to inference attacks [12]. A potential mitigation strategy is to use local differential privacy as described next.

2.2 Local Differential Privacy

Differential privacy is a mathematical framework for ensuring the privacy of individuals in datasets, which was first proposed in [11]. It allows data to be analyzed without revealing sensitive information about any particular user in the dataset, thereby providing a strong guarantee of the individual's privacy. In the traditional model of differential privacy, also known as centralized differential privacy (CDP), there exists a trusted aggregator or curator which collects and holds the sensitive data of all the individuals. It is responsible for protecting their privacy by adding a measured amount of noise, i.e., perturbing the data by infusing sufficient noise to the output. Effectively, the goal is to mask the contribution of individual data elements and yet preserve overall analysis accuracy. The aggregator, however, has to first collect original data of the users before releasing the perturbed aggregated information publicly [40].

An issue with CDP is that there is an implicit but complete trust on the data curator, which may not always be the situation in real world. Even the largest and most reputable companies cannot guarantee their customers' privacy and may fall victim to data breaches. To address the problem of trusting a central authority for managing user data, the mechanism of Local Differential Privacy was proposed. In LDP, a user's data is perturbed locally before it is sent to the aggregator. The original data is only accessible to the owner, which provides a much stronger privacy guarantee for the user. In the LDP model, a curator holds a perturbed version of the data and not the original data. Also, all the training and any form of querying is performed on this perturbed dataset only. Thus, LDP protects against data disclosure even to the untrusted curators and relieves the burden on the trusted data curators to keep data secure [40]. LDP is often specified in terms of what is known as ϵ-Local Differential Privacy. As defined in [39], a randomized mechanism $\mathcal{M}$ is said to satisfy ϵ-Local Differential Privacy if and only if for any pair of input values v and $v' \in D$ and for any possible output $S \subseteq Range(\mathcal{M})$,

$$P(\mathcal{M}(v) \in S) \leq e^{\epsilon} P(\mathcal{M}(v') \in S)$$

Here $\epsilon > 0$ is the privacy budget - smaller the value of ϵ, stricter is the protection with lower data availability, and vice versa. Federated Learning is a natural application domain for LDP and has been investigated from different aspects and application domains [14, 24, 33, 35, 42].

2.3 Permissioned Blockchains and HyperLedger Fabric

In a blockchain, which is a decentralized append-only ledger, blocks are added by miners after combining several transactions. Each node in a blockchain network operates without a need to trust other nodes. Once a block has been added to the blockchain ledger, immutability of the data within the block inhibits any attempts to alter the added information. Blockchains are primarily categorized into two types: Permissionless blockchains and Permissioned blockchains. In this paper, we use a permissioned blockchain, namely, HyperLedger Fabric, wherein only authorized nodes are permitted to participate in the transactions. However, even if a node is authorized, it is granted visibility to only the content made available to it by means of a suitable access control mechanism [2].

Hosted by the Linux Foundation [3], HLF is a modular and extensible open-source system specifically designed for deploying and operating permissioned blockchains. Within the HLF ecosystem, several essential components contribute to its architecture and functionality. Those of relevance to our work include the concept of *World State*, which refers to the current state of the ledger. It represents the latest values of all committed transactions. *Organizations* in HLF represent entities that participate in the network and interact with the blockchain. Each organization maintains its set of peers and can have specific roles and permissions within the network. *Peers* in HLF maintain the ledger and handle the state updates. They endorse, validate and commit transactions, playing a crucial role in the consensus process. A *chaincode* in HLF is essentially similar to the concept of smart contracts, representing the business logic governing transactions within the network. It is run on the HLF distributed ledger network and facilitates execution of transactions. Chaincodes provide the rules and agreements that govern how data can be accessed, updated and validated within a HyperLedger Fabric network. An *asset* in HLF refers to any digital or physical entity that has value and is owned or controlled by an organization or an individual. *Endorsement* in HLF is a crucial step in the transaction validation process. Transactions must be endorsed by a specified number of endorsing peers before being committed to the ledger. Endorsement policies can be configured to ensure that transactions meet specific criteria, enhancing the security and reliability of the network.

3   Proposed Approach

In this section, we outline our proposed approach named BTLF (*Blockchain-based TBI-LDP-FL framework*) to address the various limitations faced by FL with the integration of permissioned Blockchains, Local Differential Privacy and Token Based Incentive mechanisms. Figure 1 shows a high-level flow diagram of BTLF. Once the HyperLedger Fabric blockchain network is initialized in Step 1, the clients can start sending their model weights to the ledger after applying local differential privacy. These are replicated to all the peer nodes (Step 2). The server function aggregates the data and awards tokens to the participating clients for that round of federated learning. Updated global parameters are next written back to the ledger (Step 3). Clients that are interested in reading the weights obtained from this round, make a request through their associated nodes. The corresponding node reads the data from the ledger and shares the same with their clients and deducts tokens for data utilization (Step 4). The client updates its local model with the new global weights and checks for improvement in training accuracy (Step 5). The next round of FL can now start.

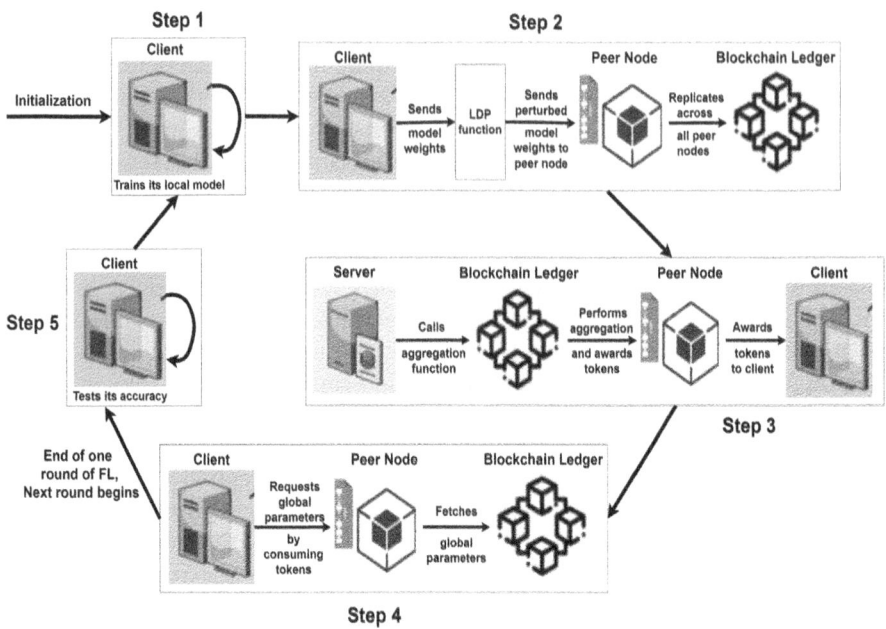

Fig. 1. Overview of the BTLF framework

Algorithm 1 Pseudo code for Client

1: $(m, w) \leftarrow \text{init}()$
2: **while** till convergence or maxRounds reached **do**
3:     $w \leftarrow \text{trainModel}(m, w)$
4:     $w' \leftarrow \text{LDP}(w, \epsilon)$ ▷ perturbing the parameters of local model
5:     $\text{putParameters}(w', \epsilon)$
6:     $w \leftarrow \text{globalParameters}()$
7: **end while**

3.1   Blockchain-Based LDP-FL

Before the FL process is started, the participating organizations join in to achieve consortium and create the HLF network. In order to participate in the FL process, a client must complete registration and authorization within the permissioned blockchain network. Thereafter, it follows the pseudo code given in Algorithm 1. Upon initiation, each client within the blockchain network initializes a DNN model m with randomized parameters as shown in Line 1 of Algorithm 1. Subsequently, this model undergoes training using the available local dataset for multiple epochs, resulting in the updating of model parameters. The updated model parameters are then subjected to perturbation, with respect to the prescribed privacy factor ϵ of the LDP technique. Following this, the clients submit their perturbed weights to the peers of the blockchain network, to be securely written onto the blockchain ledger. The training is done entirely off chain on the local system of the client, and only the perturbed model weights are uploaded to the ledger via a chaincode.

In this paper, we use the LDP mechanism introduced in [33]. This algorithm, when applied to the weights W of a model, produces a perturbed tuple W^* by introducing randomness to each dimension of W. For every weight $w \in W$, assuming $w \in [c - r, c + r]$ where c is the center and r is the radius of w's range, the following LDP mechanism to perturb w is used:

$$
w^* = \begin{cases} c + r \cdot \dfrac{e^\epsilon + 1}{e^\epsilon - 1} & \text{with probability} & \dfrac{(w - c)(e^\epsilon - 1) + r(e^\epsilon + 1)}{2r(e^\epsilon + 1)}, \\[4mm] c - r \cdot \dfrac{e^\epsilon + 1}{e^\epsilon - 1} & \text{with probability} & \dfrac{-(w - c)(e^\epsilon - 1) + r(e^\epsilon + 1)}{2r(e^\epsilon + 1)}. \end{cases}
$$

$$(1)$$

From Eq. 1, it can be observed that, with a given privacy budget ϵ, the larger the range r, the higher is the range of perturbed weights w^*, i.e., greater is the noise in the perturbed weights. In BTLF, based on prior knowledge, all the clients and the server agree on the same weight range represented by (c, r) at the beginning, i.e., while initializing the weights. In addition, each client chooses its value of ϵ. When sending its local parameters to the ledger, the client perturbs them using the LDP function given by Eq. 1. For any weight $w \in [c - r, c + r]$, where c is the center and r is the radius of w's range, it has been proven in [33] that the mechanism satisfies ϵ-LDP w.r.t. $[c - r, c + r]$. Both ϵ and r have an impact on the privacy level. The degree of concealment of the individual data in a crowd of data is determined by ϵ, and r determines the size of the crowd. Note that, in the traditional LDP approach, all the clients uniformly use the same ϵ for perturbing the model parameters. However, the exploration of distinct privacy factors in a multi-privacy framework remains an active area of research. By incorporating multiple privacy factors, BTLF facilitates better incentivization and increases fairness among the participating clients.

After writing the perturbed weights in the ledger, clients can query the peers in the blockchain network to access the global model parameters as shown in Line 6 of Algorithm 1. Clients with access to the global model parameters then update their local model parameters and iteratively fine-tune their models using respective datasets for multiple epochs. This process continues until convergence is achieved or a predefined number of rounds (maxRounds, which is same for all the clients) is reached.

The server must also complete registration and authorization within the blockchain network. Algorithm 2 shows the pseudo code of the chaincode function invoked by the server after authorization. In Lines 2–4, the server chaincode accesses the model parameters of a random subset of num clients from the complete set of clients to maintain fairness across the network. To achieve this, it takes help of `getClientList` which returns a list of all clients with data for the corresponding round as shown in Line 2. Now, the list is randomly reordered using a seed and the first num clients are selected. Further, the client selection and parameter aggregation steps are not done by the server but instead are performed by the chaincode. This ensures fairness even when the server happens to

---

Algorithm 2 Pseudo code of chaincode invoked by server

1: **function** GETROUNDDATA(num int, seed int)
2:     clientList ← getClientList($round$)
3:     clientList ← shuffle($clientList, seed$)
4:     clientList ← clientList[: num]
5:     μ_u ← getParameters($clientList$)
6:     $aggregate$ ← **0**
7:     **for** $\mu_u[i]$ in μ_u **do**
8:         $aggregate$ ← $aggregate + \mu_u[i]$
9:     **end for**
10:     μ_{fed} ← $\frac{aggregate}{num}$
11:     issueTokens(clientList)
12:     putParameters(μ_{fed})
13: **end function**

---

be malicious. The parameters are aggregated[1] as the average of the local model weights obtained from each of the clients as shown in Lines 6–10 of Algorithm 2. Equation 2 describes this calculation for a single round. The aggregated parameters are subsequently updated on the ledger, accessible to all participating clients (Line 12). This chaincode will be invoked by the server either till convergence or a predefined number of rounds is reached.

$$\mu_{fed} = \frac{\sum_{i=1}^{i=num} \mu_u[i]}{num} \tag{2}$$

3.2   Integrating Token-Based Incentivization with LDP-FL

Augmenting the blockchain-based LDP-FL setup with a token-based incentivization mechanism enhances its practicality. The peers in the blockchain network hold copies of the ledger and the clients can access the updated global parameters only if they possess a sufficient number of tokens. This controlled access is facilitated by the permissioned blockchain, wherein read and write permissions can be governed by the availability of tokens. In order to achieve this, those clients which have been polled by the server will be awarded tokens for that round (Algorithm 2 - Line11) that they can later use to fetch the updated global parameters written to the ledger by the server, thus refining their local model.

Our proposed incentivization mechanism employs a simple linear function based on the privacy budget ϵ to reward clients who share greater amounts of information. Since keeping ϵ too low results in almost total obfuscation of weights and conversely a very high ϵ implies negligible privacy, ϵ is considered to lie between ϵ_{min} and ϵ_{max}. We award those clients having $\epsilon = \epsilon_{min}$ with 0.5 tokens and those having $\epsilon = \epsilon_{max}$ with 1 token. All other values of ϵ are allocated tokens between these two extremes. The following equation demonstrates this,

---

[1] Aggregation of model weights as mentioned in [25].

where T is the number of tokens awarded:

$$T = 0.5 + \frac{\epsilon - \epsilon_{min}}{2(\epsilon_{max} - \epsilon_{min})} \tag{3}$$

Participants with lower values of ϵ receive fewer tokens, while those with higher ϵ values receive a proportionately greater token reward. This encourages clients to share their data more freely in order to reap the benefits of federated learning. The cost of fetching global weights for each client has been kept as 1 token. When initializing the entire setup, we allocate a few tokens to each client, to ensure some form of stability in the initial stages so that the clients have adequate tokens to fetch global parameters. Gradually, they will become solely dependent on their privacy budget ϵ to obtain tokens for consumption.

4 Implementation Details

In this section, we give a detailed description of our implementation of the BTLF framework using HyperLedger Fabric.

4.1 Chaincode Implementation

If a client or the server's organization is not yet enrolled in the HLF network upon logging in, the enrollment process is first initiated. After the organization enrollment is completed, the client or server can proceed to enroll and register as a member or participant within that specific organization. The server and client backend codes are developed using Node.js and the chaincodes are written in Golang. The chaincode asset definition is as follows (Appendix A - Listing 1.1):

- **Layer struct:** Stores the weights and biases for a particular layer in the neural network model of the client.
- **NeuralNetworkModel struct:** Stores an array of Layer structs, each representing a layer of the neural network model.
- **ClientData struct:**
    - **ClientID:** Stores the Common Name (CN), which is a field in X.509 certificates used for client authentication. It serves as an identifier for a client and is also utilized as a key to store the data of a client.
    - **Data:** It is a NeuralNetworkModel struct, storing the current state of the model for the particular client.
    - **Tokens:** Stores the currently available tokens for this specific client.
    - **Round:** Stores the last round in which model parameters were pushed.
    - **RoundSeen:** A list of all the rounds in which the client fetched the global model parameters from the server.
    - **Epsilon:** Stores the value of the privacy budget ϵ set by the client.

There are several chaincode functions, of which the client has the ability to invoke the `PutData` and `GetResult` (Appendix A - Listing 1.2) for writing local model parameters to the ledger and for getting the global parameters from the ledger, respectively. On the other hand, the `GetRoundData` chaincode function is exclusive to the server for parameter aggregation, which also uses to select the random num of clients `SelectSubSet` (Appendix A - Listing 1.3).

4.2   Client and Server Workflow

After registration, each client is assigned a number of random samples from each of the possible classes in the ML model. Upon initialization, every client's neural network model is equipped with randomly generated weights and biases. The ledger is also initialized for the client by calling the `InitLedger()` function of the chaincode. In each round, the client i chooses a random value of privacy budget and trains the local model using the current model parameters and the local training data. The updated parameters are then perturbed by calling the LDP function along with a privacy budget. Now, the obfuscated weights are written to the ledger by invoking the `PutData()` function of the chaincode. This transaction is endorsed by the endorsers in accordance with the endorsement policy and gets added to the ledger along with the client's round number R_i.

Given that the system operates asynchronously, issues such as node failure or network disturbances may cause a client to fail in updating local parameters, hence its round number may fall behind. To address this, the round R_i of the client i is updated as $\max(R_i + 1, R_{\text{server}} + 1)$, where R_i and R_{server} denote the latest rounds in which the client and server respectively wrote their parameters into the ledger. Subsequently, the server invokes the `GetRoundData(`num`,`$seed$`)` chaincode function, which returns a random subset of num clients' local parameters for the round. The polled clients are rewarded with tokens according to the token-based incentivization function (Eq. 3). Finally, all the local model parameters are aggregated and written to the ledger.

Once the global parameters have been updated, the clients may request these parameters from the ledger if they have a sufficient number of tokens. Then the global parameters are retrieved using the `GetResult(`$serverCN$`,`$round$`)` chaincode function, which fetches the parameters for $round$. Upon retrieval, the available tokens decrease by the cost of reading from the ledger (currently kept as 1 token). If the client requests global parameters for a round it has previously requested, the cost to read that round's global parameters will be zero. The local parameters are then updated with the global model parameters. Consequently, if the client had adequate tokens, the local model will be the updated global model itself. Otherwise it will only be the locally trained model. At the end of the round, each client tests the model on its own training dataset and computes the accuracy. The process then repeats, starting a new round.

5   Experimental Results

We first describe our dataset and experimental setup followed by detailed results.

5.1   Model Architecture and Dataset Preparation

While the proposed architecture for BTLF is ML model and dataset agnostic, for our experiments, we have used a Convolutional Neural Network (CNN) model for image classification. The classification task considered is to identify

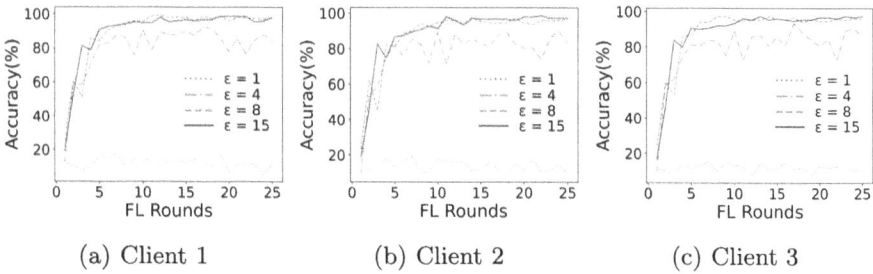

(a) Client 1   (b) Client 2   (c) Client 3

Fig. 2. Model accuracy progression over Federated Learning rounds. Each subplot shows the individual accuracy trajectory of a client model trained with 5 epochs and 75 samples per digit across different ϵ values

the handwritten digits in the MNIST dataset [19], where each image is a gray scale 28×28 pixel representation of a digit (0–9). The first layer in the model is a convolutional layer with 16 filters and a 3×3 convolutional kernel. Rectified Linear Unit (ReLU) activation is then applied. Next, a 2×2 max pooling layer is employed to down-sample spatial dimensions, before passing to the third layer. This is again a convolutional layer with 32 filters, a 3×3 kernel and ReLU activation. Another 2×2 max pooling operation is applied before flattening the output from the convolutional layers into a 1D array. Now a dense layer is added with 128 units and ReLU activation. Finally, the output layer is applied comprising 10 neurons, representing the 10 possible classes (digits 0–9). Softmax activation is employed to obtain the class probability distribution. The Adam optimizer was chosen as the optimization algorithm, allowing adaptive learning rate modifications during training. Categorical cross entropy was used as the loss function since it is suitable for multi-class classification tasks with one-hot encoded labels. Model performance was evaluated using the accuracy metric.

5.2 Detailed Results

Since there are several design parameters that can affect the accuracy of BTLF, we carried out a number of experiments. In each, some of the parameters are kept fixed while varying the rest. In Fig. 2, we make each client run local training on its dataset with the current set of weights for five epochs. The updated weights are written to the ledger once these epochs are completed. There are 75 samples for each digit used in training. We show the impact of variation in the number of FL rounds on accuracy for different values of the privacy parameter ϵ. Figures 2(a)–(c) depict the results for three representative client nodes (Client 1 - Client 3). From the figures, it is observed that while $\epsilon = 1$ leads to poor training, as the value of ϵ is increased, accuracy goes up significantly with more number of federated learning rounds. The same experiments were repeated by setting the number of epochs to 10 and the corresponding results are shown in Fig. 3(a)–(c). While a trend similar to that in Fig. 2 is seen, it is observed that for the same

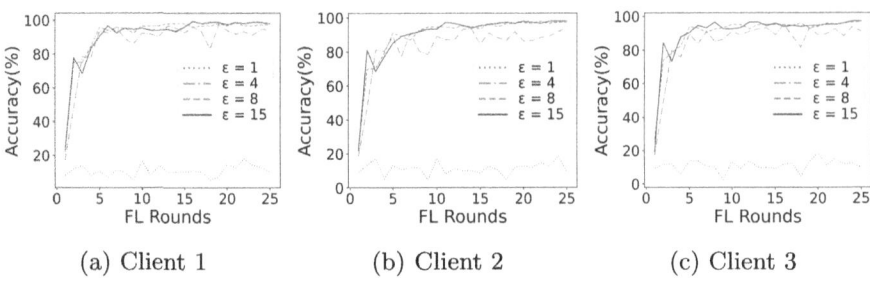

(a) Client 1   (b) Client 2   (c) Client 3

Fig. 3. Model accuracy progression over Federated Learning rounds. Each subplot shows the individual accuracy trajectory of a client model trained with 10 epochs and 75 samples per digit across different ϵ values

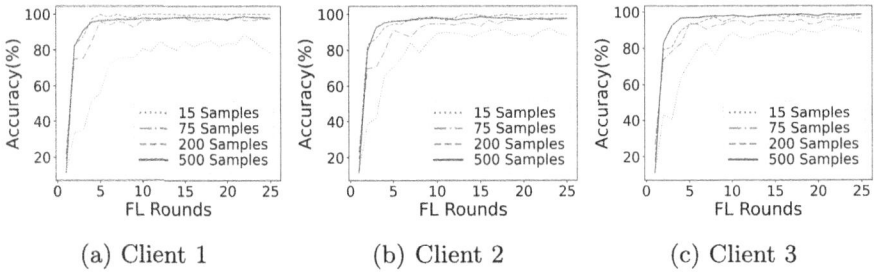

(a) Client 1   (b) Client 2   (c) Client 3

Fig. 4. Variation of accuracy with training dataset size for individual clients, each trained with 10 epochs and a fixed privacy budget ($\epsilon = 8$)

number of FL rounds and the value of ϵ, a higher accuracy is obtained for each client when the number of epochs used for local training is more.

We next study the impact of sample size on accuracy. From the previous two figures, it is observed that an epoch size of 10 gives satisfactory performance, and an ϵ value of 8 leads to convergence in a reasonable number of rounds. Hence, we keep these values fixed in this experiment. Figure 4 shows the variation of accuracy for different dataset sizes in terms of number of samples. From Figs. 4(a)–(c), it is observed that the sample size has a significant impact on the accuracy upto a certain extent. Beyond a sample size of 75, there is not much improvement. Also, for lower sample sizes, accuracy reaches a steady value for higher number of rounds. For all the three clients depicted in the figures, the trend is similar, implying that these are steady and reliable observations.

In the experiments so far, we used the same value of ϵ for all the clients. We next vary its value for different clients. In Fig. 5(a), the ϵ values are 8, 10 and 12 for Clients 1, 2 and 3, respectively. In Fig. 5(b), the values are 5, 10 and 15, while in Fig. 5(c), the values are 1, 10 and 20. The intent is to study the impact of the absolute as well as relative variation in the value of ϵ across clients. From the plots, it is seen that ϵ has a strong impact on accuracy. As is seen from the figures, when one of the clients (Client 1) has $\epsilon = 1$, the accuracy drops

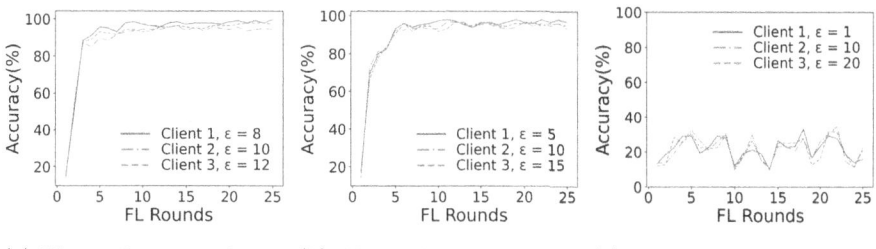

(a) Three clients with ϵ as 8, 10, 12

(b) Three clients with ϵ as 5, 10, 15

(c) Three clients with ϵ as 1, 10, 20

Fig. 5. Variation of accuracy with number of FL rounds with different clients using different privacy budgets.

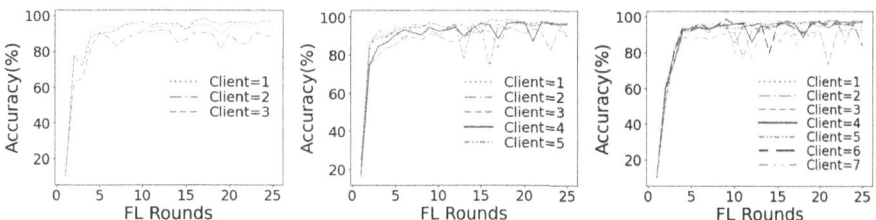

(a) Three clients varying ϵ across rounds

(b) Five clients varying ϵ across rounds

(c) Seven clients varying ϵ across rounds

Fig. 6. End-to-end working of the complete BTLF system with random values of privacy budget (ϵ) used by each client in every round.

significantly. Since lower ϵ implies higher data obfuscation and hence, higher privacy, a value of 1 implies Client 1 completely perturbs its parameters while sharing. This in turn affects the global parameters to such an extent that the entire federated learning process fails.

In the final set of experiments, we not only let the clients have different privacy settings, they are also allowed to dynamically change the value of ϵ across rounds. The effect of token based incentivization is also depicted in these results. From the previous set of experiments, it was observed that ϵ values between 5 to 15 give satisfactory performance. Hence, each client chooses a random value between 5 and 15 in each round while perturbing its data. Further, considering the results in the earlier figures, we keep the number of training epochs fixed at 10 and sample size to 75. With these settings, the results are shown in Fig. 6. Note that, for these experiments, we show the results for not only three clients (Fig. 6(a)) but also for five (Fig. 6(b)) and seven clients (Fig. 6(c)). The reason is to show the end-to-end working of the complete BTLF system. It is observed from the figures that, as the clients acquire tokens and later use those for reading the updated weights, the accuracy increases. However, if they obfuscate the data too much by choosing lower ϵ values, accuracy suffers. This is reflected in the dip in their accuracy values in some of the rounds. It needs to be emphasized here that the number of clients used in the above experiments is limited by the resource availability in a single computer with 16 GB RAM, and is not an inherent limitation of the proposed architecture.

6 Threat Model and Discussions

In this section, we identify some of the assumptions made in the BTLF framework (Refer to Fig. 1) in light of the considered threat model. We also discuss how the assumptions may be relaxed by considering alternative architectures.

We have considered the clients to be separate from the nodes. As a result, a client must trust that the peer node it is connected to will serve it correctly and not behave maliciously. For example, a malicious node may deduct tokens without delivering the global parameters to the client. The assumed threat model is that such type of behavior will not happen in a collaborative environment. This assumption can be relaxed by setting up the HLF network in such a way that each client is also a peer node in the network. Thus, the clients will not need to put implicit trust on a separate entity. Two other assumptions about the threat model are as follows.

i Since each peer node maintains a replicated copy of the blockchain ledger, they can retrieve consolidated global parameters directly from the ledger. If the peer node intends to behave maliciously, the global parameters can be accessed without any additional token costs to the clients.

ii The nodes/clients themselves can aggregate the individual local parameters of other nodes as the parameters are written to a ledger, without needing to depend on the server for aggregation. This way, clients can get most recent global parameters without using its available tokens.

These assumptions can be addressed by making appropriate modifications to the BTLF architecture. The first one can be handled through encryption. In the modified scheme, the aggregated global parameters will be encrypted by the server using a session key, which is a symmetric key generated for a particular round. The encrypted parameters are written onto the ledger instead of the plaintext parameters. A private data collection is also set up between each server-client pair, which is then used by the server for sharing the session key. When a client has sufficient number of tokens and makes a request for a particular round's session key by invoking the chaincode, tokens are deducted and the request is considered for handling by the server. Since this request is stored on the ledger, all the other peers will get to know about the request and can verify the same. The server then sends the session key to the client for the requested round through the private data collection set up in advance, thus preventing other clients from accessing the key. No unauthorized client can read the global parameters for that round. Thus, the clients cannot fetch the aggregated global parameters from the ledger bypassing the server. They must pay tokens in order to obtain the session key for decrypting the parameters.

In addition, only the symmetric session key is shared with the clients via the private data collection instead of the entire set of global parameters. Hence, it does not cause unnecessary load on the blockchain network.

For the second assumption mentioned above, we argue that if a client aggregates the local parameters of the other clients, it will need to spend its own computational power, which is more expensive than simply spending its tokens. This can be thought of as a setting where instead of every client aggregating everybody else's and also its own parameters, a single server is using its computing power. Each client effectively spends only a fraction of its available computing power. Thus, the proposed approach discourages clients from attempting to aggregate the parameters themselves.

7   Related Work

Some of the work in recent literature have explored decentralized federated learning by integrating FL with blockchain-based decentralized execution [7,10,13,16,17,21,26,28,29,31]. While they mainly focus on secured decentralized training of the FL model through multiple clients [7,10,28], a few have also considered specific use cases for decentralized machine learning [13,26,29]. A recent research by Hai et al. introduces an innovative integration of FL and blockchain technology in developing a medical record recommendation system [13]. In contrast, Pokhrel and Chai present FL with blockchain for autonomous vehicles along with automobile design challenges [31]. In yet another domain-specific application, Liu et al. discuss a secure FL framework for 5G networks, where both effective learning and security have a significant role [22].

In recent years, there has been some research towards privacy-preserving federated learning. For example, Bhowmick et al. introduce a method that protects against the reconstruction of parameters, thus enhancing privacy in FL [5]. Hao et al. focus on efficiency and confidentiality in federated deep learning [14]. Although existing work has explored secured and privacy-preserved decentralized model training through blockchain-based FL, they primarily focus on model privacy rather than data privacy for individual clients participating in the federated training procedure. There have also been attempts to leverage differential privacy (DP) to protect data. The work in [1] introduces the training of deep learning models with DP. Further literature incorporates various forms of DP with FL [27,30,34,37], yet these do not combine FL with blockchain technology.

While efforts have been made to employ blockchain in the context of federated learning [36], limited research has focused on incorporating LDP in this domain. Critical design choices necessitate careful evaluation, including determining whether the same nodes perform all three crucial operations: model parameter calculation, LDP-related computation, and distributed ledger updating. Notably, in domains like the Internet of Things (IoT), where end devices may lack adequate computational power, allocating tasks to edge nodes has significant implications for device privacy. Among the various application-specific domains, Li et al. consider FL for segmenting brain tumors [20]. In another scenario, Lu et al. integrate differential privacy with federated learning in mobile edge computing [24]. They emphasize the importance of privacy-preserving approaches in urban settings. Likewise, the focus of the work of Zhao et al. is LDP-based FL for the Internet of Things [42]. In a more general setting, the work of Truex et al. gives a formal privacy guarantee for LDP in federated learning [35].

Kong et al. propose an incentivization mechanism that does not monetize data [18]. Instead, model performance is used as the reward, i.e., those making more significant contributions can access more accurate models. It has been shown that clients will benefit by sharing as much data as they possess to participate in federated learning under this incentive mechanism. Some blockchain-based FL techniques have been proposed using specific cryptocurrencies as incentives [23]. However, the potential legal implications of such currencies can impact participants' willingness to engage. Xu et al. model incentivization in federated learning with differential privacy in industrial IoT as a Stackelberg game with the aggregating server as the leader and the client nodes as followers [39]. The server tries to maximize the utility of its available, total budget by appropriately rewarding the clients for data sharing. A systematic and comprehensive survey on incentivization in federated learning can be found in [38]. In contrast to these, our token-based incentive mechanism circumvents the identified challenges. Aligning the incentive quantum with the LDP privacy factor ϵ fosters fairness and encourages active participation in the FL process.

8   Conclusions

In this paper, we have proposed a novel approach for Locally Differentially Private Federated Learning in a permissioned blockchain setting coupled with Token-based Incentivization. We have shown how the central server typical of FL setup can be replaced by smart contracts, thereby preventing single-point failures and mitigating risks associated with potentially malicious servers or clients. By utilizing a chaincode to randomly sample a subset of nodes, it is guaranteed that transactions are endorsed by all endorsers, adhering to the endorsement policy and there is no biased selection of clients by a potentially malicious server. We highlighted the asynchronous nature of the system and explained how updating the client's round number in relation to the server's round number and the client's previous round number helps reduce the challenges posed by this non-synchronicity. Future work in this field would involve use of other LDP models and possibly more sophisticated incentivization schemes. The intricacies of achieving consortium among the organizations may be looked into as well. We would also like to implement and experiment with the improvements proposed in Sect. 6 to check for additional overhead in terms of computation and latency.

Acknowledgements. This work is partially funded by SERB, under the Dept. of Science and Technology, Govt. of India, through Research Grant Order Number CRG/2022/002565.

A   Appendix

Listing 1.1. Chaincode Data Structures

```
type Layer struct {
        Weights interface {}
        Biases []float64
}

type NeuralNetworkModel struct {
        Layers[] Layer
}

type ClientData struct {
    ClientID string
    Data NeuralNetworkModel
    Tokens float64
    Round  int
    RoundSeen []int
    Epsilon float64
}
```

Listing 1.2. Function Declarations of Chaincodes Invoked by Client

```
func GetResult(serverCN string, round int) {}
func PutData(data string, epsilon float64) {}
```

Listing 1.3. Function Declarations of Chaincodes Invoked by Server

```
func GetRoundData(num int, seed int) {}
func SelectSubSet(num int, seed int) {}
```

References

1. Abadi, M., et al.: Deep learning with differential privacy. In: Proceedings of the ACM SIGSAC Conference on Computer and Communications Security, pp. 308—318 (2016)
2. Akhil Vasishta, M.V., Palanisamy, B., Sural, S.: Decentralized authorization using hyperledger fabric. In: IEEE International Conference on Blockchain (Blockchain), pp. 238–243 (2022)
3. Androulaki, E., et al.: Hyperledger fabric: a distributed operating system for permissioned blockchains. In: Proceedings of the Thirteenth EuroSys Conference, pp. 1–15 (2018)
4. Bagdasaryan, E., et al.: How to backdoor federated learning. In: Proceedings of the Twenty Third International Conference on Artificial Intelligence and Statistics, pp. 2938–2948 (2020)

5. Bhowmick, A., Duchi, J., Freudiger, J., Kapoor, G., Rogers, R.: Protection against reconstruction and its applications in private federated learning. arXiv preprint arXiv:1812.00984 (2018)
6. Bonawitz, K., et al.: Towards federated learning at scale: system design. Proc. Mach. Learn. Syst. **1**, 374–388 (2019)
7. Chakraborty, S., Chakraborty, S.: Proof of federated training: accountable cross-network model training and inference. In: IEEE International Conference on Blockchain and Cryptocurrency, pp. 1–9 (2022)
8. Chamikara, M.A.P., et al.: Local differential privacy for federated learning. In: European Symposium on Research in Computer Security, pp. 195–216 (2022)
9. Cormode, G., et al.: Privacy at scale: local differential privacy in practice. In: International Conference on Management of Data, pp. 1655–1658 (2018)
10. Desai, H.B., Ozdayi, M.S., Kantarcioglu, M.: BlockFLA: accountable federated learning via hybrid blockchain architecture. In: Eleventh ACM Conference on Data and Application Security and Privacy, pp. 101–112 (2021)
11. Dwork, C., McSherry, F., Nissim, K., Smith, A.: Calibrating noise to sensitivity in private data analysis. In: Theory of Cryptography, pp. 265–284 (2006)
12. Gu, Y., Bai, Y., Xu, S.: CS-MIA: membership inference attack based on prediction confidence series in federated learning. J. Inf. Secur. Appl. **67**, 2214–2226 (2022)
13. Hai, T., et al.: BVFLEMR: an integrated federated learning and blockchain technology for cloud-based medical records recommendation system. J. Cloud Comput. **11**, 22 (2022)
14. Hao, M., et al.: Towards efficient and privacy-preserving federated deep learning. In: IEEE International Conference on Communications, pp. 1–6 (2019)
15. Hard, A., et al.: Federated learning for mobile keyboard prediction. ArXiv, vol. 1, pp. 1–12 (2018)
16. Issa, W., et al.: Blockchain-based federated learning for securing internet of things: a comprehensive survey. ACM Comput. Surv. **55**(9), 1–43 (2023)
17. Khan, J.A., Wang, W., Ozbay, K.: FLOATING: federated learning for optimized automated trajectory information storing on blockchain. In: IEEE International Conference on Blockchain and Cryptocurrency, pp. 1–4 (2023)
18. Kong, S., Li, Y., Zhou, H.: Incentivizing federated learning. arXiv preprint arXiv:cs.cv (2022)
19. LeCun, Y., Bottou, L., Bengio, Y., Haffner, P.: Gradient-based learning applied to document recognition. Proc. IEEE **86**(11), 2278–2324 (1998)
20. Li, W., et al.: Privacy-preserving federated brain tumour segmentation. In: Machine Learning in Medical Imaging, pp. 133–141 (2019)
21. Li, Y., et al.: A blockchain-based decentralized federated learning framework with committee consensus. IEEE Network **35**(1), 234–241 (2020)
22. Liu, Y., et al.: A secure federated learning framework for 5G networks. IEEE Wirel. Commun. **27**(4), 24–31 (2020)
23. Liu, Y., et al.: Fedcoin: a peer-to-peer payment system for federated learning. In: Federated Learning: Privacy and Incentive, pp. 125–138 (2020)
24. Lu, Y., et al.: Differentially private asynchronous federated learning for mobile edge computing in urban informatics. IEEE Trans. Industr. Inf. **16**(3), 2134–2143 (2020)
25. McMahan, B., et al.: Communication-efficient learning of deep networks from decentralized data. In: 20th International Conference on Artificial Intelligence and Statistics, pp. 1273–1282 (2017)

26. Meese, C., et al.: BFRT: blockchained federated learning for real-time traffic flow prediction. In: 22nd IEEE International Symposium on Cluster, Cloud and Internet Computing, pp. 317–326 (2022)
27. Naseri, M., Hayes, J., Cristofaro, E.D.: Local and central differential privacy for robustness and privacy in federated learning. In: Network and Distributed Systems Security Symposium, pp. 1–18 (2022)
28. Nguyen, T., et al.: Blockchain-based secure client selection in federated learning. In: IEEE International Conference on Blockchain and Cryptocurrency, pp. 1–9 (2022)
29. Ouyang, L., et al.: Artificial identification: a novel privacy framework for federated learning based on blockchain. IEEE Trans. Comput. Soc. Syst. **10**(6), 3576-2585 (2023)
30. Padala, M., Damle, S., Gujar, S.: Federated learning meets fairness and differential privacy. In: Neural Information Processing, pp. 692–699 (2021)
31. Pokhrel, S.R., Choi, J.: Federated learning with blockchain for autonomous vehicles: analysis and design challenges. IEEE Trans. Commun. **68**(8), 4734–4746 (2020)
32. Sattler, F., Wiedemann, S., Müller, K.R., Samek, W.: Robust and communication-efficient federated learning from non-i.i.d. data. IEEE Trans. Neural Netw. Learn. Syst. **31**(9), 3400–3413 (2020)
33. Sun, L., Qian, J., Chen, X.: LDP-FL: practical private aggregation in federated learning with local differential privacy. In: Thirtieth International Joint Conference on Artificial Intelligence, pp. 1571–1578 (2021)
34. Triastcyn, A., Faltings, B.: Federated learning with Bayesian differential privacy. In: IEEE International Conference on Big Data, pp. 2587–2596 (2019)
35. Truex, S., et al.: LDP-fed: federated learning with local differential privacy. In: Third ACM International Workshop on Edge Systems, Analytics and Networking, pp. 61–66 (2020)
36. Wang, Z., Hu, Q.: Blockchain-based federated learning: a comprehensive survey. arXiv preprint arXiv:2110.02182 (2021)
37. Wei, K., et al.: Federated learning with differential privacy: algorithms and performance analysis. IEEE Trans. Inf. Forensics Secur. **15**, 3454–3469 (2020)
38. Witt, L., et al.: Decentral and incentivized federated learning frameworks: a systematic literature review. IEEE Internet Things J. **10**(4), 3642–3663 (2023)
39. Xu, Y., et al.: Incentive mechanism for differentially private federated learning in industrial internet of things. IEEE Trans. Industr. Inf. **18**(10), 6927–6939 (2022)
40. Yang, M., et al.: Local differential privacy and its applications: a comprehensive survey. arXiv preprint arXiv:2008.03686 (2020)
41. Yang, Q., Liu, Y., Chen, T., Tong, Y.: Federated machine learning: concept and applications. ACM Trans. Intell. Syst. Technol. **10**(2), 1–19 (2019)
42. Zhao, Y., et al.: Local differential privacy-based federated learning for internet of things. IEEE Internet Things J. **8**(11), 8836–8853 (2020)

Does Differential Privacy Prevent Backdoor Attacks in Practice?

Fereshteh Razmi[1(✉)], Jian Lou[2], and Li Xiong[1]

[1] Emory University, Atlanta, GA 30322, USA
{frazmim,lxiong}@emory.edu
[2] Zhejiang University, Hangzhou 310027, Zhejiang, China
jian.lou@zju.edu.cn

Abstract. Differential Privacy (DP) was originally developed to protect privacy. However, it has recently been utilized to secure machine learning (ML) models from poisoning attacks, with DP-SGD receiving substantial attention. Nevertheless, a thorough investigation is required to assess the effectiveness of different DP techniques in preventing backdoor attacks in practice. In this paper, we investigate the effectiveness of DP-SGD and, for the first time, examine PATE and Label-DP in the context of backdoor attacks. We also explore the role of different components of DP algorithms in defending against backdoor attacks and will show that PATE is effective against these attacks due to the bagging structure of the teacher models it employs. Our experiments reveal that hyper-parameters and the number of backdoors in the training dataset impact the success of DP algorithms. We also conclude that while Label-DP algorithms generally offer weaker privacy protection, accurate hyper-parameter tuning can make them more effective than DP methods in defending against backdoor attacks while maintaining model accuracy.

Keywords: Differential Privacy · Backdoor Attack · Security

1 Introduction

Deep neural networks are vulnerable to backdoor attacks. The goal of a backdoor adversary is to misclassify the prediction of the target model on samples that contain a special pattern (trigger), while maintaining the inference performance on normal samples. To achieve this goal, backdoor attacks typically manipulate a small portion of training data with carefully designed triggers that lead to the mismatch between training features and labels [19]. Many studies have proposed countermeasures against this powerful attack or the more general data poisoning attacks. The most common approach in these studies is discovering abnormalities in model statistics or training data [7,8,31,35,36].

This work was funded by National Institutes of Health (NIH) R01LM013712, and National Science Foundation (NSF) CNS-2124104, CNS-2125530.

© IFIP International Federation for Information Processing 2024
Published by Springer Nature Switzerland AG 2024
A. L. Ferrara and R. Krishnan (Eds.): DBSec 2024, LNCS 14901, pp. 320–340, 2024.
https://doi.org/10.1007/978-3-031-65172-4_20

Differential privacy (DP) [12] is a fundamental concept of data privacy, guaranteeing that the inclusion or exclusion of individual data points doesn't significantly impact the outcome of any analysis. A common method to achieve DP in a deep learning model is by introducing calibrated randomness during the training process such as DP-SGD (Differentially Private Stochastic Gradient Descent), which adds noise to the gradients during the training. An alternative approach is PATE (Private Aggregation of Teacher Ensembles), which involves training multiple teacher models on disjoint subsets of the data and then using their aggregated outputs with added noise to train a student model with auxiliary data. Label differential privacy [14,34] is a variant of differential privacy that ensures that the learning process (and the resulting model) cannot reveal whether any individual's label was used or not. As the success of backdoor attacks relies on the influence of the triggered samples on the model, it is intuitive that the model might be more robust to backdoor attacks if the influence of each training sample is bounded. This concept of limiting the influence of individual samples aligns with the principles of DP. Thus a recent promising area of research focuses on using DP to build robust models against backdoor and poisoning attacks. This is accomplished by introducing randomness to the model through DP techniques, making it less sensitive to input.

There are a few works exploring this area in theory [5,25]. A few others have obtained experimental results either under a centralized setting using DP-SGD [4,10,17,40] or under the federated learning setting [24,27,28]. These studies provide some evidence that models trained with DP-SGD mitigate poisoning attacks, but they fall short of a comprehensive investigation and do not explore the power of other state-of-the-art DP models against backdoor attacks.

This paper aims to bridge the theory and practice and provide a comprehensive and in-depth understanding of whether and, more importantly, how various DP models and methods defend against backdoor attacks in practice given the theoretical promise and preliminary evidence in the literature. We study both the standard DP class of algorithms and the Label-DP variant, and compare them in their defense power against backdoor attacks. PATE and Label-DP are being examined for the first time against these types of attacks. We evaluate their performance empirically on two widely used datasets in the domain of backdoor attacks and differential privacy. To summarize, we make the following contributions:

1. **Comparative study of DP approaches against backdoor attacks, including standard DP-SGD approach and the less-studied PATE approach.** Existing studies use DP-SGD for training DP models to defend against poisoning or backdoor attacks. In this work, we explore the other well-known DP algorithm PATE against backdoor attacks in order to understand whether different DP algorithms (gradient perturbation vs. aggregation perturbation) have different powers against backdoors. We show that both of these classical DP approaches can provide robust models for backdoor attacks. Also, we will demonstrate that the ensemble structure of the PATE inherently makes it suitable against backdoors.

2. **A deeper understanding of the impact of noise and other parameters of DP approaches on backdoor attacks.** The effectiveness of DP approaches is affected by parameters other than noise. We explore the origin of these algorithms' resilience by examining whether randomness is the sole factor or if the other parameters have an impact. We empirically show that the randomness (privacy budget) contributes to mitigating the backdoor attack success rate, which is compatible with the theoretical results in the literature [39]. However, we demonstrate that the impact of other parameters can be significant on the outcome, especially for PATE, e.g., the threshold used to aggregate the teacher models' outputs.

3. **Comparative study of Label-DP approaches against backdoor attacks.** Label-DP protects the privacy of the labels of the training data by ensuring the output model is indistinguishable with respect to the label of a training sample. We study the Label-DP class of algorithms for the first time against backdoor attacks using two algorithms ALIBI [26] and LP-2ST [14]. We hypothesize that Label-DP also provides robustness against backdoor attacks while maintaining better utility than DP based on two observations. First, since Label-DP ensures the indistinguishability of labels, we expect a model with Label-DP to break the association between the backdoor triggers and their assigned target class (label). Second, Label-DP methods typically converge faster than standard DP algorithms while maintaining higher model utility. This is because indistinguishability is required only for the labels, rather than for both the features and labels, which necessitates less noise to achieve the same level of privacy.

   Our evaluations confirm that Label-DP makes the model more immune to backdoor attacks while preserving model accuracy. We show that Label-DP is superior to DP approaches in terms of convergence speed. Furthermore, we demonstrate that it can achieve better robustness accuracy trade-offs under certain settings. For instance, with a lower percentage of backdoors, ALIBI can eliminate the negative impact of the attack while achieving the highest accuracy among all approaches. For stronger attacks with higher percentage of backdoors, LP-2ST outperforms other approaches when the privacy budget is low.

2   Preliminaries

2.1   Backdoor Attacks

Backdoor attacks are a category of attacks that involve attaching a small patch to a portion of a base class of the training dataset along with flipping their labels to a specified target class. After the model has been trained using these backdoor samples, it would be vulnerable to the presence of the patch in the inputs. So as the next step of the attack, the attacker attaches the same patch to some desired test samples of the base class and passes it to the backdoored model, so that this combination of the base class pattern plus the patch pattern misleads the model to misclassify the sample as the target class. This form of backdoor attacks, initially introduced by Gu et al. [15], is a powerful attack that has gained much attention. Some other works tried to make some other type of backdoor attacks that are less detectable or employ them in other domains including videos [32,41].

2.2 Differential Privacy and Label Differential Privacy

Differential Privacy (DP). DP is a privacy-preserving notion that makes an observer unable to tell if particular information contributes to the outcome [13]. In the context of machine learning, a DP method should not reveal whether a training sample has been utilized in the training process.

Let X and Y be the feature and label domain, respectively. Also, let the training dataset consists of n samples from a domain $U = (X \times Y)_n$. Given sample x, we have a classification task for the model M to predict y. A randomized training algorithm $\mathcal{M} : U \rightarrow R$ is (ε, δ)-DP if for any two adjacent datasets $D, D' \in U$ differing on at most one sample, it holds that:

$$\forall S \subset R, P[M(D) \in S] \leq e^{\varepsilon} P[M(D') \in S] + \delta. \tag{1}$$

A smaller ε guarantees stronger privacy but typically leads to a lower utility or accuracy of the model due to the randomization in the training. Using a DP property called **group privacy**, this definition can be extended to two datasets differing in k examples where k denotes more than one data point [11]. This is achievable by a linear increase in the privacy cost.

Label Differential Privacy (Label-DP). Label DP is an extension of DP that considers the labels as the only sensitive part of the training data that requires to be kept secret. So in contrast to (ε, δ)-DP which defines privacy for datasets D and D' differing on at most one sample, (ε, δ)-Label-DP considers D and D' differing on **the label** of at most one sample. Therefore, Label-DP can be seen as a relaxation of DP algorithms that guarantees only the privacy of the labels. One of the applications of Label-DP is recommendation systems where the user's profile or search queries are public, but the history of the user rating is sensitive.

2.3 DP and Label-DP Algorithms for Deep Learning

In this section, we explore the main methods for achieving DP (DP-SGD, PATE) and Label-DP (LP-MST and ALIBI) respectively, with Table 1 showing the critical parameters of the first two algorithms.

DP-SGD [1] is the most widely used algorithm for building DP models. DP-SGD restricts the privacy loss in each iteration of SGD (Stochastic Gradient Descent), by updating model in two steps: 1) clipping the L2 norm of the gradients, and 2) inserting calibrated Gaussian noise into those clipped gradients.

Table 1. Parameters of the DP algorithms

Method	Parameters
DP-SGD	1. **Noise multiplier**: Added randomness to the model's clipped gradients to provide DP 2. **Upper bound of the clipping norm** (*Cnorm*): Bound to clip the L2-norm of the gradients to control their sensitivity to the noise
PATE	1. **Threshold** T : Queries exceeding this minimum teachers' aggregation are selected for training the student model 2. **Selection noise with variance** σ_1: Gaussian noise added to the aggregator's votes before applying threshold to enforce privacy 3. **Result noise with variance** σ_2: Noise added to the selected queries after applying threshold to guarantee DP 4. **Number of teacher models** 5. **Number of queries**

PATE [29] provides privacy through a teacher-student structure. First, an ensemble of teachers is trained on disjoint subsets of the private data. Then, given an unlabeled public dataset, a student model queries the teacher ensemble and uses their noisy aggregated vote as the label. The number of queries is restricted. Plus, their response is based on a noisy aggregation without access to any specific private data point. However, access to a public dataset forces a strong assumption on PATE compared to DP-SGD.

PATE was originally introduced with Laplacian noise [29]. Then it was revised to improve the utility and privacy trade-off through a more confident aggregated teacher consensus, called Confident-GNMax [30]. In this paper, we adopt the Confident-GNMax version of the PATE framework, which is based on Gaussian noise.

Label Private Multi-Stage Training (LP-MST) [14] is a work regarding differential privacy that achieves Label-DP for deep learning. It leverages a modified version of the Randomized Response (RR) algorithm to add noise to the labels [38]. RR outputs the actual class of a sample or randomly replaces it with one of the other classes. However, the randomness deteriorates the utility.

Ghazi et al. [14] modify the RR algorithm to compensate for the utility, by iteratively training the model on disjoint subsets of the dataset. Then they use the trained model from the previous stage to get the top-K predictions and limit the RR algorithm to those predictions. Similar to the main paper, we report our results on LP-2ST with two training stages.

Additive Laplace Noise Coupled with Bayesian Inference (ALIBI) [26] is another Label-DP method in ML that has been recently proposed. It first adds Laplacian noise to one-hot labels, then uses these soft new labels to train the model while preserving Label-DP. Since post-processing does not affect differential privacy, Bayesian post-processing de-noises the soft labels iteratively during each step of SGD. The combination of additive Laplacian noise and iterative Bayesian inference increases the utility.

3   Related Work

DP has recently been highlighted for providing robust models to alleviate the negative impact of poisoning attacks. The rationale is that according to the definition of DP and group privacy, DP models are less sensitive to the impact of one or a group of poisoned data. In this section, we go through the literature to investigate where and how differentially private approaches used to defend against backdoor and poisoning attacks. We then identify the gaps in the literature, formulate those as research questions, and try to answer them and assess the results empirically.

There are two lines of work in the literature that consider the defensive power of DP methods on poisoning attacks; theoretical and practical studies.

Ma et al. [25] theoretically prove the robustness of DP models and provide a theoretical bound. They assume a training dataset D and an attacker with full knowledge creates some poisoned dataset $\tilde{D}$ from D. The poisoned model $\theta_{\tilde{D},b}$ is parameterized through the poisoned data $\tilde{D}$ and noise parameter b of the DP model. The attacker's objective loss $C : \Theta \to R$ aims to misclassify some targets or disrupt the overall classifier's functionality. Assuming the attacker does not know the exact realization of the noise, the attack is reduced to:

$$\min_{\tilde{D}} \quad J(\tilde{D}) = E_b\left[C(\theta_{\tilde{D},b})\right] \tag{2}$$

Given k poisoned data, the authors utilize the property of differential privacy in Eq. (1) and conclude:

$$J(\tilde{D}) \geq e^{-sign(C).k\varepsilon} J(D) \tag{3}$$

According to Eq. (3) the attacker is unable to change $J(\tilde{D})$ arbitrarily because it is lower bounded by 0 if C is positive (for example, in case of Mean Squared Error) or it is unbounded from below if C is negative.

This paper provides insight into how DP methods may provide a natural immunity against data poisoning attacks. However, it has two limitations. First, the lower bound of $J(\tilde{D})$ is loose. Second, this paper implements and evaluates its theoretical findings on general attack loss functions and DP frameworks. Thus, the specific impact of Eq. (3) on SOTA deep learning models (e.g. DP-SGD) and practical attacks (e.g. backdoor attacks) remains neglected.

To overcome the second limitation, a parallel set of works has employed DP-SGD as a practical usage of DP in deep learning to achieve protection against poisoning attacks [4,10,40]. Hong et al. [17] was one of the first works that considered DP-SGD against backdoor and other poisoning attacks. However, their primary motive was not originated from the fact that DP-SGD is a private algorithm and Equation (1). Instead, they observed that during the training on a poisoned dataset, the gradients computed on poisoned samples have a higher magnitude and different orientation than those computed on clean samples. Hence they leveraged DP-SGD to offset the behavior of the model's gradients on both clean and poisoned data through the randomness of the gradients. Their results show some degree of protection against specific poisoning attacks, but their outcome is not promising on backdoor (insertion) attacks. Later, Jagielski and Oprea claimed that DP itself can not serve as a defense against poisoning attacks [18]. They argued that it is possible that the robustness of DP-SGD stems from some parameters other than noise.

4   Research Questions

The existing studies on DP-SGD are inconclusive, and there are no studies on other state-of-the-art DP approaches as a potential defense. It motivates us to extend current works by conducting more comprehensive experiments on DP-SGD and introducing other DP methods as a defense. Based on this primary motivation, we pose some research questions in this section and elaborate their significance. Then in the following sections, we will try to address them empirically.

Question 1. *Is DP-SGD a successful protective algorithm against backdoor attacks? Can PATE, as another main DP approach, mitigate backdoor attacks?* Current studies have differing views on whether DP, particularly DP-SGD, can defend against backdoor attacks. It opens the door for a more comprehensive study of DP-SGD. It's not clear whether the robustness is achieved by the randomization introduced by DP methods in general or by other algorithm-specific parameters of DP-SGD. Additionally, this outcome can emphasize the gap between DP's theoretical and practical results against poisoning data.

So in this work, we first explore DP-SGD to understand why there is no consensus in the literature on DP-SGD as a defensive algorithm. Then for the first time, we explore PATE as a DP method against backdoor attacks to demonstrate if it confirms DP models' robustness. We examine the effectiveness of these algorithms by analyzing their hyper-parameters, even those that do not contribute to the randomness for DP. With this investigation, we hope to determine whether these algorithms are effective defense mechanism solely because they are DP.

Question 2. *Can other DP notions, such as Label-DP, also provide robustness and even better accuracy and robustness trade-off? How do different DP notions and algorithms compare in the trade-off?*
Answering the research question 1, leads us to two other major challenges with regard to DP-SGD and PATE. The first challenge is their prohibitive training time. Training an ensemble of teachers in PATE is heavily costly. Also, DP-SGD requires computation of per-sample gradient norms, which is extremely slow. The other issue with the DP algorithms is the trade-off between the privacy budget and the utility, which means decreasing the privacy budget (i.e.,

achieving stronger DP) is accompanied by a drop in models' accuracy. We will show that lower privacy budgets usually lead to a lower attack success rate (ASR), which is necessary to defeat attacks. We call this simultaneous reduction in accuracy and ASR the *Accuracy-ASR trade-off*. We will define the criteria for attack success rate in Sect. 5. To address these challenges, we conduct a comparison between Label-DP and other DP algorithms by varying DP budgets and attack strengths.

5   Experimental Setup

Datasets and Models. We evaluate each DP model on two datasets: MNIST [23] and CIFAR-10 [22]. We study end-to-end training and fine-tuning since both are common practices in modern machine learning. We use the same CNN architecture as [2] with two convolutional layers for MNIST and train it from scratch. Also, for CIFAR-10, as [37] suggests, we use ResNet50 [16] pretrained on ImageNet as a feature extractor and fine-tune its classification head.

Corresponding to each DP algorithm's specification, we find an optimizer and a learning rate using a grid search algorithm to ensure the training process achieves the highest accuracy. In addition, data augmentation reduces the effectiveness of all of the attacks [21,33], leading to a bias in our results. Therefore, we skip the data augmentation in our experiments. More details on the training process can be found in the appendix.

Attack and Threat Model. All the DP models are in white-box settings. The backdoors are made based on the triggers introduced in BadNets [15]. To generate backdoors, we first randomly select two classes as base and target class. Then, we randomly select half of the samples from the base class, attach a 4×4 trigger patch to their bottom right corner and assign the target class as their labels [4]. We poison 50% base class to ensure the number of backdoors is high enough, and sufficient clean samples are left in the base class. Under this condition, the model learns both clean and backdoor data points.

Evaluation Metrics. Attack success rate (**ASR**) is the metric to evaluate the success of the backdoor attacks. According to the definition of the backdoor attacks in Sect. 2.1, ASR indicates the number of test samples from base class that are patched with the backdoor trigger and misclassified as the target class. Thus, a defense method is considered more successful if it leads to a lower ASR.

The second defensive purpose is to maintain high **accuracy** for the clean test data. The original accuracy of our CIFAR-10 vanilla model over the clean test data is 91.24% and the backdoor ASR is 98.1%. The MNIST model's initial accuracy and ASR are 98.92% and 100%, respectively.

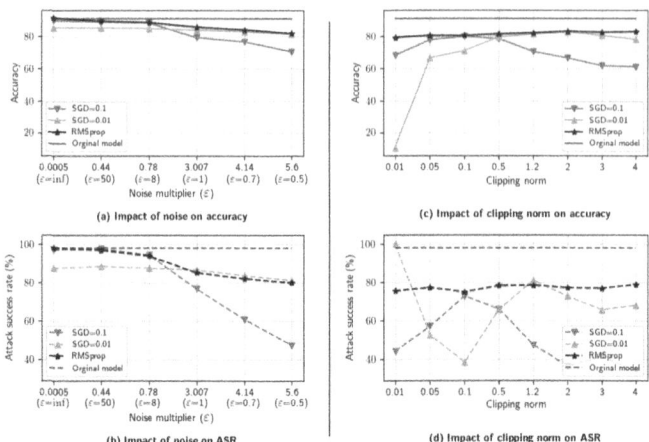

Fig. 1. Effectiveness of DP-SGD against backdoor attacks, w.r.t the noise multiplier, clipping norm, and the optimizer.

Experimental Roadmap. This subsection provides an overview of the experiments in the forthcoming sections. In Sect. 6, we analyze two DP algorithms, DP-SGD and PATE, by assessing the impact of their privacy budget and other hyper-parameters on the attack success rate. This analysis helps us clarify the underlying reason for their defensive power. At the same time, we will show their resulting accuracy and attack success rate. Then, in Sect. 7, we compare all the DP and Label-DP algorithms in various circumstances to witness which one is prominent and whether the outcome alters in a different situation. Due to space constraints, we could not include all of our experiments. Please refer to the appendix for our findings on the exploration of parameters for Label-DP algorithms and the training procedure.

6   DP Against Backdoors

This section investigates DP-SGD and PATE, against backdoor attacks. For each algorithm, we will evaluate their key hyper-parameters (introduced in Table 1) on CIFAR-10 dataset and show that some of them have a critical impact on the accuracy and ASR. The results of the MNIST dataset are very similar. So to be concise, we skip their reports here but use them to conduct the experiments in the subsequent sections.

6.1   DP-SGD vs. Backdoors

SGD is the dominant optimizer in practice paired with the DP-SGD algorithm, especially in defeating poisoning attacks [1,5,17,18]. So we consider different optimizers and learning rates to depict the sensitivity of DP-SGD performance to these factors: RMSProp, SGD with a learning rate of 0.1, and SGD with a learning rate of 0.01. Based on the size of the dataset, we set the DP-SGD algorithm as $(\varepsilon, 10^{-5})$-DP and report ε as the privacy budget [30].

Figures 1a and 1b show the impact of the noise multiplier by fixing the clipping norm to 1.2 (typical for CIFAR-10). Interestingly, the rate of the accuracy drop to the ASR drop differs for each optimizer. However, in general, higher noise levels reduce both accuracy and ASR simultaneously. This suggests that SGD can resist backdoor attacks more effectively by paying a slightly higher utility cost.

Figures 1c and 1d illustrate the impact of different clipping norms on the accuracy (top) and ASR (bottom) using a fixed noise of 5.6. In contrast to RMSProp, for SGD optimizers, the choice of learning rate creates two different patterns of ASR with respect to the clipping norm. This reveals how SGD training without an adaptive learning rate can be affected by the norm of the gradients. Therefore, while the clipping norm significantly impacts the model utility and robustness, it is difficult to optimally adjust it when the defender is unaware of the attack specifications.

According to [6], the impact of the clipping norm on accuracy is not monotonic, which is manifested as a non-monotonic pattern of accuracy and ASR in Figs. 1c and 1d. Regarding the different pattern of ASR on the left side of Fig. 1d with SGD-0.01, we speculate that the small learning rate accompanied by a high noise and small clipping norm can hardly learn the normal images' manifold, and instead it retains the repetitive and striking patterns of the backdoor triggers.

Conclusion (Q1): In our evaluations, DP-SGD was successful in mitigating the impact of backdoor attacks. However, the noise multiplier, clipping norm and training parameters determine the extent of this success. As a result, differences in these parameters contribute to the varying results reported in previous studies on the effectiveness of DP-SGD as a defense mechanism.

6.2   PATE vs. Backdoors

In this section, we evaluate the robustness of PATE against backdoor attacks and the impact of different parameters including the number of teachers, number of queries, threshold, selection noise, and result noise. The results are shown in Fig. 2. Whenever noises or threshold are not evaluated, we fix their values to 0. In the case of the number of queries and number of teachers, the default values are fixed to 10000 and 200, respectively. For training PATE, we assume 1/5 (i.e. 10000 samples) of the training data is publicly available for training the student model, and the rest is private. In the original PATE paper [29], the number of queries is set to as low as 1000. However by doing so, we naturally remove a large fraction of poisoned data and make the comparison between different DP methods unfair. Therefore, we keep the default number of queries at 10000 and in the next sections, to compare the models, we analyze the impact of both noise and the number of queries on the PATE's utility and privacy budget.

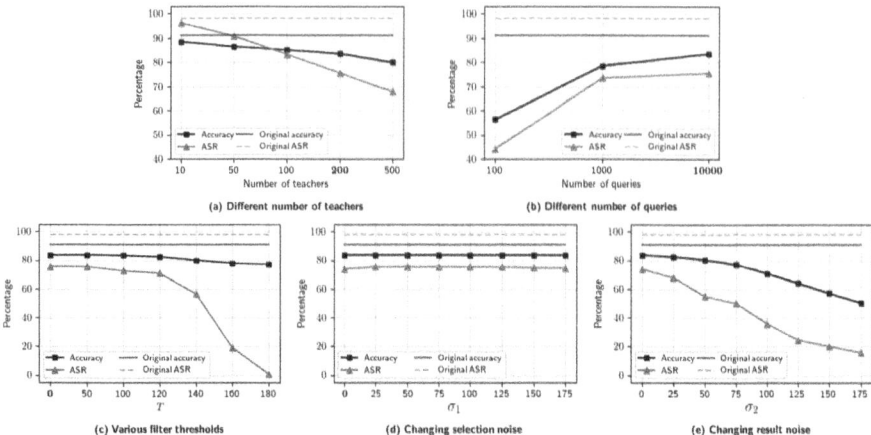

Fig. 2. The impact of number of teachers, number of queries, threshold, selection noise and result noise on the student model's accuracy and ASR from left to right and top to bottom, respectively).

Figures 2a and 2b show that the number of teachers and the number of queries impact the accuracy and ASR in opposite ways. A higher number of teachers means fewer training data and lower accuracy for each teacher, hence less accurate consensus from the aggregator. This also compromises the consensus on assigning the target class to the backdoor samples and decreases the ASR, which aligns with the literature finding that bagging can hinder the success of the backdoor attacks [3,9,20]. Furthermore, in Fig. 2b, a lower number of queries is associated with less training data for the student model and fewer backdoors, hence lower accuracy and ASR.

Figure 2c illustrates that the aggregation threshold is crucial in defeating backdoors and has minimal impact on utility loss. This finding complements previous results suggesting the use of bagging against poisoning attacks. The threshold forces the aggregation process to filter out uncertain data and backdoors, resulting in higher accuracy and lower ASR in the student model. To the best of our knowledge, this factor has not been considered in previous works as a major contributor to the effectiveness of bagging.

Figures 2d and 2e demonstrate the effect of selection noise and result noise used in selecting and randomizing queries which form the basis of DP for PATE. We found similar trends when one of the noises is fixed to a random positive value. Based on these results, to defeat ASR we need a high result noise which leads to a drop in accuracy. Since we fixed the number of queries and only varied the noise values to control privacy, the privacy budget still remains as large as $\varepsilon = 4$ at a high noise level of 175.

Conclusion (Q1): PATE is very successful in defeating backdoor attacks. It can be more successful than DP-SGD but it is highly sensitive to the algorithm parameters. Result noise (σ_2) and the number of queries which are the most

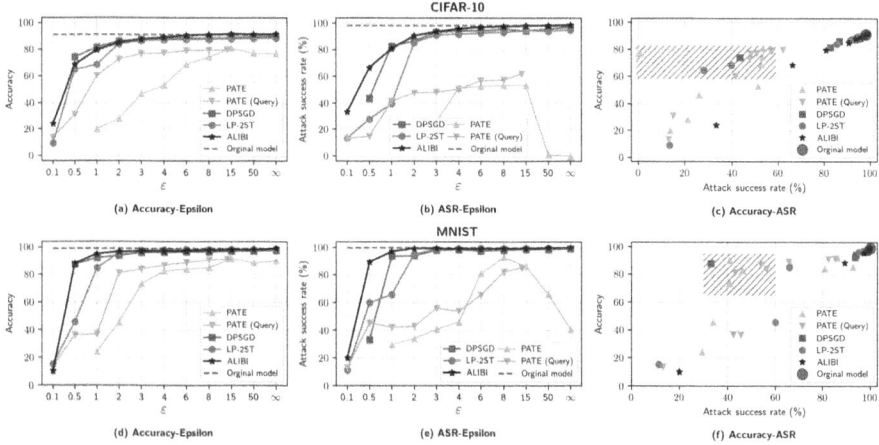

Fig. 3. The impact of epsilon on DP and Label-DP methods using MNIST (top) and CIFAR-10 dataset(bottom).

influential parameters on the privacy budget (ε) decrease the ASR but also cause a drastic decrease in the accuracy at the same time. Conversely, the best result is achieved through tuning the threshold, although it cannot provide any DP by thresholding alone.

7   Comparison of DP and Label-DP Methods

In this section, we compare all the DP and Label-DP algorithms to discover which one and under what conditions are more successful.

7.1   Privacy Budget Analysis

The ϵ in DP and Label-DP serves two different goals. So we do not directly compare the ϵ values of the two methods even though both can be reduced to label DP [14]. Instead, what we focus on is the trade-off between accuracy and ASR provided by varying ϵ of the two methods. We select the best parameters from the results in the previous section to conduct the current experiment. These best parameters lead to high accuracy and a low ASR. Wherever there is a trade-off between accuracy and ASR, we prioritize accuracy. For MNIST, we do not present those parameter selections due to the similar outcomes.

Figure 3a,b compare the accuracy and ASR of the different methods for CIFAR-10 with varying ϵ while Fig. 3c shows the trade-off of accuracy and ASR of different methods (the ideal case correspond to 100%accuracy and 0% ASR). PATE can achieve different levels of privacy by varying two factors: 1) noises (lime green plots), and 2) number of queries (orange plots). The first observation is that non-DP PATE outperforms all other results and methods (the rightmost point of the lime green plot). It indicates the power of bagging with a threshold

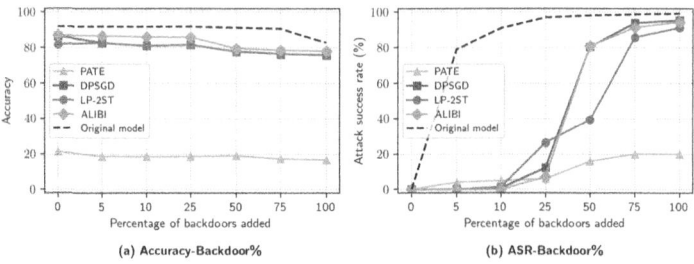

Fig. 4. The significant impact of poisoned data on DP-based defense methods. The epsilon is fixed to 1 and then all the methods are compared by varying the percentage of the training data that has been poisoned.

against backdoor attacks. LP-2ST for some ϵ values works well. For instance, $\varepsilon = 1$ has high accuracy (78%) and a significantly decreased ASR (39%). However DP-SGD gives the best results when $\varepsilon = 0.5$. For ALIBI, both accuracy and ASR drop proportionally.

Figure 3d,e,f show similar trends for MNIST. Figure 3f combines the results of the two other columns by directly comparing the accuracy and corresponding ASR. The rectangular areas with the hatched pattern in the last column consist of the most desired results with high accuracy and dropped ASR regardless of their privacy budget. This area includes different private algorithms, but mostly PATE, which indicates the dominance of PATE.

Conclusion (Q2): The DP and Label-DP techniques effectively reduce the vulnerability of backdoor attacks, albeit at the cost of decreased accuracy. If the optimal approach is determined by the accuracy-to-ASR ratio, then the superiority of each DP or Label-DP model depends on the allocated privacy budget.

7.2   Attack Strength Analysis

We discussed the hyper-parameters and the privacy budget of the algorithm as two factors that impact the immunity of the DP approaches against backdoor attacks. A third factor that should be considered when assessing the level of immunity is the strength of the attack itself. So far, we have synthesized powerful attacks by poisoning 50% of the data with backdoors. However, in practice, the attacker conceals her malicious activity by limiting the percentage of poisoned data introduced into the pipeline. Therefore we change the percentage of the backdoors in the base class to develop a range of more realistic and more powerful (but less realistic) attacks.

Figure 4 shows the accuracy and ASR with respect to the number of backdoors, when the privacy budget for all DP algorithms has been fixed to $\varepsilon = 1$. We observe that the accuracy does not drastically change with respect to the number of backdoors, yet the ASR increases as the attack becomes more powerful.

Looking at the pattern, we can see that the DP algorithms almost entirely diffuse the attack when the percentage of backdoors is sufficiently small. It should be noted that the low accuracy of PATE is a result of controlling its privacy budget by adding noise, rather than limiting the number of queries according to the reasoning we had in Sect. 6.2.

Conclusion (Q2): These results illustrate the effectiveness of DP-SGD, LP-2ST, and ALIBI against more realistic backdoor attacks (with backdoor% $\leq$ 10). For such attacks, the accuracy drops by 10%, and the attack achieves no success. This is compatible with Eq. (3) that shows that the attacker's loss limit in DP models is theoretically linked to the number of poisoned data.

7.3   Accuracy-Privacy Trade-Off

To see the accuracy when a perfect defense is desired (close to 0 ASR), we have analyzed different privacy budgets for each DP method and found the greatest ε where the ASR does not exceed 1%. This small ASR is achievable when the number of backdoors is insignificant (we set it to 10%). By doing so, we achieve the least randomness that leads to a successful defense. After removing the impact of the attack, we can have a fair comparison of accuracy and training time.

Table 2. Comparison of the highest accuracy and epsilon that DP methods can achieve while ASR=0.

	DP-SGD	PATE	ALIBI	LP-2ST
Accuracy	88.67	85.02	**89.53**	79.9
Epsilon	2	**inf**	2	0.9
Time	140 s	220 s	**59 s**	**58 s**

Table 2 highlights the best values of accuracy, privacy budget, and training time in each row. The previous findings indicate that a deterministic version of PATE, with noise removed, is the most resilient against attacks. However, when the goal is to simultaneously defend against backdoors and protect privacy, this result is not favorable for PATE. DP-SGD and ALIBI, with the same privacy budget, can achieve better accuracy than PATE.

Finally, with respect to training time, two Label-DP methods demonstrate a considerable reduction in training time, surpassing other DP techniques. It is important to note that this experiment was conducted on a CIFAR-10 fine-tuning task, where training time is negligible. However, in more complex architectures with end-to-end settings, time may become a bottleneck for PATE and DP-SGD.

Conclusion (Q2): When a perfect defense is desired, Label-DP methods offer the best efficiency and comparable or better accuracy trade-off compared to DP approaches.

8   Discussion and Conclusion

This paper posed important questions regarding the ability of DP to provide robustness against backdoor attacks in practice. In addition to DP-SGD, we explored the other commonly used DP algorithm (PATE) and two Label-DP algorithms (LP-2ST and ALIBI) for the first time. We have several main findings.

First, the noise and randomness added to the private models can indeed decrease the attack success rate of the backdoors, but at the cost of utility drop for clean input. In a nutshell, a model trained with privacy guarantee has an inherent benefit in robustness against backdoor attacks. This statement holds for all four methods mentioned above. A somewhat unexpected outcome is that PATE delivers the best results, even without the use of noise (without DP guarantee) due to the ensemble based teacher-student structure.

Second, contrary to the claims of some previous studies, DP-SGD provides good resistance against backdoors while keeping the accuracy relatively high. We also observed the same phenomenon for Label-DP algorithms. The accuracy-ASR trade-off is diverse among the DP and Label-DP methods we analyzed. One model may outperform the others depending on the privacy budget, algorithm parameters, and attack specifications. Therefore, it is possible to use DP models as defense strategies. A proper selection of the above mentioned factors can adequately balance the accuracy and ASR.

This work was an empirical study on two benchmark datasets, MNIST and CIFAR-10. It offered new empirical insights into the connection between DP and backdoor attacks in relation to existing theoretical understandings. Future research could focus on exploring the impact of Label-DP on particular type of poisoning attacks focusing on labels such as label-based flipping attacks. Additionally, given the ability of DP methods to enhance robustness, there is an opportunity to develop modified DP algorithms that offer greater protection against poisoning attacks, and simultaneously fulfill both privacy and robustness objectives.

A   Appendix

A.1   Experimental Setup Details

Training Configuration. For MNIST and CIFAR-10 datasets, we used different architectures for neural networks. For CIFAR-10, the ResNet-50 head was followed by an average pooling layer and two linear layers of size 256 and 10. For MNIST, the neural network consisted of two convolutional layers with 16 and 32 filters, each of kernel size 8 and 4 followed by max pooling layers and two final linear layers with 32 and 10 neurons. The learning rate of all four DP and Label-DP algorithms was 0.001 and the number of epochs was fixed to 50. In contrast to Label-DP algorithms where an SGD optimizer was good enough to train the model, for DP-SGD and PATE we required to use more adaptive optimizers, i.e. RMSProp and Adam, respectively.

Table 3. Parameters of the DP and Label-DP algorithms

Method	Parameters
LP-2ST	1. **Data split ratio** : The portion the training dataset split between two training stages (more in the first stage helps with accurate prior but causes underfit in the second stage)
	2. **Temperature T** : For logit z_i and calculation of prior p_i of class i, a small T in $p_i = \frac{exp(z_i/T)}{\sum_j exp(z_j/T)}$ boosts the confidence of the top classes and a large T makes the priors more uniform
	3. **Epsilon ε** : Randomness parameter that is equivalent to the privacy budget
ALIBI	1. **noise of soft training labels** : Laplacian noise with $\delta = 0$ which is applied once and determines the privacy budget

B   Label DP against Backdoors

In this section , we evaluate LP-2ST, and ALIBI as two Label-DP models. We investigate if their randomness or other related parameters can help to mitigate the backdoor attacks. To this end, Table 3 presents the various parameters involved in these algorithms.

B.1   LP-2ST vs. Backdoors

Since the Label-DP algorithms randomly change the labels, we found that the accuracy in high noise fluctuate among multiple runs. So for each experiment on LP-2ST and ALIBI, the accuracy and ASR are the averages of 10 trials. For each figure from left to right, we pick a parameter shown on the x-axis (which is chosen randomly) and apply it for the experiments in the succeeding figure. For the first two figures, we set $\varepsilon = 1$.

Figure 5a demonstrates the effect of temperature with a random data split of [80/20]. Compatible to [14], sparsifying the priors helps to improve the utility, but to our surprise, it decreases ASR. We speculate the reason is that the backdoor still has a touch of the base class. Thus the first round of LP-2ST predicts target and base classes as the backdoors' top-2 classes. The sparsified prior shifts the probabilities of these two classes far away from zero, so the algorithm selects the base class more confidently.

In Fig. 5b the training data has been partitioned for two stages. [p1/p2] on the x-axis indicates the percentage of the data in stage 1 and stage 2 of LP-2ST, respectively. When 100% of data is allocated to the first stage, it means that we are using LP-1ST with RR. There is not a clear pattern between ASR and data split. But an LP-2ST model with more data in the first stage has more enhanced priors and higher accuracy. Figure 5c compares different privacy budgets ε, which is the random factor of the RR algorithm. Naturally, more randomness helps to decrease the ASR. The results for $\varepsilon = 1$ are particularly impressive since it drops the ASR to less than 40%, while the accuracy is still roughly 80%.

Conclusion: Surprisingly, even though Label-DP only randomizes the labels, it is still successful against backdoor attacks. In this success, all parameters are involved, but noise has the major impact. LP-2ST can vividly mitigate the attack, but it is very important which ε is selected to obtain a reasonable accuracy-ASR trade-off.

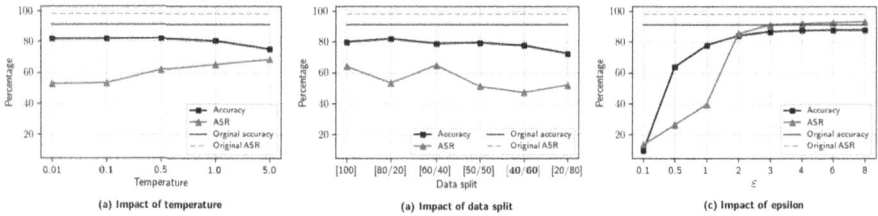

Fig. 5. The impact of temperature, data split between two stages and epsilon on LP-2ST (from left to right). Epsilon, the factor of privacy-preserving in LP-2ST, can drastically deteriorate the ASR with an acceptable utility cost (c).

B.2   ALIBI vs. Backdoors

According to Fig. 6, ALIBI with higher noise drops both accuracy and ASR proportionally. This can be justified by the fact that all the labels randomly change just once at the beginning of the training.

Conclusion: On average, ALIBI can mitigate the effect of backdoor attacks but with reduced utility costs.

B.3   Training Process

In this section, we compare the training process of DP-SGD, LP-2ST, and ALIBI on CIFAR-10. These comparisons are based on two privacy budgets $\varepsilon = \infty$

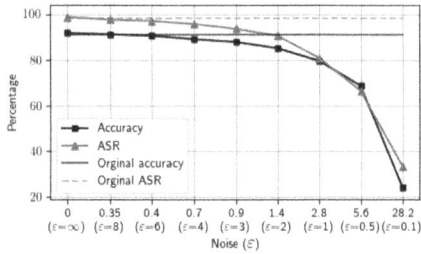

Fig. 6. Effectiveness of randomizing labels on reducing ASR in ALIBI. The noise added to one-hot labels in ALIBI impacts both accuracy and ASR proportionally.

and $\varepsilon = 1$, to provide an overview over the training process with and without randomness. For LP-2ST, we only illustrate the training of the second and final stage of the algorithm. In Fig. 7, each column demonstrates a different method, and each row indicates one of the privacy budgets. For all three differentially private methods, on the first row, with $\varepsilon = \infty$, the loss of the backdoor samples drops below the clean loss on early training epochs. It is the opposite for all three methods when $\varepsilon = 1$ on the second row. For LP-2ST the backdoor loss does not converge to the clean loss and remains higher. It is consistent with the results of LP-2ST at $\varepsilon = 1$ in Fig. 5c. For ALIBI the clean and backdoor losses change very closely. It explains the similar values for the ALIBI accuracy and ASR in Fig. 6. DP-SGD can resist the backdoor samples on early epochs. So one of the suggestions is to stop the training early to avoid backdoors from overfitting.

Conclusion: During DP training, the model underfits or suppresses the backdoor samples which results in defusing the backdoors' impact on the model. This finding confirms the results of the paper.

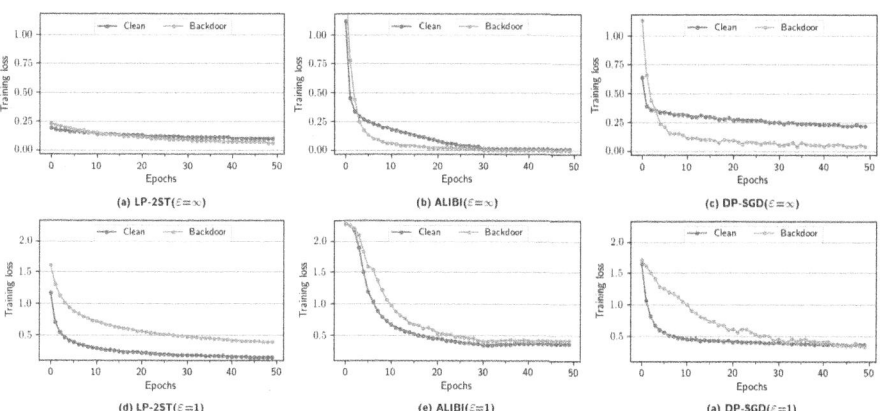

Fig. 7. An overview of the training process of LP-2ST, ALIBI and DP-SGD using $\varepsilon = \infty$ (upper) and $\varepsilon = 1$ (lower).

References

1. Abadi, M., et al.: Deep learning with differential privacy. In: Proceedings of the 2016 ACM SIGSAC Conference on Computer and Communications Security, pp. 308–318 (2016)
2. Andrew, G., Chein, S., Papernot, N.: Tensorflow privacy library (2020)

3. Biggio, B., Corona, I., Fumera, G., Giacinto, G., Roli, F.: Bagging classifiers for fighting poisoning attacks in adversarial classification tasks. In: Sansone, C., Kittler, J., Roli, F. (eds.) MCS 2011. LNCS, vol. 6713, pp. 350–359. Springer, Heidelberg (2011). https://doi.org/10.1007/978-3-642-21557-5_37

4. Borgnia, E., et al.: Strong data augmentation sanitizes poisoning and backdoor attacks without an accuracy tradeoff. In: ICASSP 2021-2021 IEEE International Conference on Acoustics, Speech and Signal Processing (ICASSP), pp. 3855–3859. IEEE (2021)

5. Borgnia, E., et al.: DP-instahide: provably defusing poisoning and backdoor attacks with differentially private data augmentations. arXiv preprint arXiv:2103.02079 (2021)

6. Bu, Z., Wang, Y.X., Zha, S., Karypis, G.: Automatic clipping: differentially private deep learning made easier and stronger. In: ICML TPDP Workshop (2022)

7. Chan, A., Ong, Y.S.: Poison as a cure: detecting and neutralizing variable-sized backdoor attacks in deep neural networks. arXiv:1911.08040 [cs] (November 2019), arXiv: 1911.08040

8. Chen, B., et al.: Detecting backdoor attacks on deep neural networks by activation clustering. In: SafeAI@AAAI (2019)

9. Chen, R., Li, Z., Li, J., Yan, J., Wu, C.: On collective robustness of bagging against data poisoning. In: International Conference on Machine Learning, pp. 3299–3319. PMLR (2022)

10. Du, M., Jia, R., Song, D.: Robust anomaly detection and backdoor attack detection via differential privacy. In: ICLR 2020 (2020)

11. Dwork, C.: Differential Privacy, vol. 2006, pp. 1–12. ICALP (2006)

12. Dwork, C.: Differential privacy. In: Bugliesi, M., Preneel, B., Sassone, V., Wegener, I. (eds.) ICALP 2006. LNCS, vol. 4052, pp. 1–12. Springer, Heidelberg (2006). https://doi.org/10.1007/11787006_1

13. Dwork, C., Kenthapadi, K., McSherry, F., Mironov, I., Naor, M.: Our data, ourselves: privacy via distributed noise generation. In: Vaudenay, S. (ed.) EUROCRYPT 2006. LNCS, vol. 4004, pp. 486–503. Springer, Heidelberg (2006). https://doi.org/10.1007/11761679_29

14. Ghazi, B., Golowich, N., Kumar, R., Manurangsi, P., Zhang, C.: Deep learning with label differential privacy. In: Advances in Neural Information Processing Systems, vol. 34 (2021)

15. Gu, T., Liu, K., Dolan-Gavitt, B., Garg, S.: Badnets: evaluating backdooring attacks on deep neural networks. IEEE Access **7**, 47230–47244 (2019)

16. He, K., Zhang, X., Ren, S., Sun, J.: Deep residual learning for image recognition. In: Proceedings of the IEEE Conference on Computer Vision and Pattern Recognition, pp. 770–778 (2016)

17. Hong, S., Chandrasekaran, V., Kaya, Y., Dumitraş, T., Papernot, N.: On the effectiveness of mitigating data poisoning attacks with gradient shaping. arXiv preprint arXiv:2002.11497 (2020)

18. Jagielski, M., Oprea, A.: Does differential privacy defeat data poisoning. In: DPML Workshop (2021)

19. Jagielski, M., Oprea, A., Biggio, B., Liu, C., Nita-Rotaru, C., Li, B.: Manipulating machine learning: poisoning attacks and countermeasures for regression learning. In: 2018 IEEE Symposium on Security and Privacy (SP), pp. 19–35. IEEE (2018)

20. Jia, J., Cao, X., Gong, N.Z.: Intrinsic certified robustness of bagging against data poisoning attacks. In: Proceedings of the AAAI Conference on Artificial Intelligence, vol. 35, pp. 7961–7969 (2021)

21. Koh, P.W., Steinhardt, J., Liang, P.: Stronger data poisoning attacks break data sanitization defenses. arXiv preprint arXiv:1811.00741 (2018)
22. Krizhevsky, A., Hinton, G., et al.: Learning multiple layers of features from tiny images (2009)
23. LeCun, Y., Bottou, L., Bengio, Y., Haffner, P.: Gradient-based learning applied to document recognition. Proc. IEEE **86**(11), 2278–2324 (1998)
24. Lu, S., Li, R., Liu, W., Chen, X.: Defense against backdoor attack in federated learning. Comput. Secur. **121**, 102819 (2022)
25. Ma, Y., Zhu, X., Hsu, J.: Data poisoning against differentially-private learners: attacks and defenses. In: Proceedings of the 28th International Joint Conference on Artificial Intelligence. IJCAI'19, pp. 4732–4738. AAAI Press (2019)
26. Malek Esmaeili, M., Mironov, I., Prasad, K., Shilov, I., Tramer, F.: Antipodes of label differential privacy: pate and alibi. In: Advances in Neural Information Processing Systems, vol. 34 (2021)
27. Miao, L., Yang, W., Hu, R., Li, L., Huang, L.: Against backdoor attacks in federated learning with differential privacy. In: ICASSP 2022 - 2022 IEEE International Conference on Acoustics, Speech and Signal Processing (ICASSP), pp. 2999–3003. IEEE (2022)
28. Naseri, M., Hayes, J., Cristofaro, E.D.: Local and central differential privacy for robustness and privacy in federated learning. In: Proceedings 2022 Network and Distributed System Security Symposium (2020)
29. Papernot, N., Abadi, M., Erlingsson, U., Goodfellow, I., Talwar, K.: Semi-supervised knowledge transfer for deep learning from private training data. arXiv preprint arXiv:1610.05755 (2016)
30. Papernot, N., Song, S., Mironov, I., Raghunathan, A., Talwar, K., Erlingsson, Ú.: Scalable private learning with pate. arXiv preprint arXiv:1802.08908 (2018)
31. Peri, N., et al.: Deep k-NN defense against clean-label data poisoning attacks. In: Bartoli, A., Fusiello, A. (eds.) ECCV 2020. LNCS, vol. 12535, pp. 55–70. Springer, Cham (2020). https://doi.org/10.1007/978-3-030-66415-2_4
32. Saha, A., Subramanya, A., Pirsiavash, H.: Hidden trigger backdoor attacks. In: Proceedings of the AAAI Conference on Artificial Intelligence, vol. 34, pp. 11957–11965 (2020)
33. Schwarzschild, A., Goldblum, M., Gupta, A., Dickerson, J.P., Goldstein, T.: Just how toxic is data poisoning? A unified benchmark for backdoor and data poisoning attacks. In: International Conference on Machine Learning, pp. 9389–9398. PMLR (2021)
34. Tang, X., et al.: Machine learning with differentially private labels: mechanisms and frameworks. In: Proceedings on Privacy Enhancing Technologies (2022)
35. Tran, B., Li, J., Madry, A.: Spectral signatures in backdoor attacks. In: Advances in Neural Information Processing Systems, vol. 31 (2018)
36. Wang, B., et al.: Neural cleanse: identifying and mitigating backdoor attacks in neural networks. In: 2019 IEEE Symposium on Security and Privacy (SP), pp. 707–723. IEEE (2019)
37. Wang, L., Zheng, J., Cao, Y., Wang, H.: Enhance pate on complex tasks with knowledge transferred from non-private data. IEEE Access **7**, 50081–50094 (2019)
38. Warner, S.L.: Randomized response: a survey technique for eliminating evasive answer bias. J. Am. Stat. Assoc. **60**(309), 63–69 (1965)
39. Weber, M., Xu, X., Karlaš, B., Zhang, C., Li, B.: Rab: provable robustness against backdoor attacks. In: 2023 IEEE Symposium on Security and Privacy (SP), pp. 1311–1328. IEEE (2023)

40. Xu, C., Wang, J., Guzmán, F., Rubinstein, B., Cohn, T.: Mitigating data poisoning in text classification with differential privacy. In: Findings of the Association for Computational Linguistics: EMNLP 2021, pp. 4348–4356. Association for Computational Linguistics, Punta Cana, Dominican Republic (2021)
41. Zhao, S., Ma, X., Zheng, X., Bailey, J., Chen, J., Jiang, Y.G.: Clean-label backdoor attacks on video recognition models. In: Proceedings of the IEEE/CVF Conference on Computer Vision and Pattern Recognition, pp. 14443–14452 (2020)

Author Index

© IFIP International Federation for Information Processing 2024
Published by Springer Nature Switzerland AG 2024
A. L. Ferrara and R. Krishnan (Eds.): DBSec 2024, LNCS 14901, pp. 341–342, 2024.
https://doi.org/10.1007/978-3-031-65172-4

GPSR Compliance

The European Union's (EU) General Product Safety Regulation (GPSR) is a set of rules that requires consumer products to be safe and our obligations to ensure this.

If you have any concerns about our products, you can contact us on ProductSafety@springernature.com

In case Publisher is established outside the EU, the EU authorized representative is:

Springer Nature Customer Service Center GmbH
Europaplatz 3
69115 Heidelberg, Germany

The manufacturer's authorised representative in the EU is Springer
Nature Customer Service Centre GmbH, Europaplatz 3, 69115 Heidelberg,
Germany. If you have any concerns regarding our products, please
contact ProductSafety@springernature.com

Printed and bound by CPI Group (UK) Ltd, Croydon, CR0 4YY

24/04/2026

02096358-0010